THE BEEKEEPER'S HANDBOOK

The Beekeeper's Handbook

THIRD EDITION

Diana Sammataro and Alphonse Avitabile

Illustrations by Diana Sammataro and Jan Propst

Comstock Publishing Associates a division of

Cornell University Press | Ithaca and London

First edition published 1978 by Peach Mountain Press
Second edition published 1986 by Macmillan Publishing Company
Third edition published 1998 by Cornell University Press
First printing, Cornell Paperbacks, 1998

Printed in the United States of America.

Library of Congress Cataloging-in-Publication Data

Sammataro, Diana.
Beekeeper's handbook / Diana Sammataro and Alphonse Avitabile:
illustrations by Diana Sammataro and Jan Propst. —3rd ed
p. cm.
Includes bibliographical references (p.) and index.
ISBN-13: 978-0-8014-8503-9 (pbk. : alk. paper)
ISBN-10: 0-8014-8503-7 (pbk. : alk. paper)
1. Bee culture—Handbooks, manuals, etc. I. Avitabile, Alphonse. II. Title.
SF523.S35 1998
638'.1—dc21 97-44378

Cornell University Press strives to use environmentally responsible suppliers and materials to the fullest extent possible in the publishing of its books. Such materials include vegetable-based, low-VOC inks and acid-free papers that are recycled, totally chlorine-free, or partly composed of nonwood fibers. For further information, visit our website at www.cornellpress.cornell.edu.

Paperback printing 10 9 8 7 6 5

To my father, Joseph Michael Sammataro, who has guided me in my many interests and endeavors with gentle kindness, and to my maternal grandfather, George Weber, who first introduced me to the world of bees. He and his two brothers were commercial beekeepers in Illinois and Montana early in the last century.

D. S.

To Leonard Insogna, a social scientist, who gave me insights into life and politics that have helped guide me in my daily life.

A. A.

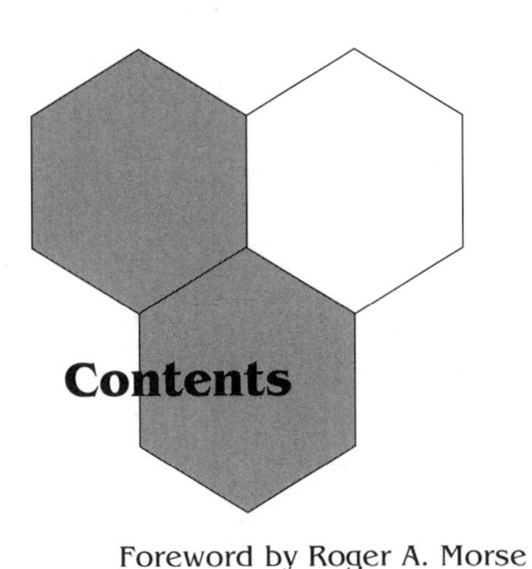

Contents

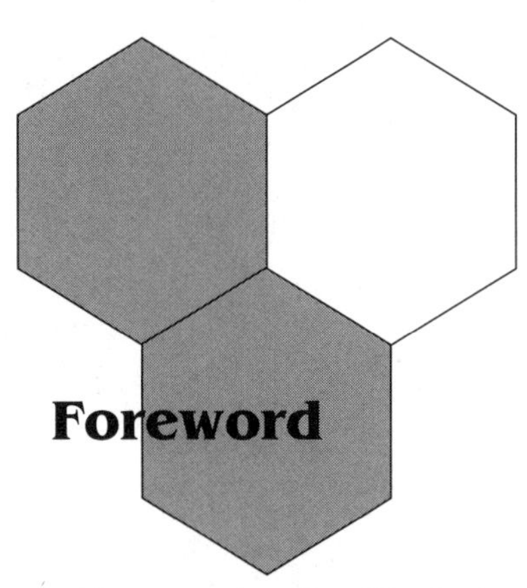

Foreword

For two decades *The Beekeeper's Handbook* has guided thousands of beginning and advanced beekeepers in the how-to's of this entertaining and profitable pastime. But in recent years the science and the art of beekeeping have changed drastically, and this new, thoroughly updated edition will enable beekeepers at all levels to keep up with those changes. This third edition brings beekeeping to the threshold of the twenty-first century, with all its challenges.

No one could do this better than authors Diana Sammataro, a noted honey bee researcher, and Alphonse Avitabile, a retired honey bee scientist and college instructor. Dr. Sammataro is also a beekeeper. She produces honey, raises queens, uses bees to pollinate crops, assembles equipment, and engages in all the other activities of beekeeping. Her intimate knowledge of honey bees is evident throughout this book. Alphonse Avitabile, also an experienced beekeeper, is a successful gardener, nurseryman, and greenhouse manager.

The popularity of the first two editions resulted from a simple premise underlying both books: there are many ways to do things right. And this latest edition, too, unlike much of the genre, presents time-tested methods and techniques, introduces the most current ideas and concepts, and lets readers choose those which best suit their individual skills, location, and requirements. Although originally designed for beginners, *The Beekeeper's Handbook* will appeal to more advanced beekeepers as well. Rather than limit the seasoned beekeeper to traditional ways of doing things, it puts forward the newest and safest methods to deal with today's problems.

With this book, beekeeping has never been easier. Simply put, it is the best of the best of beekeeping books.

Roger A. Morse

Ithaca, New York

Acknowledgments

So many people have contributed to this book since its inception over a decade ago. Jan Propst originally came up with the format, for which we are forever grateful. Thomas D. Seeley provided the photograph of a honeycomb for the cover. The following list includes folks who were generous with their time, contributions, and support. Thank you all.

Joe and Nelva Sammataro
Ruth Avitabile
Ron and Carol Conkey
John and Gwen Nystuen
Dean Breaux
M. Tom Sanford
Lynn Birny Wescott
C. Allen Dick
Ann Harman

D. S.
A. A.

THE BEEKEEPER'S HANDBOOK

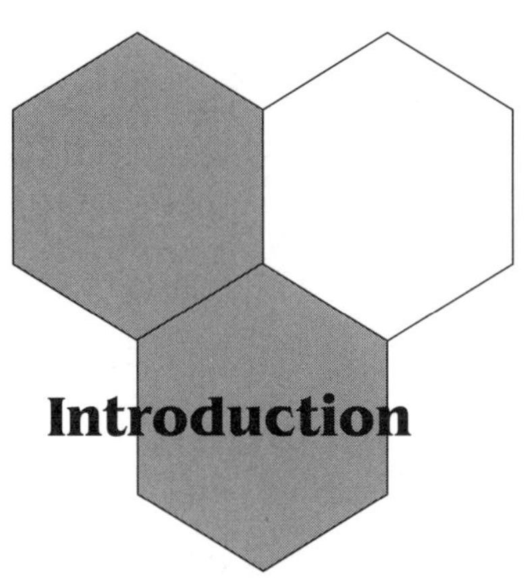

Introduction

Beekeeping is an interesting and rewarding activity if you love nature, have a fascination with the unique social organization of bees, are consumed with an active curiosity about how things work, and enjoy honey.

This handbook is designed to help you become a good beekeeper, whether you intend to start keeping bees or already have them and need a ready guide to help you accomplish the various and often complicated tasks that you need to perform in the beeyard. It is designed to assist both new and experienced beekeepers in setting up or reorganizing an apiary and in improving their style of working with and understanding bees.

The book outlines the many colony management operations you will encounter. The text presents the key elements in keeping bees, describing all the major options available to you. It also lists the advantages and disadvantages of each important technique to help you decide which one is best for you. Each section is cross-referenced to point you to more detailed information.

Numerous diagrams and illustrations accompany the text to reinforce or illuminate the descriptions given. Space is also provided at the ends of the chapters so you can keep notes on your own successes and failures.

The reference section has been updated to include as many important books, organizations, and Internet resources as possible. In addition, a glossary has been added to help new beekeepers understand the terminology of bees.

Although considered to be a "gentle art," beekeeping in reality can be physically demanding and strenuous. The typical picture of a veiled beekeeper standing beside the beehive with smoker in hand does not show the aching back, sweating brow, smoke-filled eyes, or painful stings. This handbook is intended to enable you to maximize the more interesting and enjoyable aspects of the art. Have fun, learn a lot, and share your knowledge with others!

Legal Requirements

All states have some laws that pertain to keeping honey bees and registering hives containing bees. Certain city and state laws limit the number of *hives* (the wooden boxes in which colonies of bees live) in urban areas. Because some cities may have declared bees a nuisance, local laws must be studied before an *apiary* (place where beehives are placed) is established. Most states have an apiary inspection law developed to aid beekeepers by providing means for controlling and eradicating bee diseases and pests.

General requirements usually include some of the following:

- Beekeepers may have to register hives or apiaries containing honey bees with their state Department of Agriculture.
- The director of agriculture and appointed deputies may be authorized to inspect, treat, quarantine, disinfect, and destroy any diseased hives.
- Beekeepers may have to ascertain and comply with town or county zoning ordinances that pertain to bees and bee hives.
- Transportation of bees and equipment may have to be certified by the bee inspector or other designated official.
- All beekeepers shall have bee colonies in hives containing movable frames.
- Penalties may exist for violations of applicable apiary inspection laws.

Now, because of the introduction of parasitic bee mites and the Africanized honey bee, some southern states have special laws regarding keeping bees. For specific legal requirements, check your state Department of Agriculture's apiary inspection law (see "Management of Bee Colonies" in the References).

Bee Sting Reactions

An important question that you must consider is your individual response to bee stings. Although most beekeepers never exhibit serious reactions to bee stings, after a few years some individuals can develop an allergy to bee venom, bee hairs, or other hive components.

When a person is stung, the bee's stinging apparatus pierces the flesh, and venom enters the surrounding tissues and is transported by the blood throughout the body. Fortunately for most people, a *localized* reaction results—that is, pain, reddening, itching, and swelling occurs at the sting site. Sometimes the swelling can be quite alarming, but it usually subsides over a few days. Your unique body chemistry will react in its characteristic way.

On the other hand, you may experience a more serious reaction to bee stings. This is called a *systemic* reaction, a positive sign that you are allergic to bee venom.

A systemic or general reaction takes place a few minutes after a person is stung and means that the entire body is reacting to the venom proteins. Signs of a systemic reaction may include both those of a localized reaction as well as other symptoms, such as itching of the extremities (feet, hands, tongue) or all over the body (hives), breathing difficulty, swelling away from the sting site, sneezing, abdominal pain, and loss of consciousness. This type of reaction occurs when the body is allergic to the bee venom proteins and, if not treated, could be fatal.

Anaphylactic shock reactions are rare but can occur in a very short time in highly allergic persons. Symptoms include labored breathing, confusion, vomiting, and falling blood pressure; if not treated promptly, such a reaction could lead to fainting and death.

The percentage of people that become allergic to bee venom is very small, but for those individuals, such an allergy must be considered serious. Fewer than 20 deaths per year (United States) result from bee stings, compared to heart disease (977,700 per year), auto accidents (46,000 per year) and lightning (85 per year). If there is ever any question about whether you are developing an allergy to bee stings or bee hairs, wax, or propolis, consult a physician or local allergy clinic immediately! For more information, see Chapter 5, "What to Do When Stung," "Bee Sting Reaction" in the References, and Appendix C. Also check out chapter 27 of J. M. Graham, ed., *The Hive and the Honey Bee,* published by Dadant and Sons, 1992.

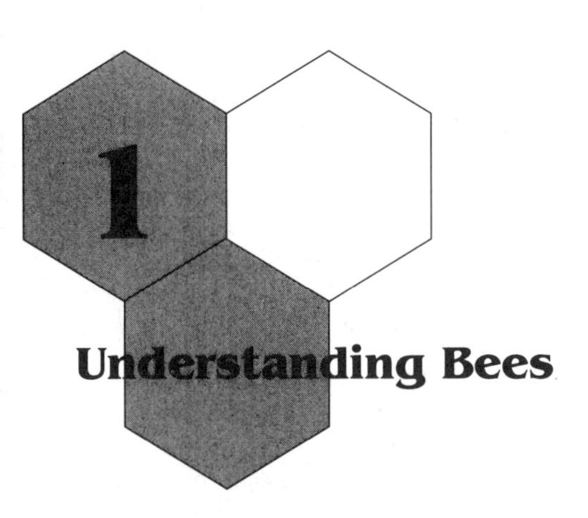

1 Understanding Bees

Bee Ancestors

Although fossil records are incomplete, insects seem to have first appeared about 300 million years ago, during the Carboniferous period. The probable ancestors of the order Hymenoptera, to which honey bees belong, evolved some 200 million years ago as predatory wasps. Fossil insects preserved in Permian rock, dating from the close of the Paleozoic era, display hymenopteran-like structures, including the membranous wings and the antlike waists.

Approximately 50 million years later, in the middle of the Mesozoic era, the hymenopterans were firmly established in the fossil records, primarily in amber, and included primitive and subsocial ants that were mostly predatory. Bees appear to have evolved from predatory sphecid wasp ancestors, about 100 million years ago (mid-Cretaceous). The switch from animal to vegetable protein and the presence of branched hairs separate wasps from bees.

During the vast periods of time that followed, the flowering plants became more specialized and more dependent on mobile pollinators. Insect visitors such as the bees were very important, and they and the plants they pollinated coevolved structures to their mutual benefit as a result of this interdependence.

It wasn't until 65 million years ago (Tertiary period) that the stinging hymenopterans became common; by this time the land was dominated by the flowering plants, or *angiosperms,* which provided plenty of pollen (a source of protein) and nectar (a source of carbohydrate). Those plants that attracted bees because of their shape, color, odor, and food offered were pollinated and therefore set seed for the next generation. In their turn, bees developed branched hairs on their bodies to trap the pollen of flowers, inflatable sacs to carry away sugary nectars, and a highly structured social order with elaborate defense and communication systems to exploit the most rewarding of floral resources.

In the order Hymenoptera, there are over 200,000 species in 10 or 11 families and about 700 genera.

The placement of the honey bee in the animal kingdom is as follows:

- **Phylum:** Arthropoda (many-jointed, segmented, chitinous invertebrates, including lobsters and crabs)
- **Class:** Hexapoda or Insecta (six-footed)
- **Order:** Hymenoptera (Hymen is the Greek god of marriage, hence the union of front and hind wings [*pteron*])
- **Suborder:** Apocrita (ants, bees, and wasps)
- **Superfamily:** Apoidea (between 8 and 10 families)
- **Family:** Apidae (characterized by food exchange, pollen baskets, storage of honey and pollen); three subfamilies (see the figure on taxonomy)
- **Tribe:** Apini (long tongues, nonparasitic, highly eusocial)
- **Genus:** *Apis* (bee, native of the Old World, probably evolved in India and Southeast Asia)
- **Species:** *mellifera* (honey-bearing)

Evolution of Social Structure

Social structure is defined by the degree of community living. The true, highly specialized, or *eusocial,* societies are those of ants, termites, and honey bees. The sophistication of the social structure of honey bees is indicated by a number of characteristics:

- Longevity of the female parent (queen) coexisting with her offspring.
- Presence of reproductive castes (two female, one male).
- Siblings assisting in the care of the brood.
- Progressive feeding of brood instead of mass feeding.
- Division of labor, whereby queen lays eggs and workers perform other functions.
- Nest and shelter construction, storage of food.
- Swarming as a reproductive process.
- Perennial nature of colony.
- Communication among colony members.

A eusocial community of honey bees consists of two female castes, a mother (*queen*) and daughters (sterile *workers*), that overlap at least two generations; *drones* are the male bees. Because hornet and

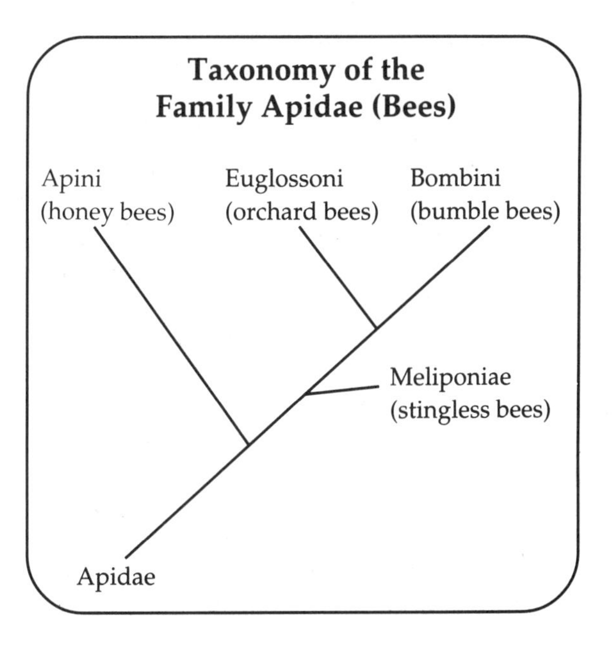

wasp colonies do not normally overwinter in temperate climates, as do honey bees, they are termed *semisocial* insects.

Most insects are solitary—they neither live together in communities nor share the labor of raising their young. The 12,000 species of insects that live in communities are the ants, termites, wasps, and bees (and include some beetles, aphids, and thrips). The origin of social insects troubled even Darwin, who could not rationalize how a special, sterile caste—the workers—could pass on their genetic information if they could not produce offspring. In other words, workers were displaying "altruism" toward their siblings at the expense of not having their own children. Over the years, there have been several ways to explain the evolution of sociality. Here are a few of the most popular theories, briefly explained—kin selection, mutualism, and parental manipulation.

Kin selection theory explains that workers are genetically more related to one another than they are to their parents because of the *haploid* (possessing only one set of chromosomes) drones. Therefore, it is more advantageous for workers (as they have no children) to rear their siblings, which are closely related to them and thus share their genetic makeup. So, helping them is like workers helping themselves, enabling their genes to pass on to the next generation.

Mutualism maintains that an individual queen benefits if others, especially if they are her sisters, help in rearing her brood. If they help one another, they will be more successful and have more offspring survive to the next generation. Such cooperation may eventually lead to members of the same species occupying a composite nest and developing some kind of communal brood care.

Parental manipulation asserts that the mother gains net survival or reproductive success by tricking or manipulating others. This theory evolved because the queen dominates her daughters by means of chemical signals called *pheromones.* Such signals reduce the reproductive potential of the daughters, forcing them to become slaves and tend their siblings instead of laying eggs.

Existing subsocial and primitively social insects incorporate some or all of these hypotheses. Evolution of social behavior makes for fascinating reading, and you should look at other works on this complex behavior for a more complete understanding of the subject (see "Social Insects" in the References).

Races of Bees

There are eight species of honey bees in the genus *Apis,* four(*) of which have been added since 1988, and more are currently under investigation. A *species* is a group of organisms that can interbreed, producing offspring that can do the same. The different honey bee species in the world, to date, are *Apis cerana, A. koschevnikovi*, A. nigrocincta*, A. dorsata, A. laboriosa*, A. florea, A. andreniformis*,* and *A. mellifera.* Among these species, one (*A. mellifera*) contains 24 races.

By races, we refer to populations of the same species (*mellifera*) that originally occupied particular geographic regions with different climates, topography, and floral resources. In these areas, bees evolved characteristics that made them unique from other species. F. Ruttner, *Biogeography and Taxonomy of Honeybees* (New York: Springer-Verlag, 1988), divided bees into four groups: (1) African, (2) Near East, (3) central Mediterranean and southeastern European, and (4) western Mediterranean and northwestern European. From the European groups came the Italian, Carniolan, and German black bees; the Near East group includes the Caucasian bees. These four races provide the raw materials from which hybrid bees are derived.

The German black bees were first brought across the Atlantic about 1630 by the early American colonists to pollinate the newly flowering orchards of imported fruits. In 1859, the first Italian queens were imported to America. This variety was quickly recognized as superior to the German, which is aggressive, with a short tongue and low resistance to bee diseases. All these negative factors led to the diminished use of the German black, and now few beekeepers in North America have these bees.

Today, the Italian honey bee is the most widely distributed bee in the Western Hemisphere. The other two popular races, the Carniolan and the Caucasian, have been subsequently brought to the United States and, with the Italians, are crossbred, interbred, and inbred for disease resistance, hardiness, and gentleness.

Importation of live adult bees into the United States was halted in 1922 because of the danger of introducing bee diseases and pests (especially the then newly discovered tracheal mite) that did not already exist here. Unfortunately, South American countries did not have such restrictions, and destructive pests are now present the New World (see "Mites" in Chapter 13).

Brazil imported African honey bees (*Apis mellifera scutellata* Lepeletier 1836) to improve breeding stock in the late 1950s. The accidental release of the volatile bee known as the *Africanized* honey bee (labeled the "killer bee" by the press) has led to its spread throughout South America north of central Argentina, Central America, and Mexico. In 1990, swarms of the Africanized bee crossed the border and have now become established in Texas, New Mexico, Arizona, and California. How many states will ultimately be occupied by these bees is presently disputed, but there is general consensus that the lower half of the United States will have to deal with them, as either year-round or summer residents (see "Africanized Bees" in Chapter 13).

Recently, the spread of the Africanized honey bee has been moderated by the parasitic bee mites, tracheal and varroa. Keep current with the bee journals for new information.

Although the most common honey bee in America is the Italian, you may be interested in experimenting with some other bee races. If you raise your own queens, uncontrolled crossbreeding could result in inferior queens. A general overview of the commonly available races of honey bees now used in the United States is given below. Good references are Graham, *The Hive and the Honey Bee,* Ruttner's book, and the January 1994 article "Races" in *Bee Culture* magazine, whose drawings by M. Yatko are reproduced below.

The Italian honey bee (*A. mellifera ligustica* Spinola 1806) originated in the Apennine Peninsula (the boot) of Italy. This race, introduced into the United States circa 1859, has several color types. Generally yellow with dark brown bands on the abdomen, the "goldens" have five bands, whereas the "leathers" have three. They are known for laying a solid brood pattern, producing lots of bees late into the fall, and making a good surplus of honey. On the other hand, they forage for shorter distances and therefore tend to rob nearby colonies. They also drift frequently because they orient by color rather than by object placement.

Italian Honey Bee

Advantages:

- Good, compact brood pattern, resulting in a strong workforce for collecting a good deal of nectar and pollen.
- Excellent foragers.
- Light color, making the queen easy to locate.
- Moderate tendency to swarm.
- Moderate propolizers, so hive furniture is not glued together too much.
- Resistant to European foulbrood disease.
- Relatively gentle and calm, thus easy to work.
- Moderate to high cleaning behavior.
- Common and easy to obtain.
- Readily builds comb; white cappings common.

Disadvantages:

- Can build much brace and burr comb; Italians have a slightly smaller cell size.
- Poor orientation to home hive; drift to other hives, spreading diseases/pests and causing uneven colony populations.
- Can be bothersome by persistently flying at beekeeper when worked.
- Short-distance foragers, thus have a tendency to rob weaker hives, creating a robbing frenzy in the apiary.
- Can be susceptible to many diseases and pests.
- Slow to build populations in spring; not good for early honeyflows.
- Brood rearing continues after main honeyflow has ceased, sometimes late into fall; bees may enter the winter period with too much brood and too little honey, which can result in starvation.

The Carniolan honey bee (*A. mellifera carnica* Pollmann 1879) originally came from Yugoslavia and Austria, where the winters were cold and the honeyflows variable. They were introduced to the United States circa 1883 and are popular in northern areas. Although they are a variety of Italian bees, Carniolans have a grayish-black-brown body with light hairs; the drones and queens are dark in color. In general, they were bred for fast buildup when the spring flow started and to shut down brood production early in the fall. They are known for their gentle disposition and low propolis and brace comb production but can swarm if not given ample expansion room. Recently, a Carniolan strain from Yugoslavia, called the Yugo bee (ARS Y-C-1), was introduced by United States Department of Agriculture (USDA) Bee Labs and is being screened for mite resistance. Some breeders are now selling Yugo queens; read the current literature to see if these bees live up to their reputation.

Advantages:

- Rapid population buildup in early spring; good for spring pollination and early nectar flows.
- Brood rearing decreases if available forage is diminished, thus conserving honey stores.
- Exceptionally gentle; less prone to sting and easier to work.
- Few brood diseases; less medication may be needed.
- Economic honey consumers, therefore overwinter on smaller honey/pollen stores.
- Little robbing instinct, as they are long-distance foragers and are object oriented.
- Can have very white wax cappings, making comb honey sections attractive to customers.
- Little brace comb and propolis, making hive manipulations easier.
- Overwinter well; queen stops laying in fall, and a small number of bees overwinter on fewer stores.
- By comparison with other races, forages earlier in the morning, on cool, wet days, and later in the evening.

Disadvantages:

- Tend to swarm unless given enough room.
- Strong broodnest depends on ample supply of

Carniolan Honey Bee

pollen; can be slow to build up in summer if pollen is not available.

- Dark queen is difficult to locate, making requeening operations slower.

The Caucasian honey bee (*A. mellifera caucasca* Gorbacher 1916) is originally from the high valleys of the central Caucasus near the Black Sea, where the climate ranges from humid subtropics to cool temperate zones. Caucasian bees are black with gray or brown spots and short gray hairs. Their tongue is the longest of the European races, which could make them superior pollinators of some crops. The drones are dark, with dark hairs on the thorax. These bees were introduced into the United States from Russia circa 1905. In general, they are gentle bees, with low swarming instincts, and good in areas of marginal forage or long honeyflows.

Advantages:

- Build strong populations, but slow to start in the spring; not good for early spring crops.
- Gentle and calm on comb, thus easier to work.
- Long tongue; can exploit more species of flowers.
- Forages at lower temperatures, earlier in the day, and on cool, wet days.
- Overwinter well, shutting down brood production in the fall, conserving stores.

Disadvantages:

- Maximum propolizers; make hive manipulations difficult unless collecting propolis for sale.
- Can have wet wax cappings over honey, thus comb honey less attractive to customers.
- Can sting persistently when aroused, making inspections difficult.
- Late starters in spring brood rearing; not good for early spring pollination.
- Dark queen is difficult to find.
- Can drift and rob.
- More susceptible to nosema disease; may require more medication.

Caucasian Honey Bee

Buckfast Hybrid Honey Bee

Hybrid Bees

In addition to these races, there are hybrid bees, which can be crosses between the races or between selected strains within a race. Some common hybrids are Starline (inbred Italians) and Buckfast. Many queen breeders have their own variations of these races, as well, to meet the needs of their customers. Before deciding on which race is best for you, call some bee breeders, talk to beekeepers, and start slowly. Keep hives of different races in separate apiaries, and remember, sometimes crossing hybrids can result in inferior bees.

The Buckfast Hybrid was produced by Brother Adam (1898–1996) of the Buckfast Abbey in the United Kingdom. He crossed many races of bees (primarily Anatolians with Italians and Carniolans) in search of a superior breed that would be tracheal mite tolerant, gentle, and productive; have high cleaning instincts and disease resistance, but would also possess good overwintering abilities.

Advantages:
- Tolerant of tracheal mites.
- Low swarming instinct.
- Minimal propolizers.
- Shuts off brood early in fall; overwinters with few bees on minimal stores.
- Gentle and calm.
- Disease resistant, including to chalkbrood.

Disadvantages:
- Build slowly in spring unless good flow is in progress; not good for early spring pollination.
- Offspring queens from hybrid mother may not be like the original queen; daughter queens or their progeny may not have desired characteristics.
- Requeening every other year may be necessary to ensure that a colony is headed by a Buckfast queen and has not been superseded.

The Starline Hybrid is a closed-population hybrid combining many Italian lines. They are known for their gentle behavior and large brood numbers, providing a large workforce to exploit many nectar resources. They are produced to perform well in a commercial setting.

Starline Hybrid Honey Bee

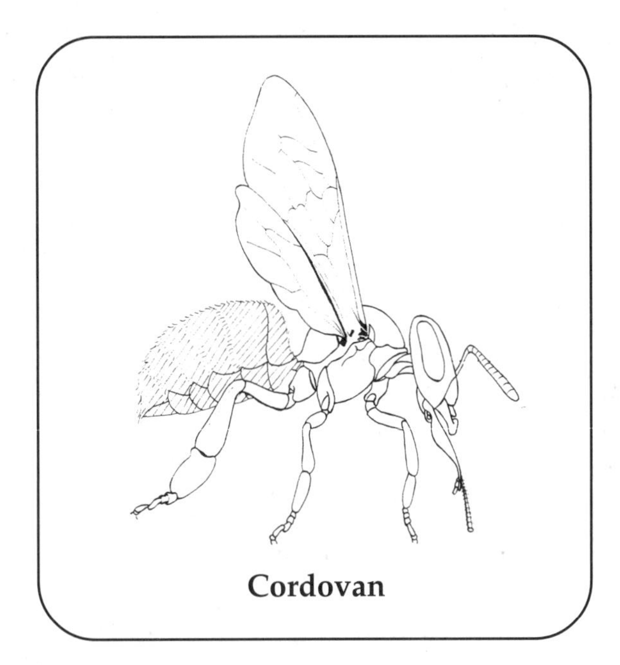

Cordovan

Advantages:
- Lots of brood, high bee populations; excellent honey producers under the right conditions.
- Fast spring buildup.
- Minimal propolizers.

Disadvantages:
- Can starve over winter because of large populations late in the fall; many commercial beekeepers overwinter their best hives in the southern states.
- Offspring queens from hybrid mother may bear little resemblance to the original queen.
- Requeening every year may be necessary.

The Italian Cordovans are a color mutation, originally produced by Dr. Bud Cale to serve as genetic markers in his Starline queen breeding program. Currently used to trace behavior and kinship relationships for research purposes, they are now also bred for hardiness, production, and disease resistance, primarily by California bee breeders. They are easily identified, because the black body color comes out red in Italians or a purple bronze in Caucasian and Carniolan bees. This latter bee is called a Purple Cordovan and is not commercially reared. Cordovans are known for their gentle behavior and pretty color—excellent for showcase observation hives.

A final note: These races evolved to take advantage of conditions in their geographic area of origin. It may not be possible to duplicate these conditions. But if the climate, timing of major honey flowers, and other factors in your area are similar to those of a particular race's homeland, try them. When trying some new races, go slowly (three to eight colonies) and observe how they compare with one another (queens vary within a race) and with the other races in your apiary.

External Structure of a Bee

The honey bee has three main body parts: a *head*, a *thorax*, and an *abdomen* (see the figure on external anatomy). Located on the head are five eyes, the antennae, and such feeding structures as the tongue (*proboscis*) and the jaws (*mandibles*). The proboscis is for lapping and sucking fluids (such as water, nectar, and honey), and the mandibles are used for manipulating pollen and shaping the beeswax.

The thorax, or middle section of the bee, contains the muscles that control the two pairs of wings; other muscles control the three pairs of legs. The legs have specialized structures and hairs on them that assist the bee in cleaning itself and in collecting and carrying pollen. The armor-plated thorax is perforated with pairs of holes called *spiracles*, which are part of the breathing or respiratory system. The first pair, called the *prothoracic spiracles*, is the site where tracheal mites are found (see the section on tracheal mites in Chapter 13). The second pair of spiracles is

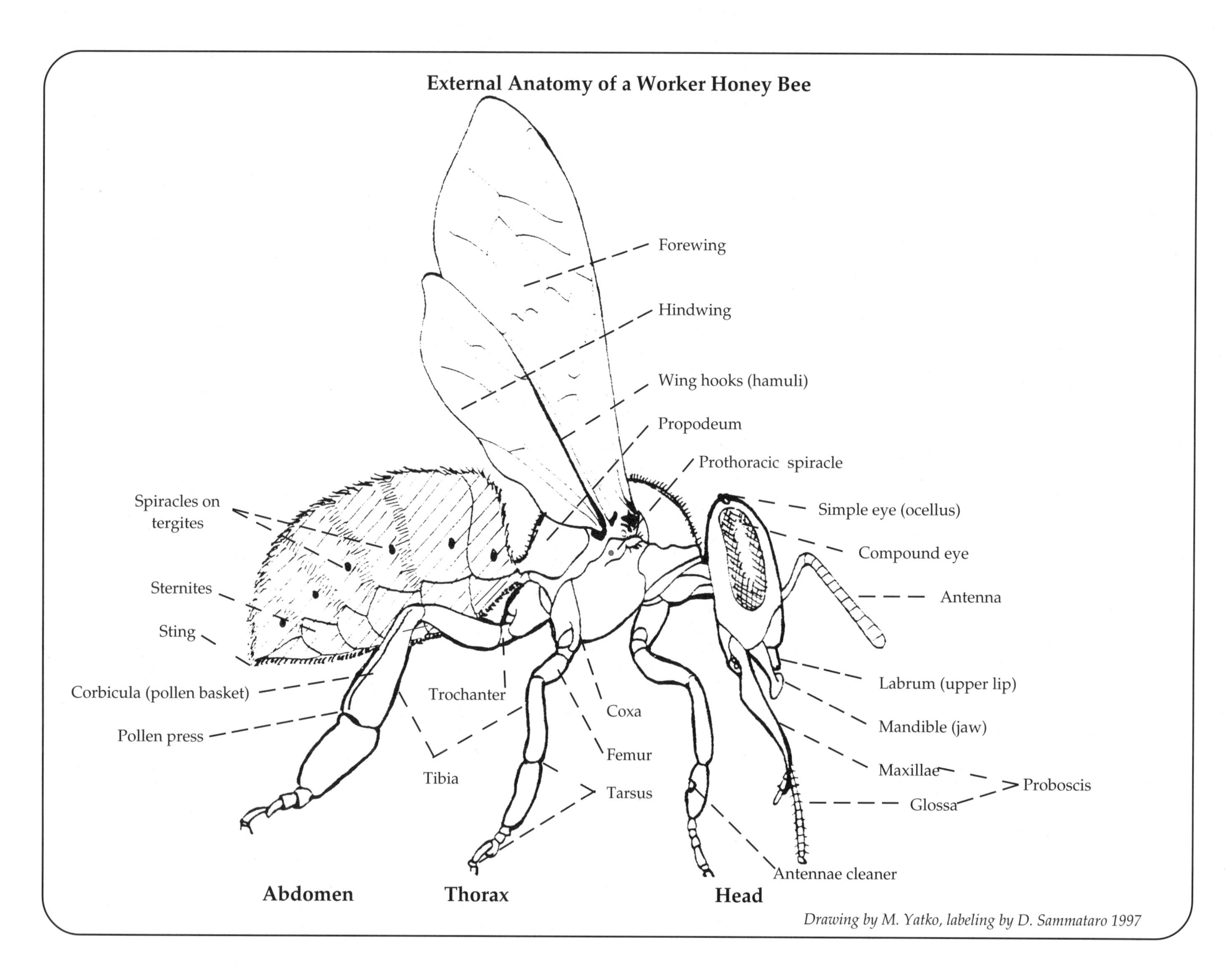
External Anatomy of a Worker Honey Bee
Forewing
Hindwing
Wing hooks (hamuli)
Propodeum
Prothoracic spiracle
Spiracles on tergites
Simple eye (ocellus)
Compound eye
Antenna
Sternites
Sting
Corbicula (pollen basket)
Trochanter
Labrum (upper lip)
Coxa
Mandible (jaw)
Pollen press
Femur
Maxillae
Proboscis
Tibia
Tarsus
Glossa
Antennae cleaner
Abdomen
Thorax
Head
Drawing by M. Yatko, labeling by D. Sammataro 1997

nonfunctional, and the last pair, although located on the thorax, is really on the first abdominal segment, called the *propodeum.*

The abdomen is the longest part of the bee and contains important organs. It is armor-plated with hard, scalelike segments called *tergites* (top segments) and *sternites* (bottom segments) that protect the bee and keep it from drying out. The abdomen is also perforated with seven more pairs of spiracles. The bee's sting, found only on the female castes, is located on the tip of the abdomen. Wax-secreting glands, on the underside of the abdomen, and the scent gland, just above the sting, are important abdominal glands in worker bees.

The queen's abdomen contains ovaries for egg production, a storage sac for drone semen, many glands that produce pheromones and a sting, but no wax glands (see Appendix A for more information on internal anatomy and Appendix B on pheromones). The drone's abdomen contains the male reproductive organs but has no wax glands and no sting. Sometimes a drone can be found with both male and female parts; these rare *gynandromorphs* may actually be able to sting you!

Bee Vision

Bees have five eyes—three simple (ocelli) and two compound. The compound eyes are composed of thousands of individual light-sensitive cells called *ommatidia.* It is with the compound eyes that bees perceive color, light, and directional information from the sun's ultraviolet rays.

The color range of bee vision has been shown to include the violet, blue, blue-green, yellow, and orange colors as well as ultraviolet light, which is invisible to humans. Because they compete with one another for available pollinators, flowers that depend on bee pollination are within these color ranges. Those plants which succeeded in attracting bees with their color, nectar, and pollen gained an edge over other plants during their evolutionary development.

The structures and arrangement of the ommatidia permit polarized light to pass through certain parts of each ommatidium at any given instance. The sun's position and the bee's direction are the factors determining which section of the ommatidia will receive full, partial, or no light, the effect of which on the compound eye is a pattern of light, dark, or shaded regions. This pattern serves as a "compass" to the bee, giving directional information. The bee is able to monitor these shifting patterns continually as it flies and, if necessary, adjust its course.

Pollen-Collecting Structures

The hind legs of worker bees are specialized for collecting and carrying pollen. An inner segment of the hind leg is covered with numerous hairs, forming the pollen combs. Bees actively collect pollen by scraping it off flowers with their jaws and legs; as the pollen is removed, a small amount of nectar is added to make it sticky. Additional pollen adheres to the bee's body by static electricity as it is being collected. The collected pollen is then transferred by the bee to areas on its body where it can be reached and removed by the pollen combs.

Removal of the pollen from the pollen combs is accomplished by rubbing the legs together so that the pollen is squeezed from the inside to the outside of the legs. The pollen will eventually be deposited into a depression called the *pollen basket.* When the baskets are full, the bee returns to the hive, backs into a cell, and deposits the pollen pellets. The hive bees will pack in the pollen solidly, ultimately capping it with honey for winter stores.

The Sting

Stinging insects, or aculates, belong to the order Hymenoptera, which includes both social and solitary bees and wasps. The more aggressive species of stinging insects are the hornets and the yellow jackets (both of the Vespidae family); less aggressive are the bumble bees (Bombidae) and the honey bees (Apidae).

The venoms of these stinging insects are not chemically alike. Thus a beekeeper who is allergic to yellow jacket venom will not necessarily develop an allergy to honey bee venom or to the venom of other stinging insects. The stinging mechanism is a modification of the egg-laying organ (*ovipositor*) of female insects. Queens generally use their stings only to dispatch rival queens. The entire stinging apparatus consists of a poison sac (sometimes called the *acid gland*), an alkali (or Dufour) gland, associated alarm substances, and the mechanical equipment (muscles and hardened plates) of the sting (see Appendix C for information on sting reaction and anatomy).

The recurved barbs of the sting's two lancets catch in the victim's skin, and as the bee pulls away, the entire sting structure, including the venom sac, is ripped out of the bee's body. Muscle pumps near the base of this now-detached sac force more venom into the wound for about a minute. The alarm odors are released at the sting site, inducing other workers to sting. To minimize the amount of venom received, it is important to remove the sting *promptly* by scraping or flicking it off with your fingernail. The sting also releases alarm odors, which have the effect of tagging the victim. Apply smoke to any sting site to mask the alarm odors. Bees usually die shortly after stinging but may occasionally live for hours or even days (for more information on the compounds in bee venom, see "Bee Venom" in Chapter 12).

The Worker

There are three types of bees in a colony, divided into the female castes (workers and queens) and the

Average Development Time for a European Honey Bee

	Egg[a]	Larva[a]	Pupa[a]	Total	Adult life span	Weight
Queen	(fertilized) 3 days	4.6 days	7.5 days	15–17 days	2–5 years	178–292 mg[c]
Worker	(fertilized) 3 days	6.0 days	12.0 days	19–22 days	15–38 days summer 140–320 days winter[b]	81–151 mg
Drone	(unfertilized) 3 days	6.3 days	14.5 days	24–25 days	4–8 weeks	196–225 mg

[a] Average time between metamorphic stages, at 93°F (33.9°C).
[b] Workers in winter have well-developed hypopharyngeal glands and more fat bodies, which may enable them to live longer. Egg dimensions: 1.3–1.8 mm long; 48–144 hours to hatch, average 72 hours.
[c] Conversion: 1 mg = 0.000035 ounce, 1 mm = 0.004 inch. Weights of emerging adults vary depending on cell size, number of nurse bees, colony population, food availability and type, and season of the year.

males (drones) (see Appendix A for information on morphology of the different castes). The most numerous members of a bee colony are the workers, a sterile female caste incapable of laying fertile eggs; in a normal hive they reach a peak population of 40,000 or more by midsummer. The workers are smaller than the drones and have shorter abdomens than the queen.

Life Stages of a Bee

Under normal circumstances, the queen lays all the eggs in a hive. If the queen is lost and the bees are unable to rear a new one, workers will sometimes lay unfertilized eggs. Workers and queen bees hatch from fertilized eggs, and drones, from unfertilized eggs; a *fertilized* egg is one formed by the union of a sperm and an egg. The egg (whether fertilized or not) consists of a nucleus and a large yolk reservoir. As the nucleus divides to form cells, it is the yolk that provides the building blocks for the new cells. These rapidly multiplying cells eventually form a layer or *blastoderm* on the inner surface. As the lower layer of cells thickens, they become the *germ band,* which marks the beginning of the embryo. As the embryo grows, it will gradually differentiate into the various organs and tissues that make up a bee.

The egg is incubated in the nursery region (called the *broodnest*) of the comb at 91.4–96.8°F (33–36°C), where the embryo will develop through four stages: *egg, larva, pupa,* and *adult.* This kind of development is called *complete metamorphosis* and is shared by most common insects, such as the butterfly, beetle, and fly.

Eggs lose about 30 percent of their weight during incubation and after 48 to 144 hours, temperature dependent, hatch into larvae. All honey bee eggs hatch not by rupturing the shell (*chorion*), as in most insects, but by gradual dissolution of the membrane during hatching, a characteristic unique to honey bees. Genetics and race of bees will dictate how well the egg hatches and the brood survives. It is essential for bees to have survival strategies that involve queens laying a large number of eggs per day as well as colony numbers sufficient to incubate those eggs, feed the hatching larvae, and build adequate space for the new population of bees.

When death occurs during any of these metamorphic stages, adult bees will clean the dead material, sometimes removing it or eating it, the latter especially during a time of dearth. In either case, they are performing a crucial hygienic function, which is an important genetic trait.

Once hatched, the larva (a white, wormlike grub) becomes an eating machine, with a huge digestion system consisting of a mouth, spiracles, midgut, hindgut, salivary and silk glands, and (closed) excretory tubes. You can see larvae in their cells, a white C-shaped worm lying in the bottom of the cell (see the table detailing development time and the figure on developmental stages). Each larva is fed between 150 and 800 times per day and will gain around 900 times the egg weight by the fifth day. About 33 percent of the larval weight (dry weight) is made up of fat bodies, organs that are utilized in the pupal stage. To grow this fast, the larva has to molt six times because the skin is unable to expand sufficiently to accommodate the rapidly growing insect. Four molts take place during the first four days, a fifth one as a prepupa, and the final one just before the bee emerges.

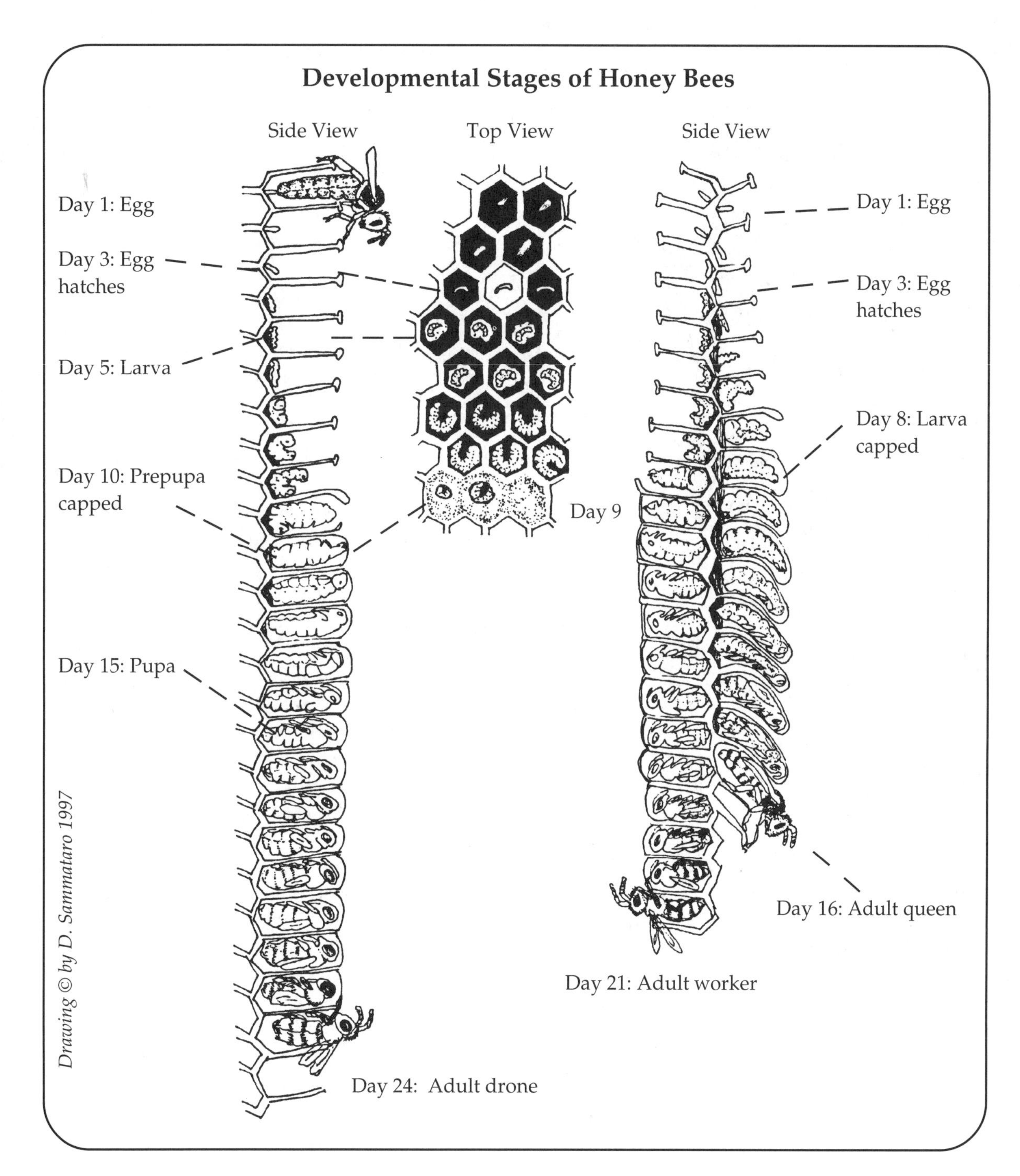

Two different diets are fed to larvae destined to become worker bees. First, the larvae are lavishly or mass-fed a diet of brood food (sometimes called *worker jelly*), which consists of 60–80 percent *clear fluid* produced by the hypopharyngeal glands of nurse bees and 20–40 percent *milky fluid* produced by their mandibular glands. On the third day, the diet is switched to the clear component only, then to pollen and honey on the remaining days. At this time, the larvae are fed progressively, or only as needed. This diet change, which contains fewer proteins, lipids, minerals, vitamins, and sugars, and the switch from mass to progressive feeding, appears to be responsible for the differentiation into worker bees. Other larvae, hatching from fertilized eggs and mass-fed only royal jelly, develop into queen bees.

At the end of the fifth day, the cell, which until this time is called *uncapped* or *open brood,* is capped with a waxlike cover and is now called *capped* or *sealed brood.* The larva molts into a prepupa, defecates, and spins a cocoon with silk produced from the thoracic salivary glands. Hive temperatures of about 95°F (35°C) are necessary for normal development; if lower, the development time can be delayed by several days.

During the next stage, called the pupal stage, massive internal and external morphological changes occur. Recognizable parts of the bee form—the legs, wings, and abdomen—and all the internal organs and muscles develop. Most internal changes happen in eight to nine days. The pupae are full of fat bodies carried over from the larval stage. Fat bodies are cell-like organs that serve as food storage reserves of lipids (waxes, oils, fatty acids, and steroids in cells) and glycogen (a stored form of glucose). Fat bodies also provide essential compounds called amino acids, which assist in hardening the cuticle of young bees as well as helping to synthesize proteins (long-chained molecules that are the foundation of all living organisms). In addition, they contain mitochondria (small organs that capture and convert food energy) and enzymes (proteins that act as catalysts), such as lipase (an enzyme that digests fat for energy).

The pupal skin, or *cuticle,* gradually darkens, and after a final molt, the adult is ready to emerge. Because the cuticle is so soft, this bee, called a teneral bee, stays inside her cell three to four hours to harden before emerging. At the end of the twelfth day, the emerging bee chews away the cell cap, which is reused by other bees for other brood cappings. Teneral bees can't sting and are relatively soft, with thoracic hairs that are light in color and matted down. She soon begins the first of many tasks she will perform during her life span. Over the next few days, glandular development and environmental conditions rule bee activities (see Chapter 2).

The young bees are still full of fat bodies and must ingest pollen proteins after the first few hours of emergence. Without protein, their life span will be shorter, and glandular development will be impaired. They will need protein until they are five days old, and will also beg nutritious brood food from other nurse bees.

The worker bee's age and the needs of the colony dictate the work she is to do for the rest of her life. Generally, workers from one to three weeks of age remain within the hive, where they:

- Rest
- Feed and clean larvae and their cells
- Tend the queen (feed, groom, help spread queen pheromones)
- Clean the cells and the hive
- Build new comb and cap cells containing honey, pollen, and brood
- Guard the entrance and other areas of the hive
- Patrol the hive, looking for intruders
- Help to heat or cool the hive as needed
- Accept nectar from foragers, storing and curing it
- Pack pollen
- Take brief orientation flights to familiarize themselves with landmarks near the hive (also called "play flights")

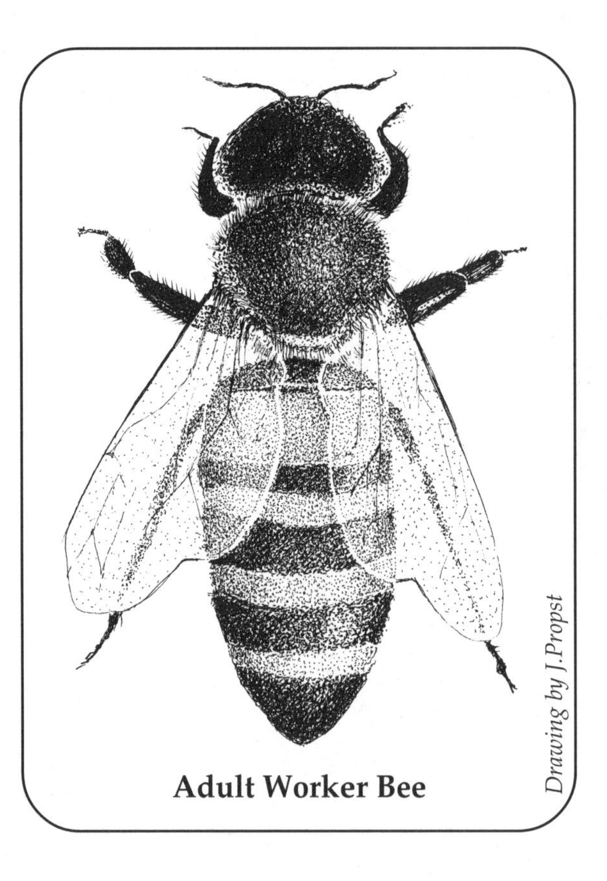

Adult Worker Bee

After about three weeks of hive duties, the glands that produce the larval food and wax have begun to atrophy. These workers then move away from the warm broodnest (where the eggs, larvae, and pupae are) onto broodless combs. Here they come in contact with returning foragers and are eventually recruited to food sources.

As foragers, they will usually collect honeydew, pollen, nectar, water, or propolis. Foraging activities take a heavy toll on workers, most of whom die after about three weeks of outside duties. During the winter, however, many workers survive for several months. For a complete breakdown of worker activities, see Chapter 2.

The Queen

Bee colonies are usually monogynous—that is, they have only one egg producer, the queen. The queen is the longest bee in the colony; her wasplike, slender abdomen, usually without color bands, distinguishes her from both workers and drones. Any larva that hatches from a fertilized egg is a potential queen. Thus, worker bees can raise a new queen from larvae up to three days old, either when their old queen has been accidentally lost or when she is injured or too old to perform her duties.

The ability to find the queen is important if her presence in the colony needs to be confirmed or if you wish to replace her. Requeening, or replacing an old queen with a new one, is successfully accomplished when the existing queen is located and removed from the colony. New beekeepers need to gain the facility to find the queen among the workers and drones; once you have accomplished this, you are truly a beekeeper.

Any larva hatching from a fertilized egg is a female bee. This fact simplifies the raising of queens for commercial purposes and gives worker bees a wide latitude in selecting larvae to become new queens (see Chapter 10).

The pathway the female larvae follow is directly connected to the food they receive during their larval life. Worker bees often initiate queen rearing by constructing special cup-shaped cells. These *queen cups* are usually located on the lower edges of combs; after the queen has deposited an egg in them, they are then called *queen cells.* The presence of cups or larvae in queen cells does not necessarily lead to the production of queens.

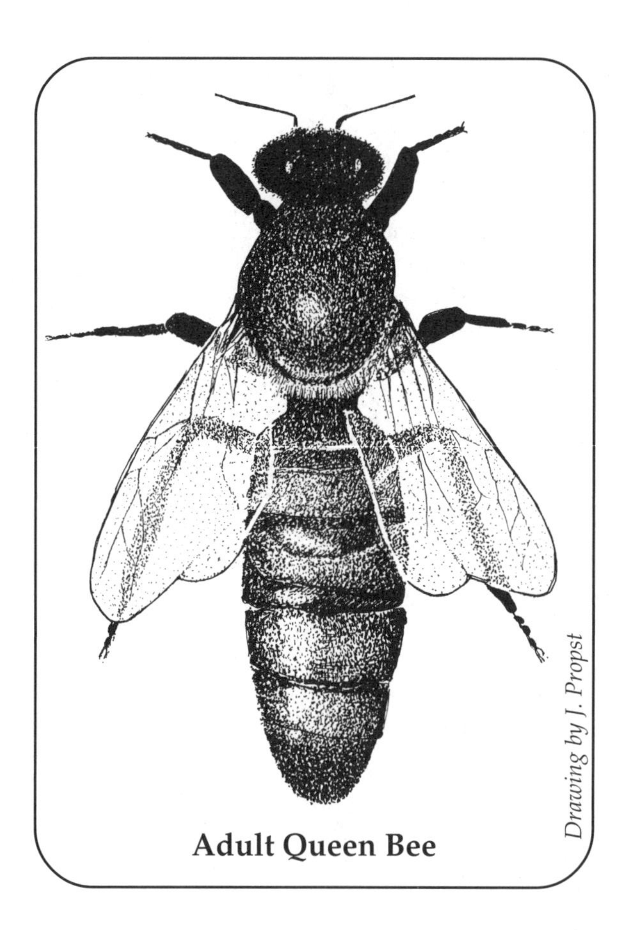

Adult Queen Bee

When queen cells are needed, workers clean the cups or worker cells are modified to become queen cells. Larvae in these cells or cups are mass-fed *royal jelly* during their entire larval development. The royal diet contains several components. For the first three days it consists of white mandibular gland secretions, after which it is fed in copious amounts as a 1:1 ratio of the white to clear components of the hypopharyngeal glands, similar to worker jelly. On the last two days of larval life there is an important addition—honey. This diet shift, with the elevated sugar content and the addition of a hormone called *juvenile hormone,* produces queen bees.

Whether cells containing queen larvae begin as cups or as worker cells, as the larvae grow, these cells are enlarged and elongated by worker bees, gradually taking on a peanutlike appearance. The openings of drone and worker cells lie horizontally but are inclined slightly upward on the comb. Cells that cradle the queens hang vertically (see the figure showing relative cell sizes).

Isn't it interesting that young worker bees play such an important role in a colony by selecting the next generation of queens? Remember, female larvae selected to be workers begin larval life on a diet similar to that of queen larvae, but after two days they are weaned from it and thereafter receive worker jelly, which consists primarily of proteins mixed with honey and pollen.

Queen Cell Production

Three conditions trigger queen rearing by honey bees: (1) the colony is making preparations for swarming, (2) the queen's physiological and behavioral activities are substandard, and (3) the queen is lost or dies. In each case, the purpose is to replace the existing or resident queen in the colony.

Relative Cell Sizes

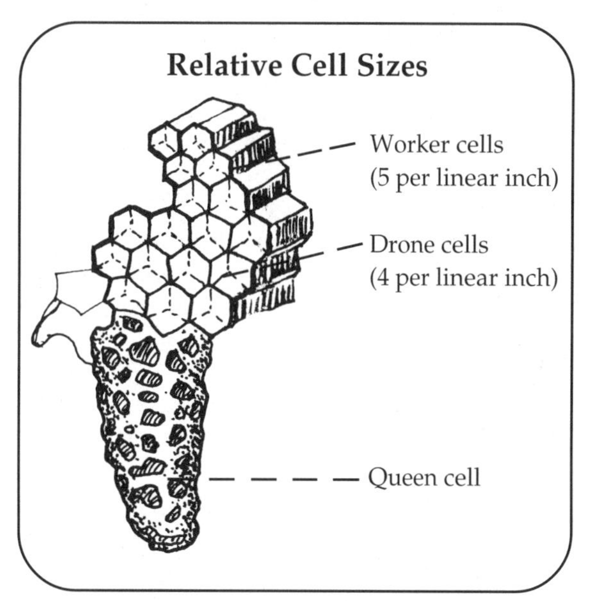

Swarming is a process whereby the existing queen will depart the colony with about half of the workers and a few hundred drones to form a new colony. Colonies preparing to swarm will begin this process by constructing a great number (from 10 to 40) of queen cups on the lower edge of combs. Newly constructed cups are light yellow in color and hang vertically from the lower edges of the honeycomb. These cups become *queen cells* once the queen deposits eggs in them. If found during the swarming season—a period when colonies are casting swarms—they are called *swarm cells.*

Another condition that leads to queen cell construction without swarming occurs when bees prepare to replace a queen that is substandard, which is called *supersedure.* Workers begin this process by either constructing queen cups or modifying existing worker cells containing young larvae. In this case, these supersedure cells are few in number and can be found throughout the brood on the face of the comb. Because the cups are constructed of old wax, these cells will be brown in color (see Chapter 10).

This replacement is triggered when either the queen's physiological or her behavioral activities or both decline—for example, her egg production slows down, her pheromone levels are reduced (usually for an aging queen), or she is injured. Worker bees are able to recognize these conditions and will rear queens to replace the resident one. After the daughter queen hatches, mates, and begins to lay eggs, she may often coexist with her ailing mother. In time, however, only the replacement queen will be found.

The last condition of queen replacement, or *emergency,* occurs when the queen is absent from the colony. This can be from natural causes, beekeeper error, or predation. Occasionally, the queen will fall off the comb during hive inspection and is unable to return to the hive or is crushed between two frames as they are being removed or replaced (a process called *rolling the queen*). In such cases, unless by good fortune there are queen cells already present, bees must turn worker cells into queen cells. Worker bees will "select" young larvae in worker cells and modify the cells into queen cells; emergency cells are found on the face of the comb, but in small numbers.

Virgin Queens

While still in the queen cells, virgin queens will often *pipe* or *quack* to one another. After emerging, a virgin queen may *toot,* or call to other virgins. She then begins to search for and partially destroy any other queen cells, leaving the workers to discard the pupae or larvae inside. Some cells may contain queens ready to emerge, in which case she will partially open these cells and sting the occupants. While performing these tasks, she may also encounter other emerged queens; fighting ensues, and ultimately only one virgin queen survives.

About six days after emerging, the queen will leave the hive on a mating flight; if weather is inclement, this flight will be delayed until more favorable weather appears. During her flight, the queen's pheromones attract male bees from drone-congregating areas, and she may mate with up to ten or more drones in succession, over a few days. When her sperm sac (*spermatheca*) is filled, she will never leave the colony again, unless accompanying a swarm (see Appendix A for information on the internal organs of the queen). Three days or so after mating, the now bigger and heavier queen will begin to lay eggs. The queen continues to lay eggs the rest of her life, pausing for a month or so late each fall. It has been reported that a good queen is able to lay over one thousand eggs a day for brief periods, provided there is enough space and worker bees to incubate and care for these eggs. Over her lifetime, a good queen will lay 200,000 eggs per year.

Genetic Traits

Because the queen mates in the open, the beekeeper has limited control over which drones will inseminate her. Those few that do mate with her may be from several apiaries or from "wild," or *feral,* colonies. However, since the appearance of parasitic bee mites, most feral colonies have now been killed off, drastically reducing the number of wild drones.

As a consequence of this random mating pattern, the queen's spermatheca may contain semen from genetically different drones. Her worker and queen progeny, therefore, will consist of individuals that are not necessarily genetically alike. The drones, hatching from unfertilized eggs (known as *parthenogenesis*), are all full brothers because the queen will lay genetically similar drone eggs whether or not she has been inseminated. Only when the queen has been instrumentally inseminated with semen from recorded drone stock (or from a single drone) will a colony's workers be of known origin.

Because the queen is the sole egg producer, she is responsible for the genetic traits of a colony. If a colony has undesirable traits, requeening should change the hive's genetic makeup and, therefore, its character.

Queens should be of superior stock to optimize desirable characteristics, such as:

- Color
- Temperament
- Industry (how early in the morning bees are out foraging)
- Production (how much honey is collected)
- Swarming tendency
- Winter hardiness
- Propolizing tendency
- Burr-comb building
- Nectar-carrying capacity
- Disease resistance
- Mite tolerance
- Longevity
- Cleanliness
- Total hive population
- Brood pattern
- Tongue length
- Pollen hoarding
- Honey hoarding
- Plant preferences
- Handling ease
- Whiteness of honey cappings
- Conservation of stores

The Drone

Because drones are larger, beginners often mistake them for the queens. They can be distinguished from queens if one knows that the abdomen of a queen tapers to a point, whereas the drone's is blunt or rounded, making him appear chunky in appearance. The number of drones per colony may be in the hundreds to thousands, usually about 15 percent of the total colony population.

Drone larvae hatch from unfertilized eggs, which under normal conditions are laid by a mated queen in hexagonal wax cells similar to, but larger than, worker cells (see the figure showing relative cell sizes). On the fourth day, drone larvae are fed a diet of modified worker jelly, which contains a larger quantity of pollen and honey.

After six and a half days of feeding, the cells of drone larvae are capped with wax. The capped drone cells are dome shaped, like a bullet's head,

Adult Drone Bee

and are readily distinguished from the slightly convex shape of the capped worker cells. Remember, capped cells lying on a horizontal plane are either worker or drone cells; those ultimately peanut shaped and suspended on a vertical plane are queen cells.

Newly emerged adult drones are fed by workers for two to three days and then will beg food from nurse bees, which contains a mixture of pollen, honey, and brood food. Older drones feed themselves from the honey stores. Adult drones have no sting (the sting is a modified female egg-laying structure) and have very short tongues (unsuitable for gathering nectar) (see Appendix A). Drones never collect food, secrete wax, or feed the young. Their sole known function is to mate with virgin or newly mated queens.

Drones first leave the hive about six days after emerging on a warm, windless, and sunny afternoon. As they get older, they fly to locations known as *drone congregating areas.* Whenever the drones in these areas detect the pheromones of a virgin or newly mated queen, they pursue her. A few succeed in mating with her, but those few that copulate die soon afterward.

Whenever there is a dearth of nectar (when no food is being collected), worker bees expel drone brood and adult drones from the colony. During the summer, you can see workers dragging drones in various stages of metamorphosis out of their cells and dropping them in front of the hive. Normally, in the fall, all adult drones and any remaining drone brood are gradually evicted from the hive. The evicted drones die of starvation and exposure. Queenless hives and those with laying workers or drone-laying or failing queens usually retain drones longer.

Drone Layers

An unmated queen can lay only unfertilized eggs. A failing queen is one that has mated but is no longer capable of normal egg-laying activity, laying all or nearly all unfertilized eggs. Failing queens may result from sperm deficiency, physiological impairment, disease, poor mating, mite infestation, or old age. Some workers of hopelessly queenless hives (unable to rear another queen) undergo ovary development and start to lay eggs. These eggs are, of course, all unfertilized. Unfertilized eggs laid by healthy, mated, unmated, or failing queens or by *laying workers* will produce mature drones, capable of mating.

Unlike a mated queen that lays unfertilized eggs in drone cells, a failing or an unmated queen will often deposit such eggs in workers cells. Laying workers also place their eggs in worker cells, but though these unfertilized eggs are laid in worker cells, they will hatch into drone larvae, and as they near the pupal stage, the cappings will have the characteristic dome shape found on regular drone cells. The presence of scattered worker cells with drone cappings indicates that the colony's egg-layer needs to be replaced.

On further inspection, you may find that each uncapped cell within a scattered brood pattern contains not one but several eggs. These eggs, instead of being deposited at the bottom of the cell as is characteristic of eggs laid by queens, adhere to the cell walls. This is a result of the worker's abdomen not being long enough to reach the cell bottom. If you find these patterns in your hive, read about what to do in "Laying Workers" in Chapter 11.

The presence of clusters of occupied drone cells in the spring, summer, and early fall in a *queenright* colony (where a healthy, mated queen is present) is a normal part of the colony cycle. Because drones attract varroa mites, many beekeepers use this fact to trap the mites. They add at least one frame of drone-sized comb in each hive body to attract female varroa mites to lay eggs. Once the drone cells are capped, the frame is frozen to kill the mites (see "Varroa Mite" in Chapter 13).

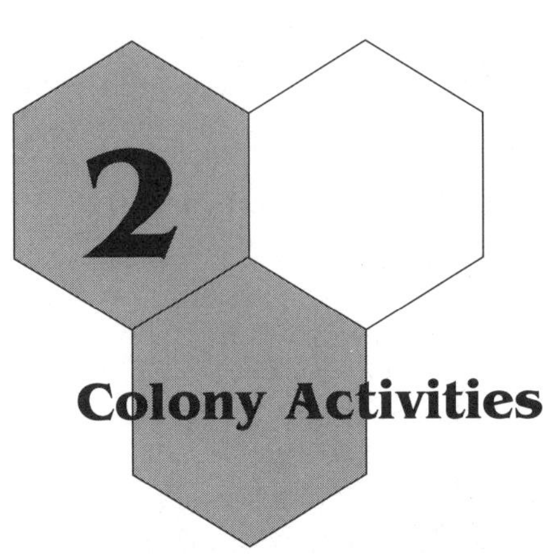

Colony Activities

As already discussed, the worker bees perform most of the tasks necessary to maintain the colony. Their activities can be divided into two categories: inside activities and outside activities.

Many factors affect how a colony behaves and the way a colony performs its duties. From reading the previous chapter, you should be able to distinguish the two female castes and the drones. Now we will talk about the worker caste, which is responsible for carrying out most of the chores associated with a bee colony (see the figure showing the cross section of a frame).

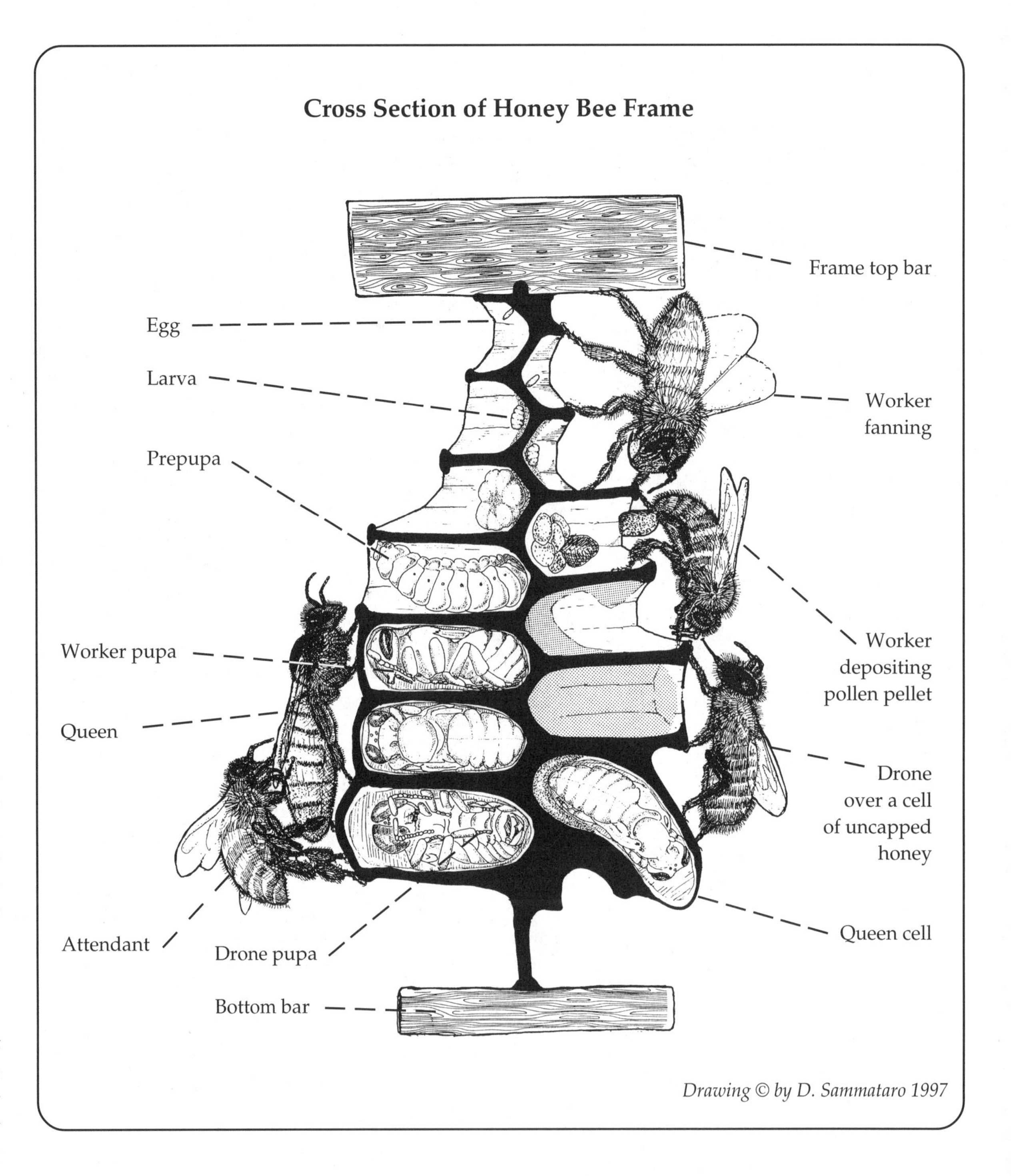

Drawing © by D. Sammataro 1997

Inside Activities

A functional knowledge of bee biology and behavior will enable you to understand and to some extent help you better manage your bee colonies. You should be able to recognize the different duties of the workers, how and when the worker bees perform them, and what to do if these duties are interrupted or are not carried out.

Division of Labor

Upon emerging, the young worker begins the first of many tasks she will perform during her life span. Over the next few days, the age of the bee, its glandular development, and the environmental conditions of the colony will rule bee activities. The organizational factors determining worker bee activities are called *age polyethism* and include nest cleaning, brood and queen tending and feeding, comb building, food handling, ventilation, guard duty, thermoregulation, nest homeostasis, and orientation and foraging flights.

Researchers long ago found a simple tool to observe polyethism in a colony. As soon as a teneral bee emerges, a dab of fast-drying paint is placed on her thorax, and the date of emergence is recorded. By using different colors on successive days and a glass observation hive, many bees can be identified by age and observed over their lifetime. As the bees age, there is a corresponding transition from one specific task to another. Under normal conditions, the full complement of tasks required for the survival of a colony are performed by bees of different ages.

When you open your colonies or observe bees in an observation hive, the ages of the various bees can be gauged by the tasks they are performing. A brief discussion of some of the more important duties is presented in this section. For a breakdown of colony work by age of bee, see also "Books on Bees and Beekeeping" in the References.

Nest Cleaning

Nest-cleaning activities include cleaning cells in preparation for egg deposition and keeping the nest free from debris and disease, as well as removing dead brood and under certain circumstances healthy brood when there is a nectar dearth or when the colony can no longer care for the brood. Another nest-cleaning activity is coating the interior hive parts and the entrance with *propolis,* a resinous substance collected from buds or the bark of trees and carried back in the pollen baskets. Often referred to as bee glue, it is a dark reddish to brown resin, sticky when warm, brittle when cold. It is used to strengthen the combs and to cover any foreign matter that cannot be removed (e.g., a dead mouse). Propolis makes it necessary to use a hive tool to pry apart the beehive.

Cell preparation is accomplished by very young workers, only a few hours old. These young bees remove nearby cocoon remains and larval feces from brood cells. The cleaned cells are then acceptable by the queen, who will lay eggs in them. Honey and pollen will also be placed in cleaned cells. Any remaining and uncleanable surface is covered with fresh wax or propolis.

Cleaning, or hygienic behavior, is a genetic trait, one that is desirable for beekeepers to perpetuate. For instance, the continual quick removal of dead brood from the cells and the bottom board is considered hygienic behavior. Colonies whose workers demonstrate good hygiene are more likely to be free from some diseases, and such behavior may also help reduce mite levels. For more information on testing for this behavior, see Chapter 10 and "Varroa Mite" in Chapter 13.

Older workers take on the task of keeping the rest of the hive clean. They can be seen:

- Removing dead or dying brood and adults from the hive. Some bee colonies also recognize varroa-killed brood and will remove it. These workers are called undertaker bees and make up about 1 percent of the worker population.
- Removing debris such as small pieces of grass and leaves as well as pieces of old comb and cappings.
- Removing granulated honey or dry sugar and moldy pollen.
- Coating the insides of the hive and wax cells with bee glue, or propolis.
- Propolizing cracks and movable hive parts, including frames, bottom board, and inner cover; some races use more propolis than others (see "Races of Bees" in Chapter 1).
- Removing healthy brood—usually drone brood and adults—when the colony is starving.

Tending to the Queen and Brood

Most of the young workers (3–16 days old) in the brood area feed uncapped larvae; these bees are often referred to as *nurse bees.* Nurse bees provide larvae with food secreted from two head glands: the hypopharyngeal and the mandibular glands. These glands reach maximum production in nurse bees. A single nurse bee will raise on average between two and three larvae during her life as a nurse. You can see these bees sticking their heads into cells for a few seconds to determine how much food is available and to feed the larvae as needed. The brood food is placed near the cell bottom, close to the larva's mouth.

In addition, these young nurse bees also feed the queen. About 6–10 workers, referred to as attendant bees, form a retinue or circle around the queen (see the illustration of the queen's retinue). Each attendant stays only from about one to three minutes and then departs. Another attendant takes her place, so the queen is always surrounded, except if she is moving quickly across the comb. The bees feed her brood food (royal jelly) directly into her mouth. The ingredients in and the amount of the food stimulate the queen's ovaries to keep producing eggs.

Another function of the attendants is to groom and clean the queen and contact her with their antennae and forelegs. In the process of these tactile activities, the attendants collect and distribute the queen pheromones. When the departing attendants contact other bees in the nest, they pass on the queen's scent adhering to their bodies, especially the antennae. On average, a departing attendant will touch 56 other nestmates in about 30 minutes with her antennae. This communicates not only that the queen is alive but also her condition (newly mated or old and failing), health, and activity. Within 24 hours, an entire colony knows if it is queenless or if the queen is failing or diseased; sometimes this takes much less time. Reduced queen pheromones may

trigger swarm preparation or queen supersedure; their absence can initiate emergency queen cell construction or the development of laying workers (see "Swarming" and "Queen Supersedure" in Chapter 11).

Comb Building

The wax comb (and surrounding cavity) is the nest and abode of the honey bee. In the wild, the comb is usually confined within a dark enclosure such as a hollow tree, cave, or opening in a house, although some nests can be found in the open. The wax for the nest is produced by workers (8–17 days old), whose wax glands are at their peak function. Wax fashioned into the hexagonal "honeycomb" cells serves multiple purposes. Such cells can be used for the egg deposition to become cradles for the young developing bees, or when free of eggs or

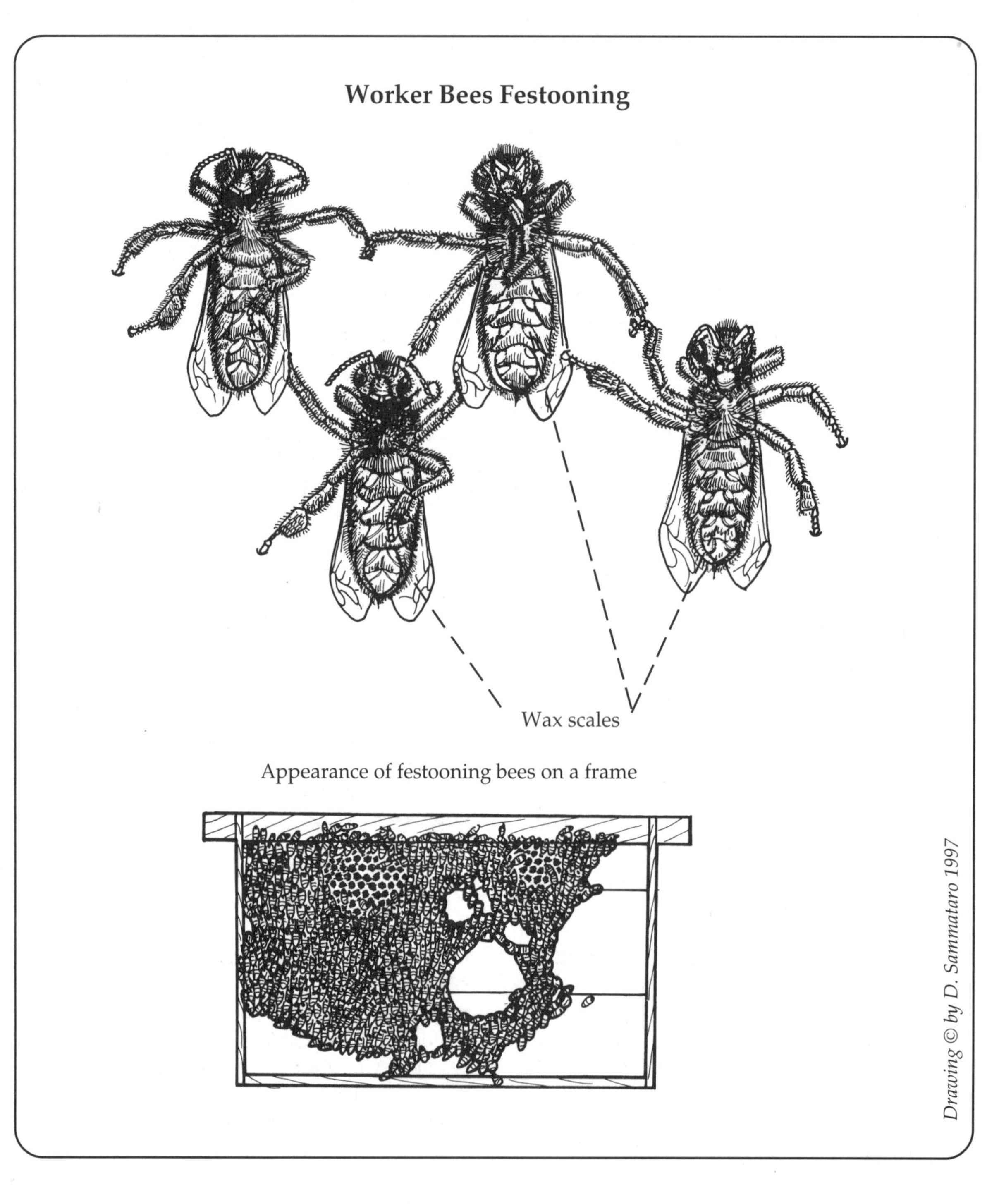

brood, they can be used for the storage of honey and pollen.

Wax fashioned into a shape that looks like an inverted cup is called a queen cup and is used to rear queens. After queens have been reared, bees usually remove these queen cells.

Beeswax, as already mentioned, is secreted from the wax glands located on the ventral or front side of the worker bees' abdomens (see the illustration of festooning and Appendix B). A wax droplet is secreted from beneath the overlapping portions of the last four abdominal segments; on contact with air, the wax hardens to a thin, oval scale, which has been described as resembling a small fish scale. The bee then removes this wax scale from the abdomen with its hind legs, passes it to the forelegs, and then to its mandibles, or jaws. The scale is then masticated, mixed with saliva, softened, and used to begin construction of or added to existing comb. If cappings are needed to cover honey cells, other workers, not secreting wax, can extend the tops of cells over the honey.

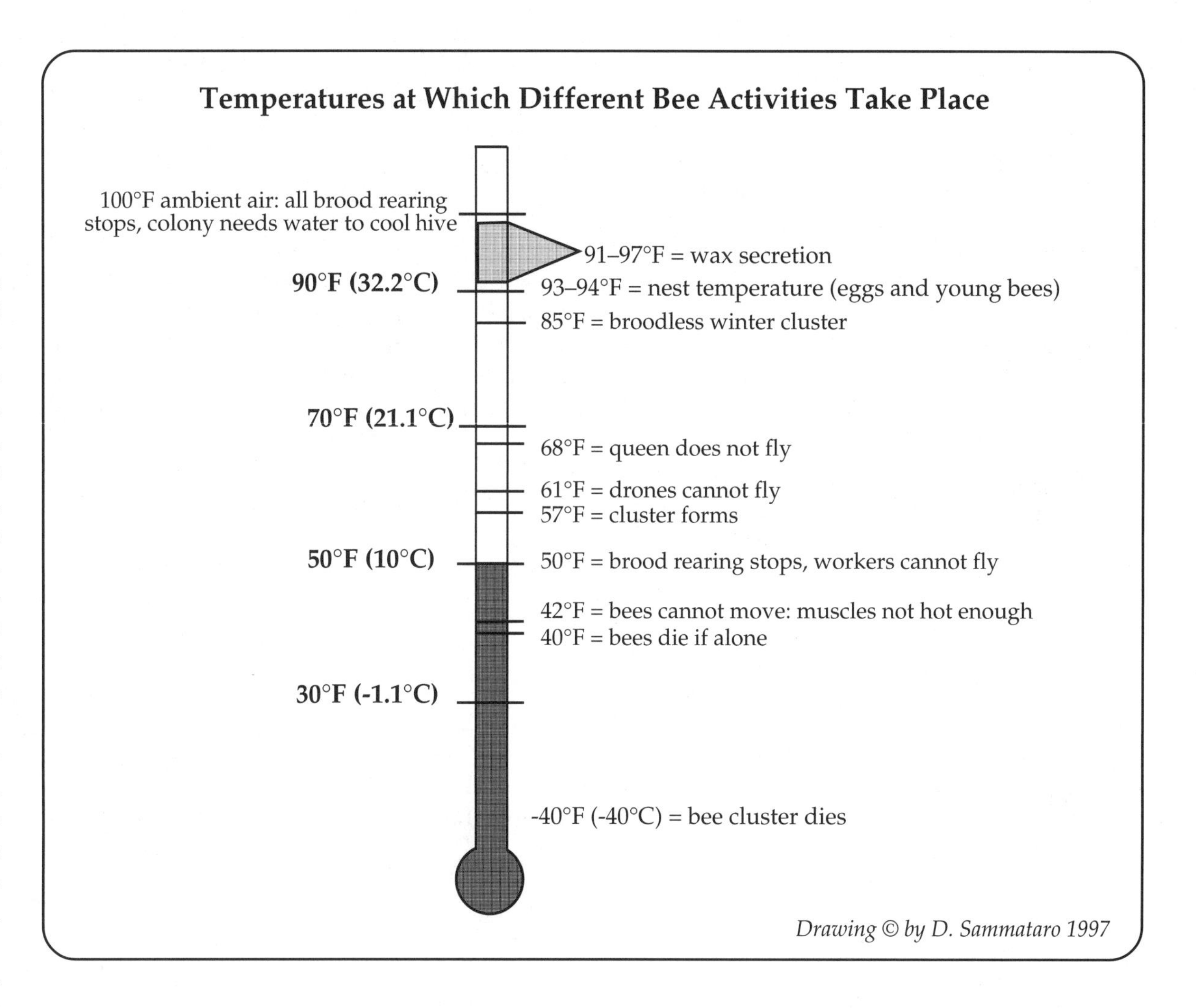

Drawing © by D. Sammataro 1997

The cells of the honeycomb do not lie on a completely horizontal plane: the openings are slanted slightly upward by 9–13°. This prevents stored materials and brood from spilling or rolling out of the cells before they are capped with wax. Each comb surface is separated from another by about 3/8 inch (9.5 mm), which is called a *bee space.*

Wax glands are stimulated to produce wax when bees gorge honey, nectar, or sugar syrup. When many bees are secreting wax, they hang in festoons or layers. If bees are seen in such a posture, called *festooning,* they are probably producing wax (see the illustration of festooning). Workers in a festoon will stay there for a time and then move off to feed brood or do other tasks, thus allowing the wax glands to recharge.

Wax secretion is stimulated by:

- High temperatures (see the figure on temperatures)
- Plentiful nectar, honey, or sugar syrup
- Ample pollen consumption

Before a colony casts a swarm, workers will gorge themselves with honey or nectar. This activity before swarming serves a dual purpose: while the swarm is clustered, the bees have a food reserve, and at the same time, this food stimulates the production of wax. When the cluster moves to a new homesite, the bees will be able to begin comb construction. Beekeepers who capture swarms make use of this knowledge and hive swarms on foundation, because they know the bees will draw it out into beautiful, new white comb.

Food Exchange, Handling, and Hive Odor

Bees within a hive exchange honey or nectar. Foragers returning from the field pass food to the hive bees, who then pass it to other bees. This food exchange, called *trophallaxis,* not only communicates what the colony is receiving from the foragers but also indicates the availability of food.

A returning forager will offer a drop of nectar to two or three house bees, who will then move off to a quiet corner to work the nectar by extending their tongues and exposing the droplet to the warm air. This helps cure the nectar into honey by evaporating some of the 80 percent water that is in nectar. As the water content is reduced, the nectar is placed in cells, where further curing takes place. This ripening is finished in the honey cells and will take from one to five days to complete, depending on the amount of humidity in the hive, water content of the nectar, ventilation, and amount of nectar being cured. Through this ripening process, the final product, honey, will have a finished water content of less than 18 percent.

House bees also process the pollen pellets deposited in cells by foragers. They pack them down with their heads, moisten them with honey and saliva (which contains enzymes), and cap the cell with honey, turning the mixture into *bee bread.* When a cell is filled with bee bread, it is covered with a layer of honey and capped with wax. Bee bread is an important food and protein source for young bees.

An additional function of food transmission is the spread of the hive's odor. Each colony has its own characteristic odor, which may aid the bees of one hive in distinguishing bees from other hives (such as robbers) and foreign queens (see "Queen Introduction" in Chapter 10). To keep foreign bees out, guard bees patrol the colony entrance and challenge any incoming bees that may be intruders. Guard bees are workers that have very high concentrations of the alarm pheromones.

The needs of a colony often dictate how fast incoming foragers are relieved of their nectar loads. If the colony is too hot, bees returning with water or more watery nectar are relieved first, forcing those foragers with nectar higher in sugar to stand around. This allows those bees with empty honey stomachs to recruit more foragers to the water source, and thus an overheating colony can collect more water to cool the hive and relieve the stress.

Ventilation

Bees can often be seen fanning their wings on the extended deck of the bottom board with their heads facing toward the hive entrance. In this position, warm air is pulled out of the hive. This fanning also takes place inside the hive as well as on the portion of the bottom board within the hive that is obscured from view. The best time to observe fanning is on warm days when abundant amounts of nectar are being brought in. Workers of all ages perform this task, but many young bees, less than 18 days old, are often fanning, especially on hot days.

Fanning circulates air through the hive and helps to:

- Regulate the hive's humidity at a constant 50 percent.
- Reduce the level of carbon dioxide (CO_2).
- Regulate brood temperature.
- Evaporate water carried into the hive to reduce internal temperatures (see the figure on temperature on p. 19).
- Evaporate excess moisture from unripened honey (nectar with a high percentage of water); as this moisture evaporates, it, too, will cool or humidify the hive.
- Keep wax from melting as temperatures climb.
- Eliminate accumulations of gases (such as CO_2).

Another type of fanning helps spread workers' pheromones. In this case, the fanning bee's abdomen is raised; a gland (Nasanoff, or scent, gland) located near the tip of the abdomen is opened; and a mixture of pheromones is released from it. These chemicals guide other bees toward the fanners; also called the "come hither" odor, it has a sweet-smelling, lemony scent.

This type of fanning is commonly seen:

- When a swarm or package of bees is emptied at the entrance of or inside a hive.
- When bees are shaken off a frame or otherwise disoriented.
- When a hive is opened that is queenless or has a virgin or newly mated queen.
- When a swarm begins cluster formation.
- As a swarm enters a natural homesite, to guide stragglers.

Guarding the Colony

Worker bees between the ages of 12 and 25 days will defend their hive by flying at and often stinging an intruder. Bees do this for only a few hours or days in their entire life. They can be recognized by their posture—standing on hind legs with the antennae held forward and the first two legs (forelegs) upraised. These guard bees often inspect incoming bees, which will not be admitted if they do not smell or behave "correctly." Stray or foreign drones, young workers, and foragers carrying a full load are generally allowed to enter. During strong nectar flows, foreign bees pass easily into a hive; during a dearth, however, guard bees closely inspect strange bees.

If a large animal approaches the colony, some of the guards will often fly out to challenge it. This defensive action should not be interpreted as "meanness" or "aggression" but rather as a defensive action. When an intruder approaches and enters or begins to open a hive, some bees raise their abdomens, begin fanning, and thereby disperse the alarm odor being released by a gland at the base of the sting. This pheromone has an odor similar to that of banana oil. It incites other bees to defend the colony. Once some of the attacking bees sting clothing or skin, some alarm odor remains at the site, tagging the victim. Thus tagged, the victim may become the target of further defensive acts as long as the odor remains on the clothing or skin.

Many factors influence the temper of a colony (see "Bee Temperament" in Chapter 5). Africanized bees, recent immigrants from South America, are known for their volatile defensive stance, which distinguishes them from European races (see "Africanized Bees" in Chapter 13).

Outside Activities

Flight

Except for occasional orientation flights, worker bees generally remain within the colony for the first three weeks of their adult lives, cleaning, feeding, building comb, ripening honey, and packing pollen. These routines are more or less discontinued at the end of the third week as bees turn to tasks that require flight. An ability to recognize the different types of flying activity will enable you to interpret activities at or near the hive entrance.

Orientation Flight. Bees on orientation flights familiarize themselves with landmarks surrounding their hive as well as void feces. These bees hover near the hive entrance for very short periods of time. A single flight will last only five minutes, with successive flights over the next few days lasting longer. This is a common sight in the late afternoon, when young drones and workers are hovering in front of the entrance.

Foraging Flight. This is the final task of a worker bee. Foraging bees fly out and away from the hive in a definite direction in search of nectar or honeydew, pollen, propolis, and water. Because their brood food and wax glands have degenerated, these bees look smaller, and the edges of their wings are often torn and ragged. The characteristic patterns of returning with pollen or flying straight into the hive or onto the extended deck of the bottom board will distinguish them from orienting bees. After all its trips, each forager will have traveled about 500 miles (800 km) before she dies (see the figure on forage area).

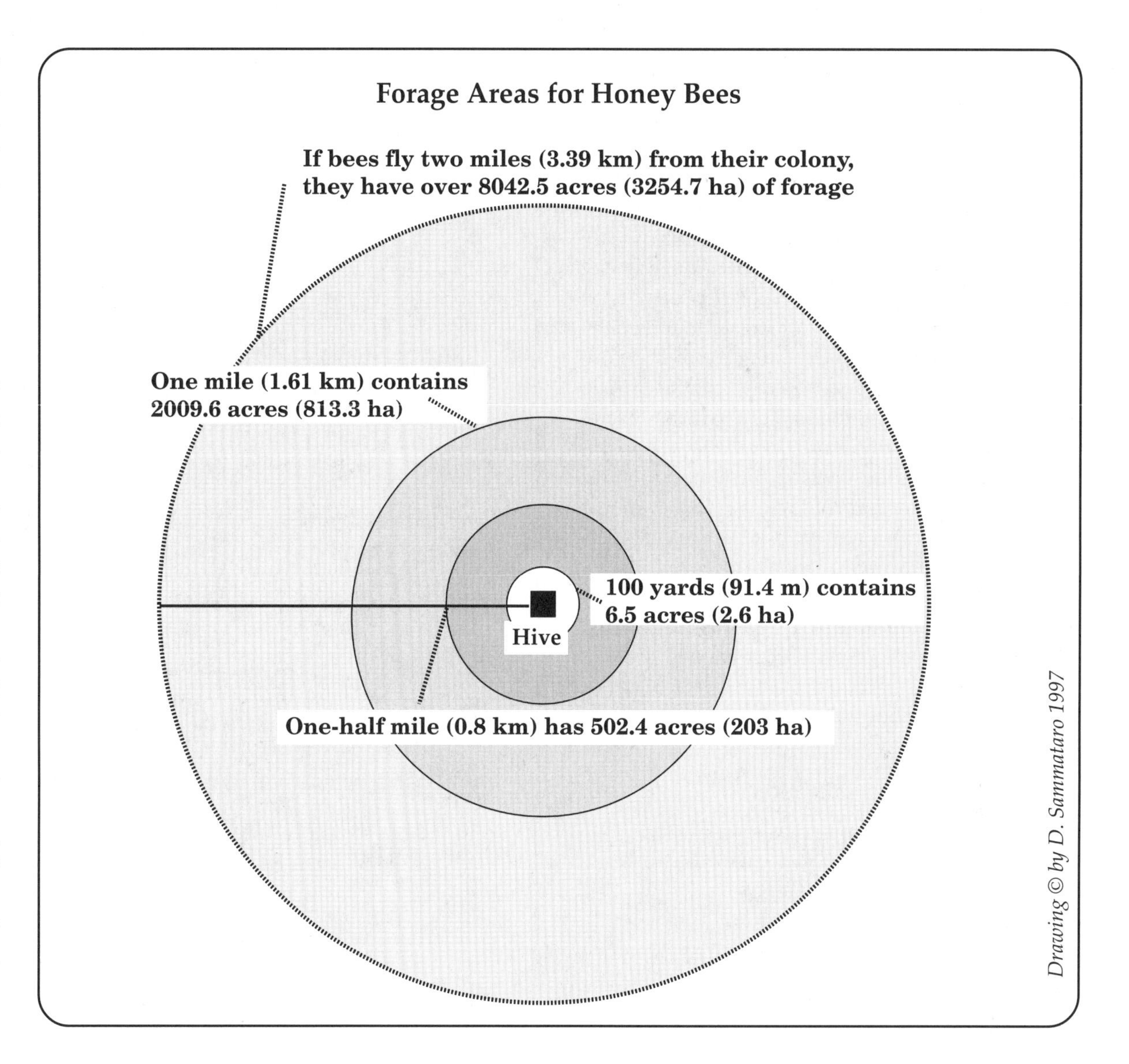

Robbing Flight. Unlike orientation flights, which are short in duration, robbing activity is a form of foraging. On first approaching a hive, the robbers sway to and fro in front of the hive to be robbed in a manner somewhat similar to a figure eight. Once the hive has been invaded, other robbing bees are "recruited" to it.

Robbing often takes place during periods of dearth (when little or no nectar is available); abandoned and weakened colonies are the first targets for robbing bees. Beekeepers who expose combs or attempt to feed weak colonies at this time may inadvertently initiate robbing. You must minimize your activities in the beeyard during a dearth or you can start a robbing frenzy (see "Robbing" in Chapter 11).

Cleansing Flight. Individual bees usually release their body wastes or fecal material outside the colony during flight. Evidence of this activity is the yellow or brownish spots or streaks found on the snow, plants, cars, or laundry hung out to dry. (The so-called yellow rain discovered in Southeast Asia during the Vietnam War turned out to be bee excrement, not chemical warfare.) When bees are confined for long periods—during the winter months, for example, or wet and cold weather and even in a package—as soon as flight is possible, the yellowish brown droppings are easily seen because of the large number of bees that defecate in a short period of time. In such circumstances, the flights are referred to as cleansing flights. Such flights usually occur in the spring, as the young bees are first able to leave and orient to the homesite.

Foraging and Communication

The gathering of water, propolis, and food for feeding larvae and for storage requires a high degree of cooperation and communication among the members of a colony. Haphazard searches for food by the older worker bees would require too much energy and time and could not be sustained over long periods without adversely affecting the well-being of the colony. Communication among the bees increases the efficiency of food-gathering activities by directing bees to known water and food sites.

A worker bee orients herself according to the following various external stimuli as she comes from and goes to collecting locations:

- The sun's position and polarized light
- Landmarks, both horizontal and vertical

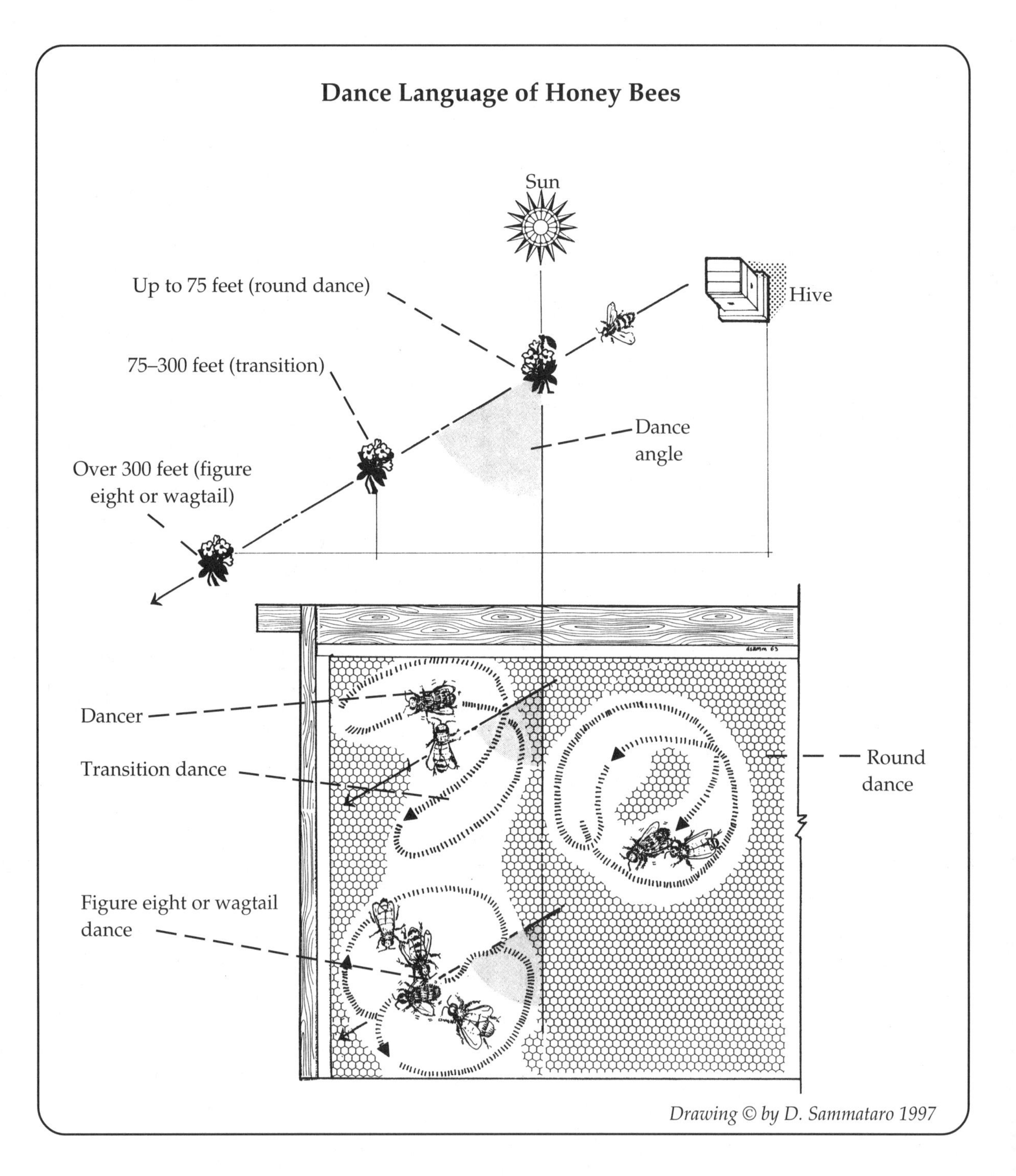

Drawing © by D. Sammataro 1997

- Ultraviolet light, which enables her to see the sun on cloudy days

A worker bee is able to inform other bees about the location of a food source through a series of body movements called *dances.* Dr. Karl von Frisch, an Austrian scientist who won the Nobel Prize for his research, and others observed that the movement of the dancer on the surface of honeycomb could be outlined as circles and figure eights. Bees near the dancer(s) can learn the distance to and often the direction in which to find food, water, and propolis. During the dancing, the dancer rubs her wings together, causing vibrations on the comb, which fine-tunes the information on distance and direction to the recruited bees following her. Other clues may include samples of the provision being sought and its associated odors. This information, as well as scent markers at the source, minimizes the time and energy needed by the recruits to locate the provisions.

Several dance configurations have been recognized, but the two most easily identified are the round dance and the wagtail dance. The round dance communicates that the provisions are no farther than 300 feet (100 m) from the hive in any direction. Bees being recruited to distances over 300 feet obtain distance and direction from the waggle, wagtail, or figure eight dance, which resembles the number eight. The wagtail dance includes three components: comb vibration, frequency of waggles, and number of waggles. The middle section of this dance is known as the straight run but is really where the bee is vibrating and waggling (see the figure on dance language). During this time, the bee transmits the waggle vibrations to the surface of the comb, where they are felt by other bees attending the dancer, especially if the dance is performed on uncapped cells. This vibration is thought to be picked up by the recruited workers via special organs on their legs. Distance is communicated by the number of straight runs in the dance per 15 seconds; the greater the number of runs, the shorter the distance to the food source.

A transition, or sickle, dance is done by some races. Workers do many other dances, and many are still being studied.

Other Activities

Washboard Movement

Beekeepers can often observe bees, usually in the early evening, on the front wall of the hive with their heads pointed toward the entrance. These bees are standing on the second and third pair of legs and seem to be scraping the surface of the hive with their mandibles and front legs, as if to clean it. As they scrape, their bodies rock back and forth in a motion similar to scrubbing clothes on a washboard. Thus it is called the *washboard movement.* The exact purpose of this activity is not currently understood, although it occurs in only very populous colonies.

Nest Homeostasis

The maintenance of temperatures and other environmental factors at a constant level inside the colony, despite external conditions, is termed *homeostasis.* Colony homeostasis is maintained by cooperative living or social behavior and is found in many insects (ants, termites, wasps, and bumble bees) and some mammals (naked mole rats). The ability of a colony to survive temperature extremes and dearth or abundant times is an obvious advantage over a solitary life, such as most other insects live.

The ability of bees to organize and execute the many tasks needed for colony survival, including nest design and construction, is indeed a marvel of the natural world.

Beekeeping Equipment

A *hive* is a structure that houses a full colony of bees (queen, workers, and drones). A man-made hive consists of a bottom board; two deep hive bodies for the broodnest and, if honey has been stored, one or more honey supers (the number depending on the abundance of incoming nectar); and an inner and outer cover (or a single migratory cover).

Some beekeepers use only standard or deep supers for their hives; others use the shallow supers for housing both the brood and the honey. If only the deep hive bodies are used, lifting off the honey will be very strenuous because a deep super full of bees, brood, wax, wooden frames, and honey can weigh over 100 pounds (45 kg).

If, on the other hand, you choose the medium supers to house your colony, the weight will be less, as one full medium super weighs about 65 pounds (29 kg). Finding the queen, however, becomes more time consuming and disruptive to the colony when using medium supers. If two medium supers are used as the broodnest, the queen can move more rapidly from frame to frame when you are looking for her than if she were on deep frames.

The number of hive bodies needed to house bees varies throughout the season; in general the broodnest occupies the equivalent of two deep supers. In regions with winter, the extra honey supers are removed, and some beekeepers overwinter their colonies in two deeps and a shallow, using the top deep and the shallow for winter stores of honey and pollen. But this can vary; other combinations are one deep and one shallow, two deeps, or sometimes even in three deeps (see "Wintering" in Chapter 8). An ample supply of provisions must remain or be supplied before the onset of cold weather.

It has been traditional to paint the hive bodies white to reflect the sun's heat in the summer months, thus keeping the colony cooler, especially in the southern part of the United States. Paint the metal top of the outer cover white to keep it cool to the touch during the hot summer months. Although white is most favored in southern climates, beekeepers in northern areas might consider painting hives darker shades to retain the heat longer. For hives located in highly visible sites where vandals or thieves may be a problem, darker colors serve to camouflage hives.

Whatever color is used, the outer sides of the wooden hive parts should be painted in order to extend the life of the equipment. Because bees produce moisture as a part of their metabolic activity, latex paint would be least likely to blister as the moisture leaks out; lead-based or other toxic paints should never be used.

To save money, you can buy off-color latex paints from hardware stores; mixing paints together, you can create a unique appearance. It is unnecessary to paint or coat any interior hive parts with any substance. Some beekeepers paint the rabbets, however, or apply a layer of petroleum jelly to reduce propolizing in this area.

Some pieces of equipment, notably inner and outer covers and bottom boards, are available in plastic, but their suitability compared with wooden equipment can be disappointing (see "Equipment" in the References). Visit with other beekeepers who have used plastic equipment.

Beekeepers in damp and humid regions immerse hive parts in wood preservatives or boiling paraffin to extend the life of their woodenware. The latter method works well but takes significant physical effort to heat the wax and can be dangerous, thus requiring great care. It is economical if you have a lot of equipment to dip (see Appendix D).

In areas where loss of beehives through theft is a concern (or if you keep many colonies in various locations), wooden hive parts could be branded with individual identification. But branding one's name or symbol on hive parts merely alerts the thief that eradication of such marks is necessary. A less expensive alternative is to write your name and address under the tin roof of an outer cover or, with a permanent marker, write your name inside your supers. The eventual cover of propolis can easily be scraped away to reveal your name. Or you can add or leave out an extra nail or otherwise individualize equipment to assist in proof of ownership in the event of theft. For just a few hives, the extra time may not be worth the effort.

Basic Hive Parts

Outer Cover

Two types of outer covers, a *telescoping* cover and a *migratory* cover, are placed on top of a beehive for a roof. The telescoping cover is usually made of wood and is covered with tin or aluminum. It "telescopes" over the rim of the inner cover and uppermost hive body. If you live in areas with long, hot summers, place some form of insulating material (newspapers work well) between the tin and the wooden roof during construction. The illustration of a beehive shows a telescoping outer cover.

The flat California or *migratory* cover does not telescope over the sides and is used in drier areas because it will not last long in extended wet weather. It is also used on migratory beehives, which are packed onto pallets and trucked. The lack of sides telescoping over the edge of the hive allows beekeepers to strap hives together to move them to different locations for pollination services or to par-

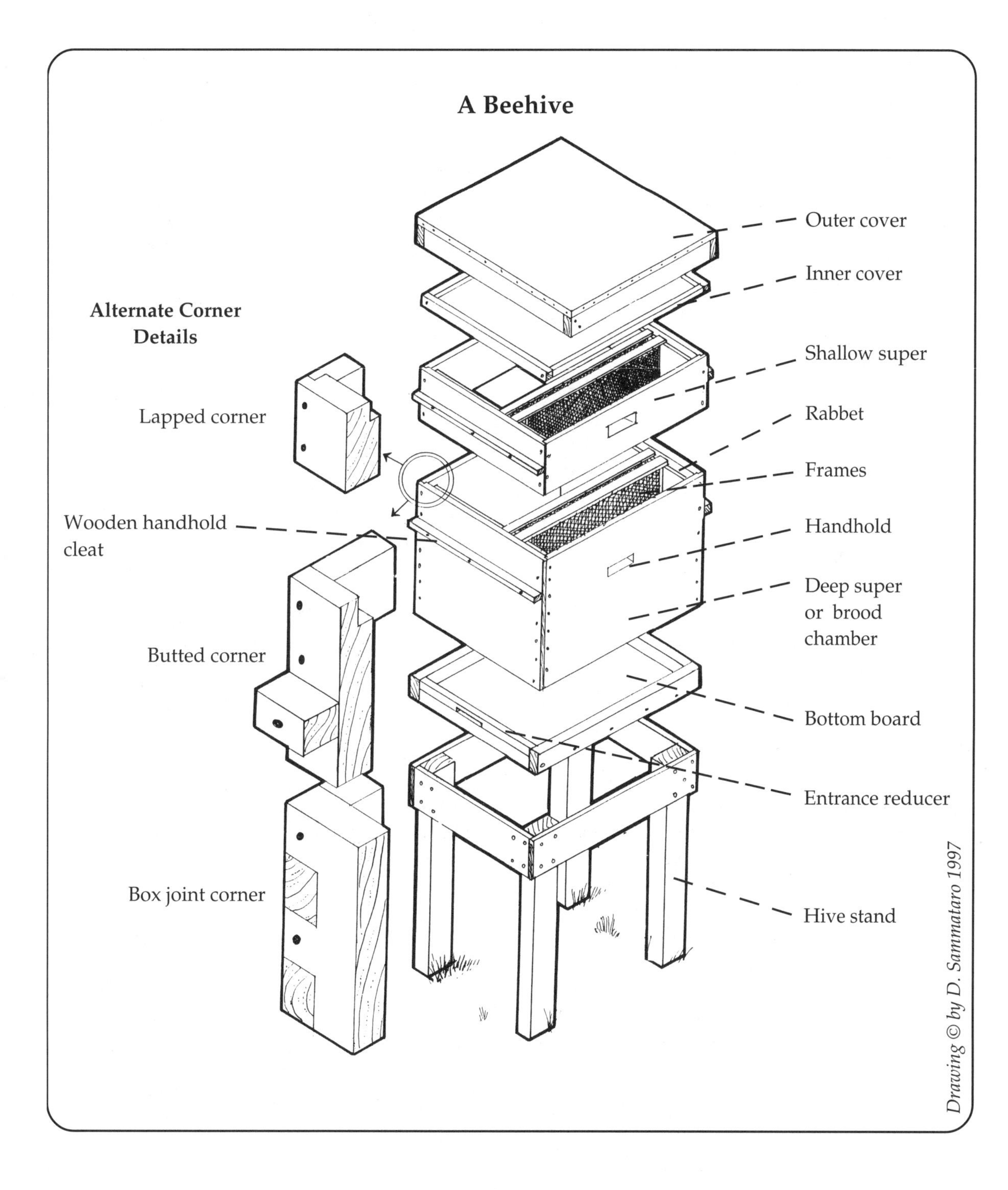

ticular or abundant honey sources. Whichever outer covers you use, place a heavy weight on top to prevent them from being blown off by strong winds. If a colony loses its roof, exposure to heat, snow, rain, and cold winds could weaken or kill it.

Inner Cover

The *inner cover* is a wooden, Masonite, or plastic board that has about a 1/2-inch (13 mm) rim on one side and an oblong hole in its center. Some inner covers come with an additional half-circle hole notched in one end rim to provide an extra entrance for bees. This hole, as well as the center opening, helps to vent moist air, especially over winter when the inner cover is turned rim-side down. If migratory outer covers are used, inner covers are not necessary.

If a device called a *bee escape* is placed in the oblong center hole, the inner cover becomes an *escape board.* By placing this below a honey super or supers, the worker bees passing through the one-way escape are unable to reenter the honey supers above and thus clear the boxes of bees; see "Removing Bees from Honey Supers" in Chapter 9 for other removal methods.

Whenever it becomes necessary to feed a colony, invert a jar or pail of syrup over the oval hole of the inner cover. Bees can collect the food without venturing outside. An empty super and the outer cover can then be placed around the feeder to enclose it (see Chapter 7).

Shallow or Honey Supers

The shallower supers, which are usually used for honey storage, come in various depths, with frames and foundation of corresponding size. In the United States, three different honey supers are currently used:

- $4\frac{13}{16}$ inches (12.2 cm), referred to as the half-depth (half a deep super), comb honey, or section super.

- $5^{11}/_{16}$ inches (14.4 cm) or shallow super.
- $6^{5}/_{8}$ inches (16.8 cm) or medium, Illinois, Dadant, or three-quarter-depth super; this can also be used as a brood super.

The different super sizes are illustrated in "Super Sizes," Chapter 9.

The number of honey supers per hive will vary during a honeyflow, depending on the amount of honey collected and the number of supers you have. Although supers are designed to hold 10 frames, this configuration is generally found only in the brood supers. Experienced beekeepers put in just eight or nine frames in honey supers so the bees will draw out wider comb. When full of honey, such thick frames are much easier to uncap. Chat with other beekeepers before experimenting.

Occasionally, you will find beekeepers using 8-frame equipment, hive bodies that are made smaller to hold only eight frames. They are lighter in overall weight and of smaller dimensions, but this size of hive furniture is difficult to find now, as most major bee supply companies no longer manufacture them.

Queen Excluder

This is a perforated zinc or plastic sheet or a wood-framed metal grill. *Queen excluders,* as the name implies, exclude the queen from the space below or above the device. Because of its small openings, only the worker bees can squeeze through; the larger drones and queens cannot. The device is usually placed on top of the broodnest to prevent the queen from entering the honey supers above. It is also used on two-queen colonies or for any other reasons that require the exclusion of the queen. Drones are often trapped by excluders and, if they die in large numbers, can clog it with their bodies. This can be prevented by leaving an escape hole in the honey supers.

Deep Super or Brood Chamber

The deep or standard super is $9^{5}/_{8}$ inches (24.4 cm) in depth. Ten full-depth frames ($9^{1}/_{8}$ inches) are used in this super. The deep super is typically used in the United States to contain brood and winter stores. The general understanding is that these deep boxes are not moved around much, so to collect honey, the shallower supers were made, to keep the weight more manageable. At one time bigger supers called the Jumbo and the Modified Dadant deep supers were common but are now hard to find. The Modified Dadant hive is longer and deeper than a standard deep; it holds 11 frames. A Jumbo hive holds 10 frames but is as deep as a Modified Dadant. Commercial beekeepers, however, tend to use only one size of box for both brood and honey production.

Bottom Board

The floor of a hive is a structure called a bottom board. Hive bodies are stacked on top of the bottom board, which should be placed on a firm foundation to keep honey-heavy hives from tipping over. Never place the bottom board directly on the ground as it will rot quickly; use a hive stand (see the illustration of the beehive). Treating it with wood preservatives may extend the life of the bottom board, but many chemicals will not only kill bees but could also contaminate honey and wax. Whatever you use, it must be nontoxic to bees, the hive environment, and humans. Some bee supply companies sell a product approved by the Food and Drug Administration. Bottom boards are a good candidate for paraffin dipping (see Appendix D for more information).

Most bottom boards have two rim heights—a short winter rim and a deeper summer rim—and are called *reversible bottom boards.* Many beekeepers, instead of reversing the bottom board (which is difficult to do), use an entrance reducer to restrict the hive opening during the winter. By reducing the entrance, weak colonies are protected from robbing bees and other insects, and cold winter weather is likewise blocked. Plastic bottom boards are available, but they sometimes buckle if the hive is very heavy, rendering them useless.

Commercial operations often use pallets, on which bottom boards are permanently attached. This system serves as a convenient way to move bees (via forklifts) and provides a solid and safe hive stand.

Hive Stand

Most bee supply companies sell an on-the-ground hive stand and alighting board combination. Although it is sufficient for some situations, this hive stand is not recommended because it will rot in a few years. It also will not protect bees from ants, mice, or other animal predators. Therefore, most beekeepers eventually make stands from materials at hand so as to keep bees higher off the ground. For more information, see "Hive Stands" in Chapter 4.

Bee Space

Bees do not space natural honeycombs at random in a wild colony or in the wooden beehive. They adhere to a strict code and do not construct comb in spaces less than about $^{3}/_{8}$ inch (9.5 mm). This fact was published by the Philadelphia minister L. L. Langstroth over 100 years ago. It was the basis on which he designed the prototype beehive used today. The $^{3}/_{8}$-inch space enables beekeepers to remove frames without having to cut the combs from the walls and covers. A $^{3}/_{8}$-inch gap separates each of the frames, the hive walls and the bottom board from parts of each frame, and the top bars from the inner cover of a hive (see the illustration of the bee space).

By utilizing this natural spacing, beekeepers as-

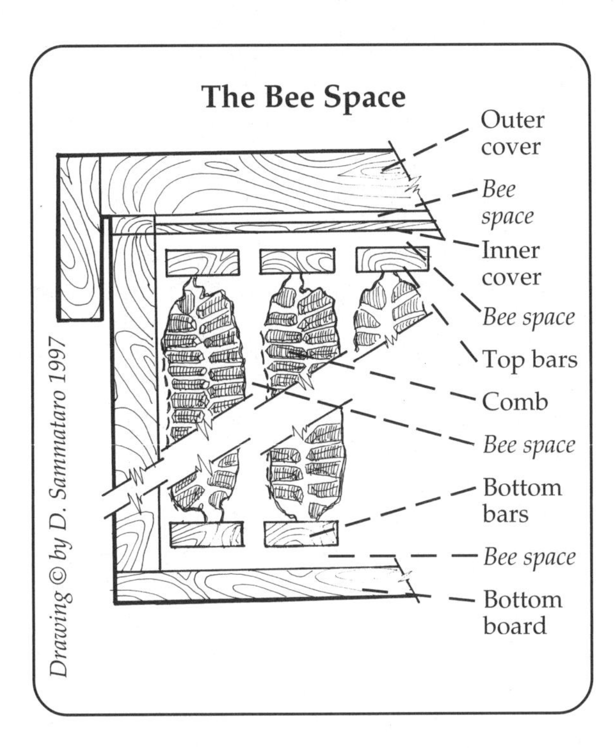

sure that the bees will not attach comb to the walls or to other sections of comb and that the frames can be easily removed. If the frames are spaced farther apart or if you neglect to return a frame to the hive after examining it, the bees will fill the gap with comb or will extend the cells of combs adjacent to this gap. This action makes it impossible to remove frames without cutting out the comb. Recent studies indicate that some races of bees leave less than 3/8 inch between combs.

Any space less than the bee space will be filled with propolis (or bee glue). This keeps cold air drafts out of the hive and kills any microbes that may live in these close spaces.

It is illegal to use beehives with fixed honeycombs, such as straw skeps or log gums. All beehives must have removable frames so bees can be inspected for diseases.

Frames

Bees in the wild attach honeycomb to the ceiling and often to the walls of a cavity. Such combs can be removed only by cutting or breaking them from their attachments. In modern beekeeping equipment, hive bodies contain *frames,* which are rectangular structures made of wood or plastic that hold wax or plastic *foundation* (see the next section) or *drawn comb* (completed comb with finished wax cells).

The vertical parts of frames, called *side bars,* are designed so that their lateral sides serve to space the frames correctly in the hive (see the figure showing frame sizes). Thus, when bees have fully fashioned foundation into honeycomb, the natural "bee space" is created between combs. This space allows the bees to move freely from comb to comb and the queen to lay her eggs.

As illustrated, frames come in all sizes and styles. The top bars of frames usually have a wedge that is removed and then nailed back once the foundation is in place. The bottom bar may be either solid, grooved, or two-pieced. Before the foundation is set, it is advisable to string the frame with wire. This will add support for the foundation, which is especially important for frames that will be placed in a honey extractor. Because the extractor spins frames at high speeds, stress is placed on the comb, which may cause it to buckle or break. The wire support will prevent this from occurring in most cases. It is also important when setting foundation to have it remain flat, or uneven or warped honeycomb will result.

A simple jig can be made to facilitate wiring frames. It is a good idea to hammer eyelets in the holes on the wooden side bars before wiring; otherwise, the taut wire strung through these holes may

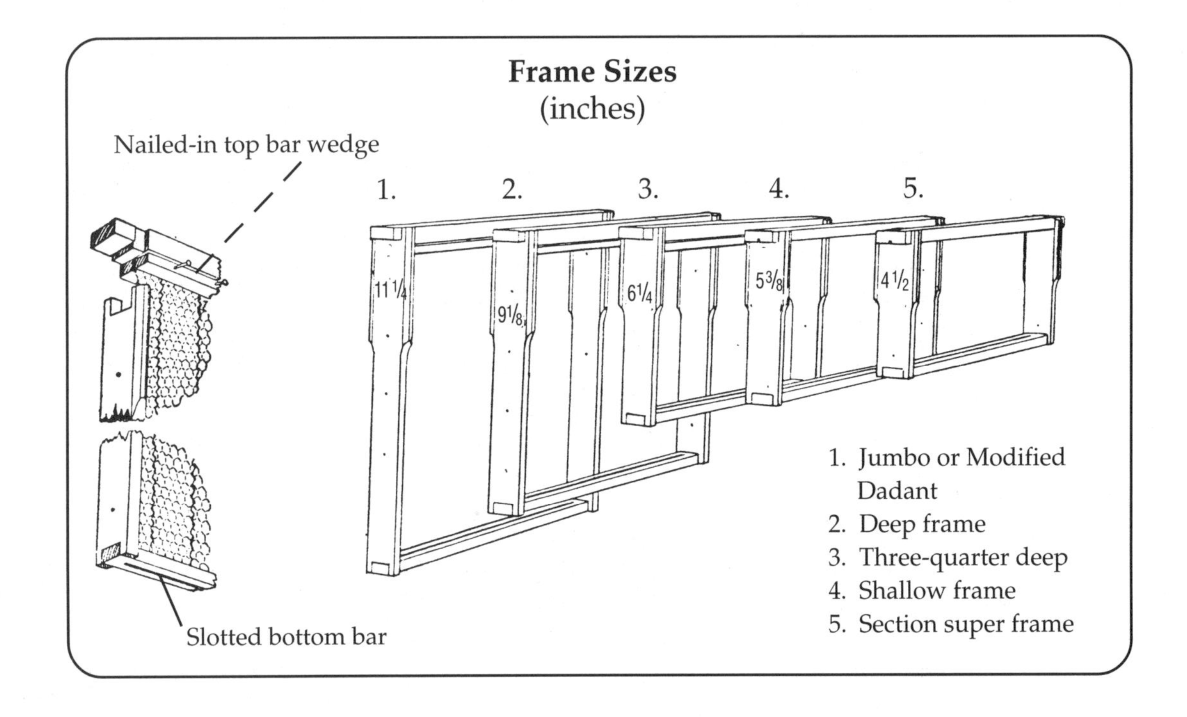

cut into the wood, which can cause the wire to slacken. The wire used for stringing frames is #28 tinned, although any thin wire that won't rust will do. Some beekeepers use 40-pound fishing line or monofilament instead of wire, in which case there is no need to embed the line in the foundation.

If you decide to use plastic-based foundation or plastic sheets sprayed with beeswax or plastic molded frames that already have the hexagonal honeycomb, wiring is not necessary. Check with your bee supply dealer or with other beekeepers in your bee club to compare notes on which is the best to use in your area (see "Foundation" in this chapter).

In wired frames, the wire is forced into the foundation by lightly heating the horizontal metal wire. Two tools can be used for this purpose. One is called a *spur embedder,* which is heated in boiling water and then rolled over the wire. The heat melts the wax around the wire; when hardened, the wax holds the wire-foundation combination in place. The other tool, called an *electric embedder,* heats the entire wire with an electric current, which melts the surrounding wax enough for the wire to become embedded.

When frames are fitted with sheets of wax foundation, start with 10 frames in the hive so the bees will draw out the foundation into even combs. Some beekeepers later remove one frame in the broodnest edge to allow easier manipulation of the remaining nine frames. To space these nine frames evenly (or eight frames in honey supers), use special built-in spacers (such as Stoller spacers) or follower boards, or merely space the frames evenly by hand. If you do not space the frames evenly, bees will fill in any larger gap with more comb.

Empty frames—those that contain no foundation or comb—should not be placed in a colony, because the comb may be attached to the bars in an unsatisfactory way, such as at right angles, which would fasten all the frames together, making their removal impossible. Some beekeepers, for reasons of economy, will put in only a small strip of foundation attached to the top bar of a frame. Although straight, even comb may be achieved by using a strip, the practice is not recommended: with deep frames, the bees may fill in the empty space with drone-sized cells or crooked comb and it will not be wired, further weakening the comb.

With the advent of the varroa mite and the difficulty in finding drone-sized foundation, beekeepers are reverting to this practice, but with a modification. They install shallow foundation in the top bar of a deep frame to encourage bees to draw out drone comb in the space below. Because varroa mites are attracted to drone brood, the infested capped drone pupae are then cut out and destroyed or frozen to reduce the level of mites (see "Varroa Mite" in Chapter 13).

A word about plastic frame-foundation combinations: some suppliers sell one-piece frames with or without fully drawn plastic combs. Some report good success with such frames, which may have a benefit in mite control, whereas others do not use them at all. If you wish to experiment, try plastic frame units on a small scale; they may or may not work very well in your area.

Foundation

Foundation is a thin sheet of pure beeswax or wax with a plastic base, embossed with either hexagonal worker or drone cells. This sheet is set in the rectangular frame and wired or pinned in place. When put into a colony gorged with honey, nectar, or syrup (a state that induces worker bees to produce wax), the bees will use the foundation as a base and draw up the walls of the cells. They do so by adding secreted wax to the base; when completely filled out, the frame is said to contain drawn comb.

The kinds and combinations of foundation are often confusing to beginning beekeepers. The three different categories are (1) 100 percent beeswax (with and without vertical wires); (2) beeswax sheets around a plastic core (with or without metal edges); and (3) plastic embossed sheets sprayed with beeswax.

Different thicknesses of wax foundation also exist, but basically, the thicker sheets of 100 percent beeswax with inlaid vertical wires are used for brood frames and in the extracting supers. The thinner 100 percent beeswax foundation is used in honey supers for producing cut-comb and section comb honey. Unwired frames should NOT be placed in the extractor. Though most foundation sold today contains worker-sized cells, some companies sell drone-sized cell foundation, which is now used to increase drone populations for varroa control or in queen mating yards (see Chapter 10).

To set the foundation firmly in the frame, especially in the brood frames, horizontal wires should first be strung across the wooden frame (see the illustration of frame parts). The embedded wires help keep the comb from sagging, keep the cells from stretching in warm temperatures, and strengthen the honey-filled frames against breakage during extracting. Distorted cells in sagging combs are unsuitable for raising brood and decrease the number of cells in which the queen can lay eggs, which results in spotty brood patterns. Such frames should be replaced with new foundation or newly drawn comb.

The plastic-core and plastic-embossed foundation sheets are easy to install and require no wiring. Drawbacks of the former include bees not drawing new comb on the plastic sheet, should the wax separate from it. Another concern with plastic sheets is how to dispose of or sterilize them if they become infected with foulbrood spores (see "American Foulbrood Disease" in Chapter 13). The best method is to irradiate such supers, but this resource may not be readily available in your area. Overall, the cost of wooden and plastic frames is about the same.

Recently, minute differences in the sizes of worker cells have become the focus of research in controlling varroa mites; keep up with the current research to learn more.

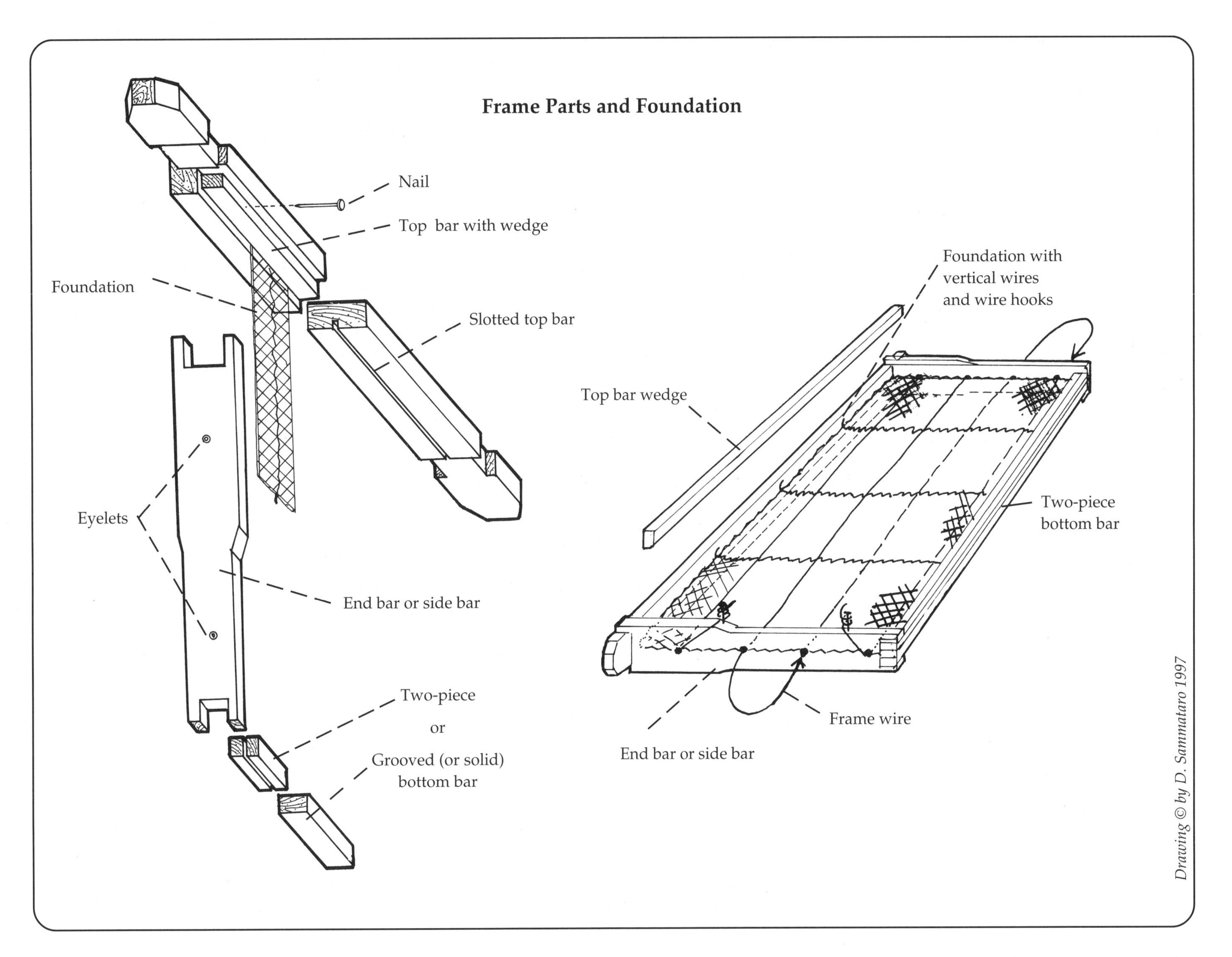
Frame Parts and Foundation
Nail
Top bar with wedge
Foundation
Slotted top bar
Eyelets
End bar or side bar
Two-piece
or
Grooved (or solid)
bottom bar
Foundation with
vertical wires
and wire hooks
Top bar wedge
Two-piece
bottom bar
Frame wire
End bar or side bar
Drawing © by D. Sammataro 1997

Foundation can be stored for a long time but should be kept away from heat or from freezing temperatures. If kept in plastic bags, the wax sheets will remain fresh and soft. Never store foundation in hot or sunny areas, as the wax will melt.

Beginner's List

Most hive parts come disassembled and may require painting. Some bee supply companies sell plastic equipment, but most hive parts are made of pine or cypress. Hive bodies made from cypress are more durable and last longer than do those made from pine. Check prices and equipment available from various bee supply houses or from your local bee club. Generally, when just starting with a few hives, it is best to buy all new equipment from one source. Hive dimensions vary just enough to cause problems when you mix and match hive furniture from different suppliers.

For one complete hive, the following are needed:

Item	Price
• 1 standard deep body, inner and outer covers, bottom board, frames	$40–60
• 1 additional deep body with frames	15–22
• 20 sheets wired foundation	20–25
• 1 large smoker	20–30
• 1 square folding veil and helmet	14–20
• 1 pound tinned embedding wire (or #40 monofilament)	4–6
• 1 spur embedder, to embed wire in foundation	3–5
• 1 10-inch hive tool	4–6
• 1 gallon exterior latex paint	16–25
• 1 medium super (6⅝ inches)	10–15
• 10 frames (6¼ inches)	10–15
• 10 sheets wired foundation, medium super	5–9
• Medication	
1. Fumidil-B: 0.5 g	9–12
2. Terramycin: one 6.4-ounce pack	4–6
3. Apistan strips: 10 strips	17–21
TOTAL	$191–277
Optional equipment	
• Gloves, plastic-coated canvas	$8–15
• Bee suit, zip-on veil	75–125
• Queen excluder	5–10
• Bee escape	1–3
• Division board feeder	3–5
• Uncapping knife, electric	58–100
• Extractor (two-frame, hand-powered)	100–300
• Jars, 12 1-pound	5–8
TOTAL	**$255–566**

The prices listed are approximate 1996–1997 prices and will vary somewhat depending on make, supplier, and so forth. Shop around to get the best deals.

For the Beginner

To the beginning beekeeper, the plethora of equipment available from the bee catalogs may prove somewhat confusing. The basic equipment listed in the beginner's list provides a starting point.

It is not a good idea to buy just one hive, because if the colony fails to develop properly or the queen becomes injured or dies, you will lose an entire year's experience at working bees and learning the art of beekeeping. From two to five hives is a better number for the beginner.

While used hive bodies and frames are less expensive than new equipment, they could be contaminated with brood diseases that are not readily apparent. If equipment is questionable, it should be sterilized. The most economical way to sterilize such equipment is to scorch it with a propane torch, but this method is not always reliable or effective. This should be done only with hive bodies, not frames.

Other, more expensive equipment, such as honey extractors, can be shared by several beekeepers on a cooperative basis. Hobbyists are cautioned not to buy every gadget on the market. When in doubt about the usefulness of a particular piece of equipment, seek the advice of other beekeepers.

Other Equipment

Slatted Racks

Slatted racks, sold by some bee suppliers, provide additional space for bees to cluster near the bottom of the colony. This piece of equipment is a wooden device that looks like a slatted board and fits on top of the bottom board. Bees are able to enter the hive normally, but they move up into the broodnest through the slats, giving them additional room under the frames to fan; it alleviates congestion in the hive during the summer. In the winter, the slatted rack allows bees to form the winter cluster lower down and frequently keeps the bottom board cleaner, because an entrance reducer is often not used, which provides the bees more space below the frames. Mice that enter the hive generally stay below the slatted rack and do not move up into the colony.

Frame Grip

When held in one hand, this spring-loaded metal or plastic tool enables you to grip the center of a frame's top bar and pull the frame out in one motion. It works best if the frame is first broken of its propolis seal. If it is not used correctly, however, you can scrape or injure clinging bees or the queen if she is on the frame. You must lift the frame slowly, carefully, and straight up out of the hive.

Bee Brush

This soft bristle brush is used to clear bees off a frame, especially frames of honey that you wish to extract. It is not a practical way to remove bees from honey frames, except on a small scale. Other things can be used for the same purpose, such as a handful of grass, feathers, or leaves.

Beewear

There is no stigma attached to wearing protective equipment; it will save you and your visitors a lot of worry if you are properly prepared to look through colonies without the fear of being stung. You will learn to gauge bee temper, handle equipment properly, and understand basic bee management by working bees. Let's look at beewear that is essential in your wardrobe.

Veil

A bee veil is a must. Although you will see photos that show beekeepers working without veils, this practice is discouraged. Stings on eyes, lips, the scalp, or inside the nose or ear canal are extremely painful; it is downright foolish to risk them. All sensible beekeepers wear veils. Veils can be purchased separately or attached to helmets.

There are several kinds of veils:

- Wire mesh veils (square, folding, or round) worn with a helmet
- Sheer veils (tulle or nylon) worn with a helmet
- Alexander-type veils, worn without a helmet, using an elastic head strap
- Veils attached to coveralls, with or without helmets
- One-piece bug bags that fit over the entire torso and arms

Check with other members of your bee club and try on different veils (and suits) to see which type you like best.

Helmets

Helmets are worn to support veils that require them; they keep the veil away from one's face. They are sold in plastic or woven mesh styles, some with ventilation louvers and others with hooks to hold the veil in place. Most have adjustable bands inside to fit your head, but none fit very well. Because they have a tendency to slip to one side or onto your face at inappropriate moments (such as when you bend over), helmets can create problems for inexperienced beekeepers.

Felt or straw hats are sometimes used as helmets but are not usually strong enough to hold up to rough use. Veils and helmets are typically worn with a coverall. If the veils are attached to the coverall with a zipper, the outfit becomes a bee suit.

Bee Suits

Bee suits, which are coveralls with zipped-on veils, come in handy if you are working with a lot of colonies. They protect you against stings and keep your clothing free of honey, wax, and propolis, which stain clothes easily and are very difficult to remove. If purchasing a bee suit, get the best you can afford, for cheaper ones tend to wear out quickly. The suits should incorporate all features discussed in this section. Homemade bee suits or those purchased from a supply house are made of a white cotton and polyester mix, with pockets and pouches to carry hive tools, matches, and the like. If you make your own, zippers on the legs make getting into the suit easier. Most commercial suits have wrist and ankle cuffs with elastic closures.

Turn up the collar before putting on the bee veil and close trouser and sleeve cuffs tightly with tape or elastic. Gauntlets made for wrists and ankle straps will keep bees from getting beneath clothing. Some beekeepers tuck their trousers into shoes or socks or fit cuffs with elastic bands for a snug fit; leg straps for trousers are also available. Make sure the pockets of purchased bee suits or coveralls do not come with slits, which are used to access your inside pants' pockets. Bees love to crawl inside the slit opening, with surprising results.

If clothing is not closed tightly, bees will crawl underneath unnoticed, and when a bee is pressed between clothing and skin, it will sting. Once a bee gets inside the clothing, you may attempt to release it, but it is easier to crush it before it stings you.

Bees are less likely to sting people wearing light-colored attire. Bees are more prone to sting dark, furry objects (such as skunks or bears) and fabric that does not "breathe" well (that is, it makes you sweat more than is usual). Wearing lightweight material is preferable, because the best part of the day for working bees is usually also the hottest part of the day. Heavy perspiration may aggravate bees; avoid heavy clothing when working with bees. Rule of thumb—fabric light in color and weight is best.

The easiest suit to use is a one-piece suit with an attached veil that requires no helmet, though you can wear a visor hat underneath to keep the sun off your face. Remember, veil mesh does NOT shade your face from sunburn. Some companies even sell jacket-length or child-size suits. Check out the bee supply catalogs and bee journals for other bee suits. Wash bee suits to keep down accumulated dirt, disease spores, and venom from stings.

Gloves

Many beekeepers disdain using bee gloves, but for the beginner it is a good idea to start with them. There should be no stigma attached to wearing gloves while working bees. With experience you will find that you will put away your gloves, at least most of the time.

Gloves sold today at bee supply outlets are made from cloth or canvas, plastic-coated canvas, leather, or some kind of plastic. Gloves that do not fit well will make handling frames more awkward and may even invite more stings than no gloves at all (yes, bees can sting even through leather). All bee supply houses carry men's sizes, but the smaller women's sizes may be harder to find (check the British supply houses; see the section on foreign suppliers in the References). Buy gloves that fit snugly.

Gloves are a great help in keeping wax, honey, and propolis off your hands, but wash them regularly, especially after working with diseased bees. One disadvantage of gloves is that they may retain the alarm odor long after bees sting them, another reason to wash gloves routinely. Most times, however, gloves are not necessary and you should not make a habit of using them—you can get stung more wearing gloves than bare handed. And never work someone else's bees with your used gloves, for they may contain foulbrood spores. Some people wear leather gloves; others, cloth. In a pinch, rubber kitchen gloves can substitute for regular bee gloves and must be used when applying Apistan strips for varroa mite control (see "Varroa Mite" in Chapter 13). Work with other beekeepers to observe how they use gloves, which type they use, and how they move around bees. You can learn much by watching others.

After gaining a bit of experience and increased confidence in working with bees, even the relatively new beekeeper may sometimes choose to work without gloves. Gauntlets that fit over the arms, keeping the hands free, are an added protection. If you wear a leather watchband, remove it during bare-handed apiary work, because such bands seem to incite bees to sting wrists.

Hardware

Hive Tools

Several types of hive tools are available from bee suppliers, even Teflon-coated tools. Again, check what others are using and how they like their tools. The scraper-nail puller type of tool is found in most hardware stores (see the illustration of hive tools) and comes in 7-inch or 10-inch lengths—buy both. Hive tools are an invaluable aid to the beekeeper when prying apart hive bodies and frames that have been propolized.

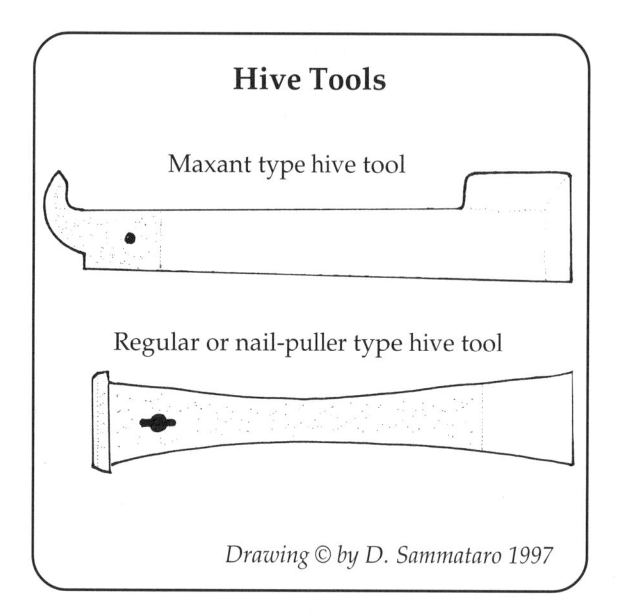

Drawing © by D. Sammataro 1997

It is a good idea to have several hive tools on hand, because they are easy to misplace. Hive tools should be periodically heat-sterilized and scoured (use steel wool) clean of excess propolis and wax. Paint hive tools bright colors (or stripes), such as neon colors, red, orange, or yellow, but not green or blue; this will help keep them from being lost in the grass. The ends should be sharpened at least once every year.

Smoker

The second most important piece of equipment you will own (besides a veil) is the smoker. The

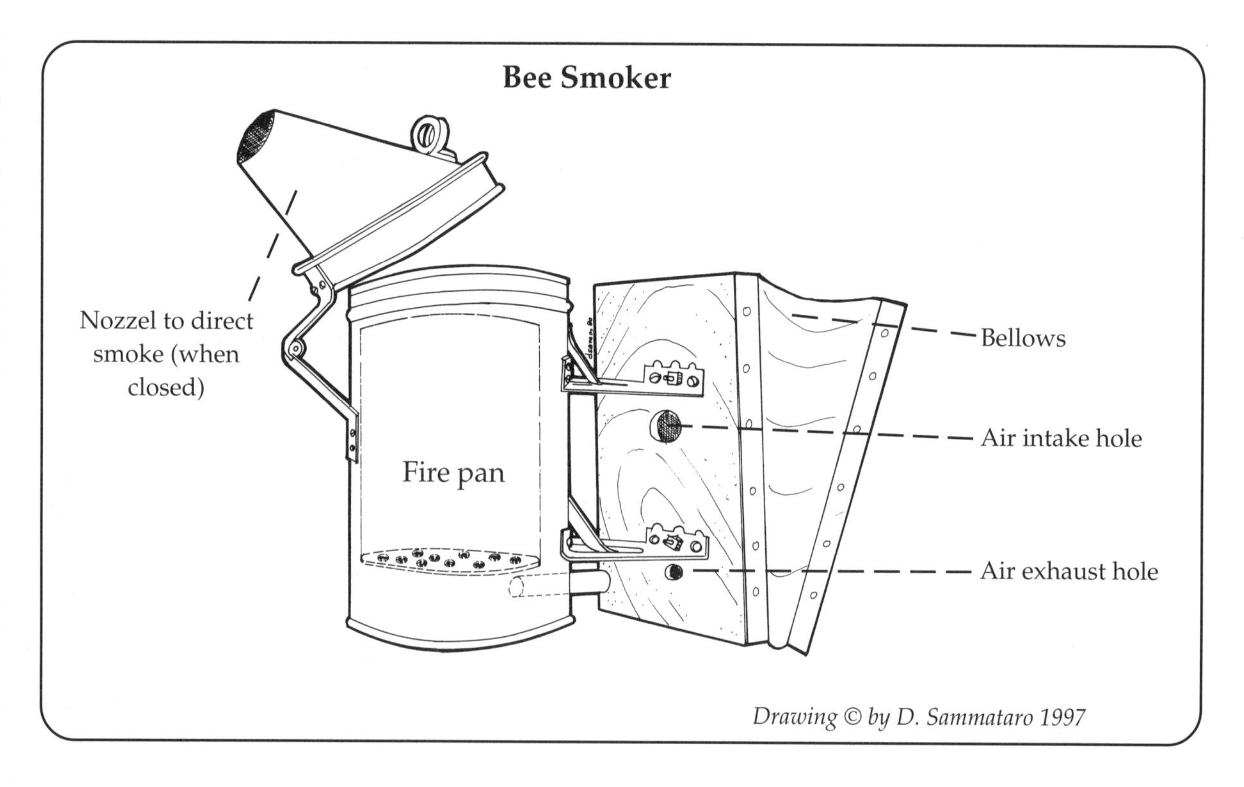

Drawing © by D. Sammataro 1997

smoker is a metal cylinder, in which a fire is lit, with attached bellows. You add the fuel and pump in fresh air by way of the bellows. Smoke blown from the nozzle (see the illustration of the smoker) is directed into the hive and between the frames to encourage bees to engorge honey. Once engorged, bees are more docile and less prone to sting.

Smokers come in a 4-inch and a 7-inch size. The small size does not hold enough fuel for more than two colonies, so when purchasing a smoker, get the 7-inch smoker if available. For long life, buy a stainless steel smoker with an attached heat shield and front hook. See Chapter 5 for more information on lighting and using your smoker.

4 Obtaining and Preparing for Bees

Whether you are a beginning beekeeper or an established beekeeper who wants to increase the number of colonies, you can obtain bees in the following ways:
- Buy package bees.
- Buy nucleus or established colonies.
- Collect wild colonies from buildings or bee trees.
- Collect swarms.

Packages

Package bees come primarily from the southern states or California and are shipped in the spring by mail or are picked up by dealers and trucked to their final destination. To order packages, look for advertisements in the bee journals and call several places; packages should be reserved in the winter months (December and January) to obtain the desired number of packages and choice of shipping dates. Request delivery three to four weeks before the dandelions and fruit trees bloom in your location.

Generally, a three-pound (about 1.35 kg) package of bees plus a mated queen will provide the ample amount of bees needed to begin a good colony. The approximate cost (1997) of such a package with one laying queen is around $40–50.

Advantages:
- Easier for beginners to work (fewer bees than in an established hive).
- More adult bees than in a nucleus (or small hive).
- Certified apparently healthy and from healthy stock.
- No brood diseases.
- Replacements are easy to obtain.
- Available in two-, three-, four-, and five-pound units (there are approximately 3,500 bees per pound), with or without queens.

Disadvantages:
- Queen could become injured as a result of stress during shipment.
- Drifting common (bees fly into other hives or become lost), especially during installation.
- Dependent on weather; if it is too cold, bees may not feed and thus may starve.
- No eggs or brood until queen starts to lay; about 21 days until new adult workers emerge.
- Must be fed heavily to draw foundation, because feeding stimulates wax glands to produce wax.
- Must be fed heavily at least until the first major honeyflow to keep bees from starving.
- Bees may not forage if weather is too cold or wet, which would delay colony development.
- Should be medicated.
- Bees could be infested with pests (see "Mites" in Chapter 13).

Nucleus and Established Hives

Nucleus hives (or *nucs*), which consist of from three to five frames, as well as established hives, both with laying queens, can be purchased from local dealers or beekeepers. Retiring beekeepers often sell complete colonies and extra equipment. If you need to have expert help, call your local bee inspector to assess the health of such colonies. You may be buying more trouble than you can afford. Before moving any bees and used equipment, check and comply with all legal requirements (see "Moving an Established Colony" in Chapter 11).

Advantages:
- Usually cared for by an experienced owner.
- Owner usually available for questions.
- Already assembled.
- Includes all ages of bees and brood when purchased during the spring and summer months.
- With established hives, surplus honey at the close of the season is almost guaranteed.

Disadvantages:
- Old equipment may harbor American foulbrood disease spores (viable for 50 years).
- Equipment could be of different types and sizes.
- Combs could be old and need replacing.
- Queen could be old or of poor quality and stock.
- Honey could be diseased.
- Large established colonies would be very populous and thus difficult for the beginner to work.
- Bees may be infested with mites.

Installing a Nuc

If you buy a nuc from a local beekeeper, make sure there is no sign of disease or mites; if in doubt contact the local bee inspector for advice. Merely install the purchased frames of bees and queen into one deep (if the frames are deep) hive body, and fill the rest of the space with foundation or clean drawn comb. Feed or medicate as needed, and provide additional room as the colony grows. This is an easy way to get a hive going, provided the queen is young, healthy, and of good breeding stock.

Other Methods

A *colony* of bees—consisting of thousands of workers, usually one queen, and sometimes drones—can be living in a building or in a tree. They can often be obtained free of charge from the property owner with appropriate permission. A *swarm* of bees is a portion of a colony between homesites.

Collecting swarms and bee trees was a common way for people to obtain bees free of charge. Today, because fewer feral, or wild, colonies exist and because of possible liability exposure, we caution you from getting bees in this way. We encourage you to catch a swarm, however, especially if it comes from your or a friend's bees. Generally, swarms are easy to collect, but as a precaution they should be treated as if diseased (or mite infested) and be given medicated sugar syrup after installation in a hive (see "Catching Swarms" in Chapter 11).

Bee colonies living in buildings are difficult to remove and can cost much in time and stings. Their removal is worth it only for the experience of getting bees out of buildings or for collecting the honey. Removing the entire colony and its combs successfully involves tearing off the outer portion or the inner portion or both portions of the building covering the colony. Leave this to the professionals.

Removing bees and combs from bee trees usually involves felling the tree and splitting it. Much of the comb and many of the bees, even the queen, are often crushed when the tree hits the ground.

Several beekeeping books and an experienced beekeeper should be consulted before deciding on any method of removing bees from trees or buildings; see "Books on Bees and Beekeeping" in the References.

Advantages:

- Interesting and educational.
- Free bees to augment weak hives, make nucs, or start new hives.
- Extra wax and honey from removed combs.

Disadvantages:

- Bees could be diseased or mite infested.
- Queen could be injured or killed.
- Could require a great deal of labor with little reward.
- Bees could be of inferior stock.
- Queen is often difficult to find and capture.
- Owner may expect beekeeper to repair dwelling.
- Beekeeper may be liable for others getting stung.
- Bees could be Africanized.

The Apiary

It is becoming increasingly difficult to obtain apiary (beeyard) sites in this day of urbanization, changed farming practices, and public awareness of "killer bees." Many who keep bees as a hobby put bees in their own backyards, much to the concern of their neighbors; but this need not be a deterrent if you practice a "good neighbor policy" as we discuss in this chapter. Maybe you can find a farmer or a friend with rural property willing to take bees in exchange for bottles of honey.

Choose a site to optimize these conditions (see the illustrations of ideal and poor sites):

- Away from areas subjected to pesticides.
- Close to fresh water: a stream or a dripping faucet or other device.
- Easy, year-round vehicle access.
- Near dependable nectar and pollen sources (within a 2-mile radius).
- On upper sides of slopes to improve air drainage away from hives.
- Away from wet bottomland.
- Along the edges of open fields.
- With a northern windbreak for winter protection and noontime summer shade to keep hives cool.
- Far from fire and flood areas.
- Near owner or neighbors (with clear commitments on both sides) to discourage vandals and thieves and to encourage visits.
- At least 2 miles from other beeyards, to diminish the spread of diseases and pests.

Identification

The name and address of the beekeeper should be posted at each *outyard* (an apiary that is not near the beekeeper's home). Information about the location of each of your yards should be kept in an accessible place in your residence, including maps, phone numbers, and other pertinent information, to help locate you should an emergency arise.

Posting your name and address in your outyards will enable bee inspectors to contact you if necessary. If the outyard is located on another person's property, request a signed statement from the owner that the hives are your property. Such paperwork may help you avoid any legal battles in the event of the property owner's death.

Good Neighbor Policy

To keep on good terms with your neighbors, landowners, or other people likely to come in contact with your bees, here are some simple rules:

- Keep your bees out of sight by planting tall shrubs in front of the hives to hide them and to force bees to fly higher than your neighbors' heads.
- Adhere to state and city zoning regulations (check with your City Hall).
- Obtain a gentle race of bees, and keep them gentle by requeening.
- Learn to keep your bees calm; if possible, work them only during sunny, warm weather or a nectarflow, and don't start a robbing incident.
- Erect a sturdy fence to keep out curious children.
- Have no more than 2–3 hives at a home property site and not more than 20–30 in an out apiary.
- Don't work hives if neighbors are outside working in their yards.

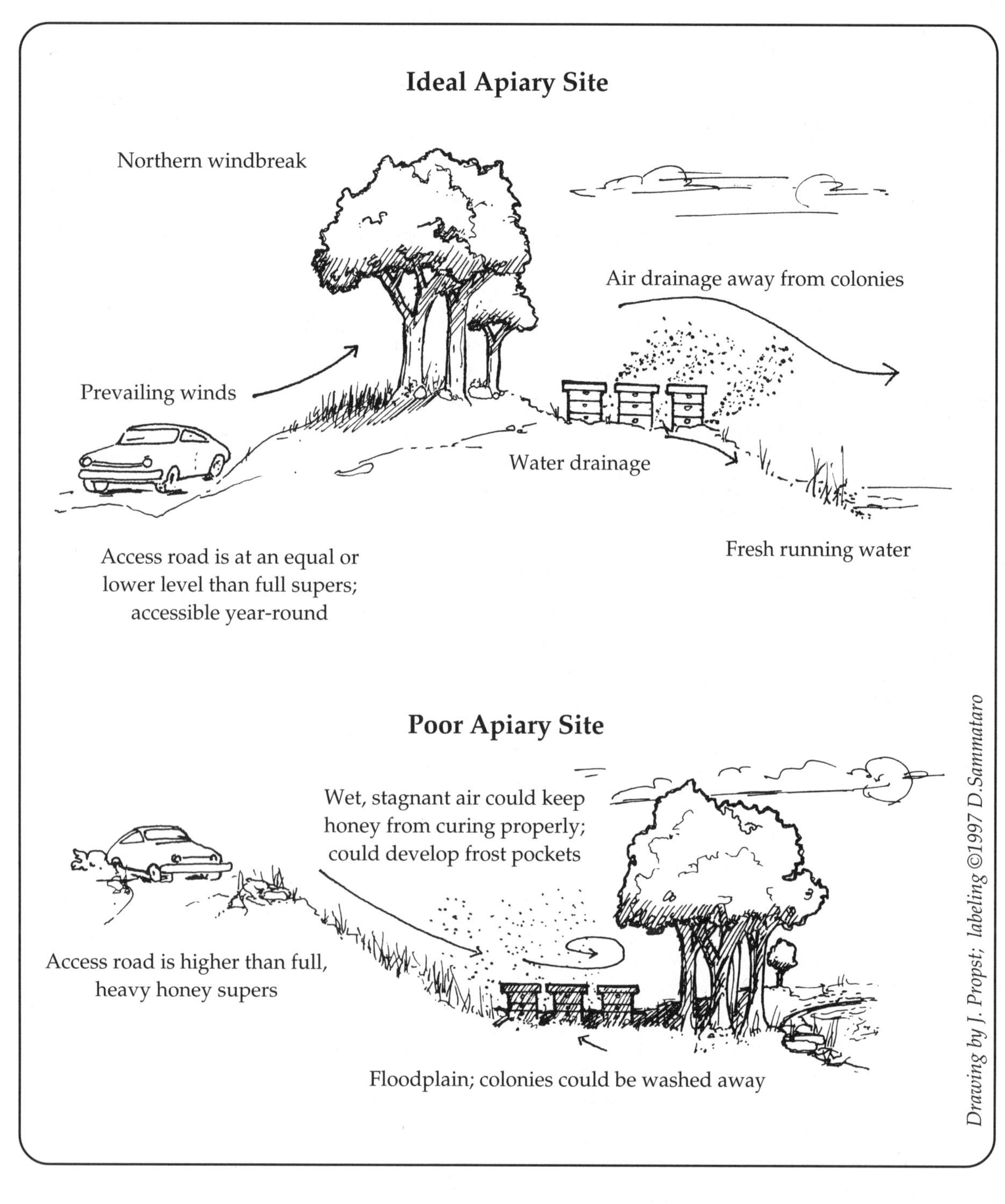

- Prevent your bees from swarming so as not to alarm your neighbors.
- Provide and maintain a water source so your bees are not in a nearby pool or birdbath.
- Give honey freely to next-door neighbors, along with a magazine or book explaining how beneficial bees are to people.

Hive Scale

A hive scale is a device that is placed under a strong colony and from which accurate records of weight gains and losses can be made. These scales can be a valuable piece of equipment. If, for example, the scale shows that the hive has grown daily heavier, it means a strong honeyflow is in progress and the hives can be *supered,* that is, have extra supers (shallows or mediums) placed on top of the broodnest. The scale formerly used to weigh hives was a farmer's grain scale, but several scales specially designed for bee colonies may be available at bee supply stores. Check back issues of bee journals too.

When a honeyflow is in progress, the hive gains weight because of the incoming nectar. You should be alerted to do certain tasks, depending on the season:

- Add frames or supers full of foundation, because the worker bees' wax glands are stimulated during honeyflows.
- Add extra supers for honey storage (supering).
- Interchange the locations of weak and strong hives (see "Prevention and Control of Swarming" in Chapter 11).
- Check hives for swarming preparation, especially during or shortly after a spring honeyflow (see "Swarming" in Chapter 11).
- Requeen (see Chapter 10).

Whenever the scale shows a hive gaining weight, the beekeeper should check and note which flowers are in bloom in order to anticipate nectar flows in future years. Occasionally, hives may gain weight when no major nectar plants are in bloom. In this instance, bees may be gathering water or honeydew, a sugary liquid excreted by insects feeding on plant sap. If, on the other hand, a scale records a continual weight loss, check the colonies to see why. The colonies may need to be fed to prevent starvation; they may be diseased, queenless, or weak because of parasitic mites; or perhaps their stores are being depleted by robber bees.

Hive Stands

The amount of bending and lifting that a beekeeper must do while working a colony can be minimized when the colony is placed on a stand about 18 inches (about 46 cm) above the ground (see the figure on hive stands). Such a stand, in addition to saving your back, will keep the hive dry, extend the life of the bottom board, help keep the entrance clear of weeds, and discourage animal pests. When working a colony on a stand, you can set the hive bodies that are temporarily removed from the hive on an empty super or extra hive stand.

Wood that is continuously wet or damp will quickly rot. Such pests as carpenter ants and termites are likely to nest in damp bottom boards (see "Minor Insect Enemies" in Chapter 13). Other pests, such as skunks and mice, have less easy access to hives that are placed on some sort of hive stand.

Some stands are constructed low to the ground in areas where dampness is not a problem. These low stands help to create a dead air space underneath the hive, thus providing extra insulation and enhancing the bees' wintering success.

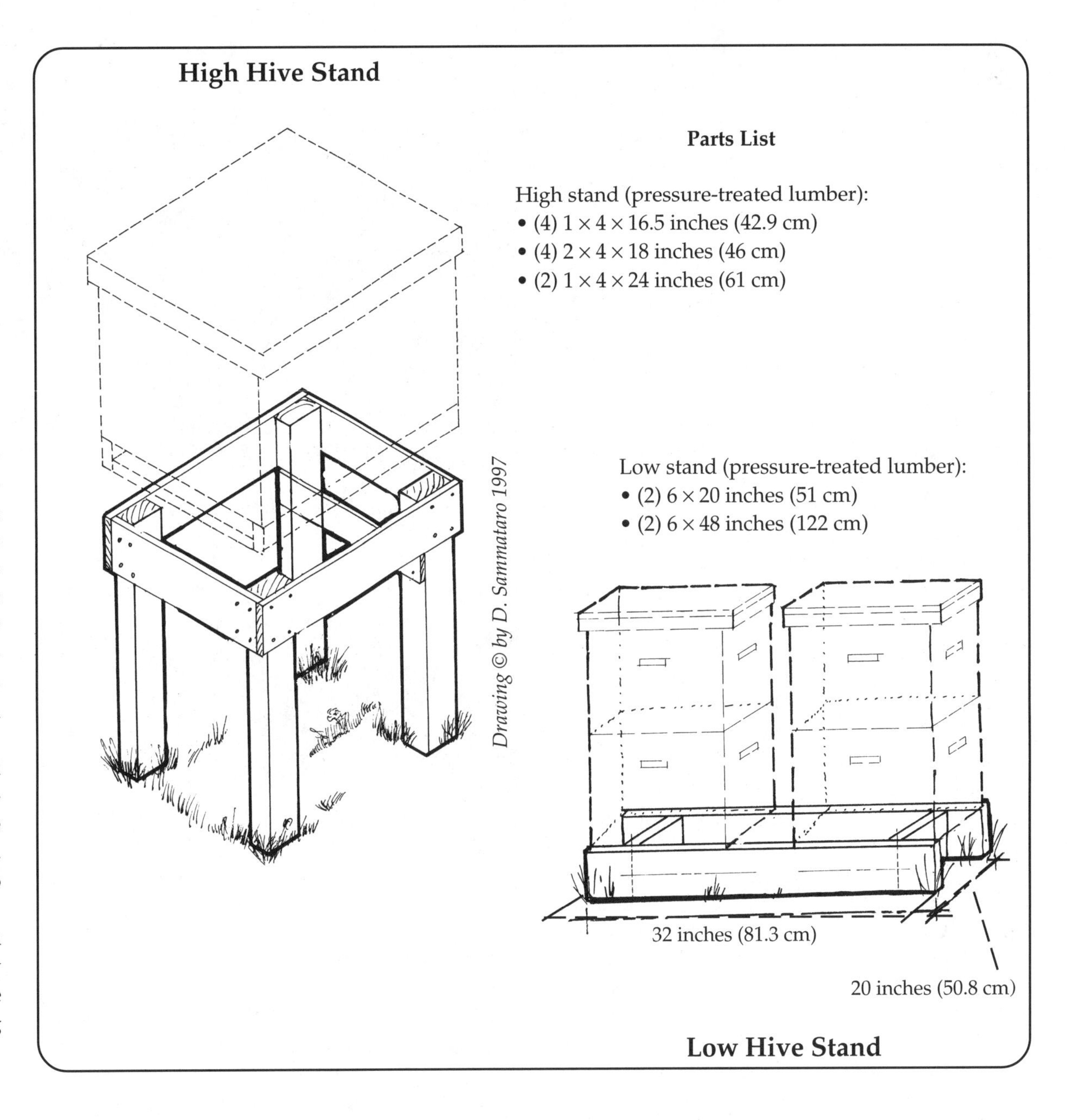

Types of Hive Stands

Hives can be kept off the ground by placing them on any one or a combination of these materials:

- Cinder blocks covered with tar paper or shingles
- Bricks or drain tiles
- Wooden railroad ties, pallets, or 2 × 4 lumber
- Wooden hive stands of durable lumber
- Permanent cement platforms or stands or even flat rocks

Hive Orientation

In most apiaries, the hives are placed in rows or paired in rows. The hives within a pair should be 6–8 inches apart (15–20 cm), and there should be 5–8 feet (1.5–2.4 m) between pairs. This spacing minimizes vibrations and jostling while working a colony but keeps the hives close enough together to make work more efficient.

When the hives are in long rows, there is a tendency for some bees to drift to the wrong hives. This drifting may be due to prevailing winds, which continually push returning bees toward the end of the rows. Drifting is undesirable because it creates colonies of unequal strength and can promote the spread of disease, mite dissemination, and robbing out of weaker colonies.

Reduce drifting by placing the hives in a horseshoe configuration (entrances facing in or out), by putting up a windbreak, or by shortening or staggering the rows (see the figure on hive orientation). If hives must be placed in a row, alternate entrances between the front and the back along the row to reduce drifting, or if this is inconvenient (because you can't drive "behind" the colonies), paint the hives or bottom boards different colors or patterns or place landmarks near the hives, such as rocks or bushes. In some areas where there is a flat horizon, bees may not be able to find their way back to a colony because of the lack of vertical landmarks. Consider erecting a snow fence or planting trees or shrubs to help the bees orient themselves.

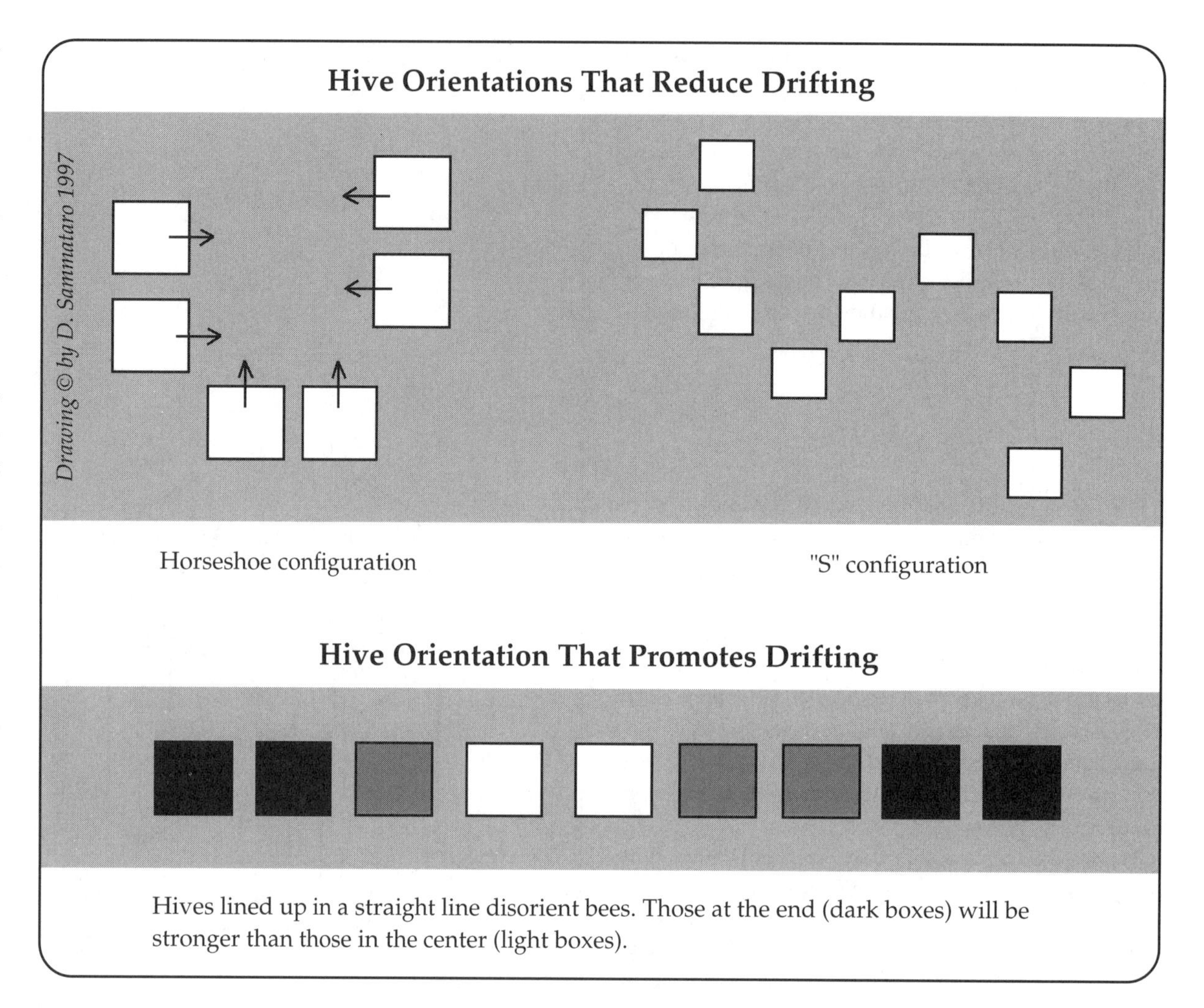

Hives lined up in a straight line disorient bees. Those at the end (dark boxes) will be stronger than those in the center (light boxes).

Record Keeping

Careful record keeping will enable you to maintain an accurate account of the condition of each hive, as well as to determine beekeeping expenses. Such records are absolutely necessary for those who desire to upgrade their stock continually. The goal that most beekeepers strive for is gentle bees that overwinter well, remain disease free, and produce a surplus of honey whenever weather and floral conditions permit (see "The Hive Diary" in Chapter 5).

A diary of the blooming times of important nectar- and pollen-producing plants will help you anticipate the time when major honeyflows and important sources of pollen will become available. This information makes it easier to plan wisely the activities necessary for successful beekeeping.

Financial records should also be kept for income

tax and loan purposes and to determine the amount of income lost or gained in a season.

Keep records of such things as the following:

- Dates of all beekeeping purchases
- Equipment bought, destroyed, stolen, or sold
- Mileage to the apiary
- Dead or stolen colonies, queen, or packages
- Pesticide loss (colonies killed by pesticides)
- Medications for the beekeeper and for the bees
- Organizational memberships; journal subscriptions
- Lectures, talks, shows, and fairs attended or entered, with associated fees
- Books and conference fees
- Equipment for selling honey (labels, bottles, and so forth)
- Amount of honey extracted, bottled, and sold
- Amount of comb honey packaged and sold

Resources

There are numerous books, pamphlets, journals, and organizations for beekeepers. In addition, talking and working with other beekeepers can be an important way to learn more about the art and science of apiculture. For the beginner and for those who wish to learn more about beekeeping, local or state groups and university extension offices usually offer workshops or seasonal meetings for beekeepers to improve techniques or to share experiences. Check the references section for books and journals.

5 Working Bees

It is important to know what you are going to look for and do before opening a hive. Be prepared so that you can keep the amount of time spent in each colony to a minimum (no more than fifteen minutes, unless you are doing some specific task that requires more time). Each time a colony is examined, the foraging activities of the worker bees are disrupted, and it may be a few hours before normal foraging resumes. During a major honeyflow, this disruption could result in a measurable drop in the quantity of honey collected.

It has been estimated that an average of 150 bees are killed every time a hive is worked. Bees that are killed or injured release their alarm pheromone, which may excite other bees to become more aggressive. Careful handling of the bees and the hive furniture can minimize the bees' release of the alarm signal and could reduce the number of stings you receive.

Avoid quick movements when working with the bees and do not jar the frames or other equipment. By proceeding slowly and gently, you allow time for the bees to move out of the way. Although killing some bees is unavoidable, the beekeeper who works slowly but precisely can keep the number of squashed bees to a minimum. Remember, work in slow motion.

When to Examine a Hive

A precise timetable for checking hives cannot be given because conditions vary from colony to colony throughout the year and some hives will require more attention than others. Here are some general guidelines explaining when to open your hives:

- In the spring, when temperatures first rise over 55°F (12.8°C); briefly check the general conditions and determine if the colony has an adequate food supply. You can do this in cooler temperatures, but don't keep the colony open too long (keep open less than 5 minutes).
- After the first fruit bloom, to check hives periodically for growth, strength, signs of swarming, and such.
- After a major honeyflow, to remove or add supers.
- Periodically after a honeyflow, to check condition of the queen and brood.
- Before the winter season sets in.

Certain changes to a hive will require that you check the colony to see how it responded. For example, check a hive:

- Fourteen days after installing a package or swarm.
- Seven days after queen introduction.
- Seven days after dividing a colony.
- Whenever pesticide damage, disease, mite infestation, queenlessness, or similar conditions are suspected.

A hive should not be examined:

- During a major honeyflow, unless absolutely necessary—for example, if disease is suspected, to requeen, or to add or take off supers.
- On a very windy or cold winter day.
- When it is raining.
- At night.

Before Going to the Apiary

The following equipment and supplies should be available before departing for the apiary. Although some of these items will not be needed during every trip to the apiary, it may be prudent to keep them near at hand, such as in the car or in the apiary shed.

- Extra hive tools
- Extra gloves, veil, and bee suit
- Extra smoker
- Matches, lighter, or propane torch
- Dry fuel for the smoker
- A squeeze bottle of water to wash sticky hands, to quench thirst, or to extinguish the smoker
- Can or jar of fresh syrup for emergency feeding and extra feeders
- Plastic spray bottle for emergency medication or for swarm capture (to dampen the bees' wings and keep them from flying off)
- Sample jars to collect bee samples for mite testing
- Extra newspaper sheets to collect comb samples for disease identification
- Extra frames, in case one has broken
- Extra hive bodies, outer covers, inner covers
- Division screen for emergency colony division (e.g., should many queen cells be present)
- Container to collect scrapings of wax or propolis, to save valuable hive products and to prevent the spread of disease
- Queen excluders
- Hive diary and extra markers, pens, and pencils
- Burlap or cotton sacking, to protect uncovered supers from robbing bees
- Newspaper to unite hives
- String, rope, or hive straps to move colonies or for other uses
- Local pest controls—live traps and bait
- Skunk/mouse guards, entrance reducers
- Hammer and nails for repairs
- Duct tape and screen to close holes and cracks
- Bee medication: Terramycin for foulbrood, grease patties for tracheal mites, and Apistan strips for varroa mites
- Extra queen cages, for those loose queens or for emergency queen introduction and marking paint (model paint)

- Apistan strips, for varroa control
- First-aid kit, including a sting kit or other medications for the beekeeper
- Pruning clippers, saw, or sickle to keep weeds under control
- Spray can of herbicide to kill poison ivy or other noxious weeds growing on your hives

The Hive Diary

Methods of keeping track of the condition of each colony vary. Some beekeepers use a system of bricks or stones placed on top of the hives in a coded manner to tell the queen's age, swarming tendencies, and the like. But because the stones can be removed by bee inspectors or others or the code can be forgotten, other methods giving more precise information should be used.

A sheet of paper stapled or tacked to the underside of the outer cover is a good place to keep records. Yet this is only a temporary solution, for bees will chew their way through the paper, and all your notes could be lost; you can stall this process by placing your notes in a plastic sandwich bag and stapling or tacking it on the cover. Another technique is to write on the inside outer cover or on the outside of the hive with a grease pencil or permanent marker, numbering your hives and noting important events.

It is best to keep some sort of a hive diary that you can fill out each time you work the colonies. Or you can purchase some inexpensive calendars and keep track of when you medicated, requeened, supered, and so on for each apiary.

Diaries are important if you plan to rear queens and select breeder queens from your best colonies. Queen characteristics should be recorded so you can tell whether a queen is up to your standards (see "Breeder Queens" in Chapter 10). In addition, by referring to the diary before going to the apiary, you will be less likely to forget any needed supplies or equipment. Every time a particular hive or group of hives is worked, note down some pertinent facts in the hive diary and add others as you go along, as follows:

- Layout of apiary, including windbreaks, water source, and compass directions
- Date
- Weather conditions (wind, temperature, humidity, and the like) the day you worked the bees
- Colony strength—that is, number of frames with sealed brood and frames covered by adult bees (do this each spring and fall)
- Characteristics of the hive—whether it is aggressive, gentle, productive
- Swarming record—how often, dates swarms found, queen source
- Manipulations made that day—reversing, supering, medicating, and so forth
- Effects of last manipulation and time elapsed—after requeening, package installation, and similar activities
- Hive weight gained or lost since the last visit
- Amount of honey harvested
- Requeening schedule—age and origin of queen
- Disease and mite record
- Wintering ability—amount of stores consumed
- Medication—what type, when, for what reason—and mite control schedule
- Number of stings received and reaction

Dressing for the Job

If you absolutely do not want to get stung (although bees can sting through leather gloves), you can lessen your chances considerably by dressing appropriately for the job, much of which depends on reading the temper of the bees and dressing accordingly (see "Bee Temperament" in this chapter). First, bees are attracted to some odors, so do not wear strong perfumes or hair sprays. Then, depending on the degree of armor you want to put on (and the outside temperature), use the checklist below to make sure you or your friends, curiosity seekers, or other visitors are adequately dressed to work or to observe you working bees (see also "Beewear" in Chapter 3).

- Bee veil, a necessity: getting stung on the face is dangerous and painful. ALWAYS WEAR A VEIL. Protect your eyes.
- Some sort of hat or helmet to keep off the sun, usually worn under the veil.
- A light shirt, bee suit, or coveralls to cover up gaps through which bees can and will crawl. A suit is best because it keeps your other clothes clean; propolis and wax stains will not come out in the wash. Avoid floppy shirt sleeves or loose material that could get trapped between hive furniture.
- Long pants, with socks over the cuffs or some other closure (duct tape, bicycle straps, or a long bee suit with elastic cuffs). The same goes for the shirt—elastic closures around the wrists are important to keep bees from crawling up your arms.
- Boots or sturdy walking shoes; many apiaries have poison ivy and other "goodies," not to mention wet grass and brambles.
- Gloves are optional but handy to have if the bees decide to get fractious. They are essential to have if you are doing work that will irritate bees, such as taking off honey or setting upright a colony turned over by bears or upset by skunks.

If you do find yourself in a stinging situation, here are some simple rules:

- Don't swat at the bee as it flies around; bees are faster than you, and swatting motions irritate them.
- If you are without a veil, push your glasses (if you wear them) close against your face to protect your eyes, and cover your mouth. Keep your head down and walk calmly into some shrubbery or forest canopy.

- If a worker lands on you or in your hair, kill her quickly before she stings you by smashing or pinching her.

Smoking

The use of smoke while working bees is essential. No hive should be opened or examined without first smoking the bees. A few periodic puffs of smoke will help keep the bees under control, but bees that are oversmoked might become irritated.

Smoke works in several ways to keep bees from stinging. When bees are smoked, they seek out and engorge honey or nectar in the hive, and bees with full stomachs are less prone to fly or sting. Also, when the hive is first opened, guard bees—which are sensitive to hive manipulations—release an alarm pheromone to alert other bees. When many bees are releasing this pheromone (when you open a hive in winter), you too can detect this alarm odor, which is similar to the odor of banana oil. The alarm pheromone causes the bees to react defensively to protect the hive from "intruders." Smoke directed from your smoker into the entrance of the hive will mask the initial release of the alarm odor of the guard bees, allowing the other bees to continue their routine hive duties rather than assume a defensive stance. Smoke may also reduce the sensitivity of the bee's receptors to the alarm pheromone; if they can't detect the pheromone, they won't release or react to it.

Smoke can also be used to drive bees away from or toward an area within the hive. Additionally, it is used to mask the alarm pheromone after you have been stung. Because the gland that releases the alarm pheromone is at the base of the sting, some of this pheromone marks the area when one is stung. Other bees that detect this signal may also sting the tagged area. Hands, clothing, and bee gloves that have been stung should also be smoked (and washed regularly) to clean off the alarm odor, dirt, honey, sweat, and disease spores.

Purchase the larger smokers: they are easier to light than the smaller ones, burn longer, and are less likely to fail when needed most. Most beekeepers carry a lidded 5-gallon pail in which they keep dried fuel, extra matches, and hive tools. Your smoker should be cleaned of soot when it builds up in the nozzle and thoroughly scrubbed with steel wool and soap after working a diseased (especially foulbrood) colony.

Lighting the Smoker

You should become thoroughly familiar with the smoker before using it at the apiary. It is a good idea to practice lighting it a few times to get the hang of it; nothing is worse than having your smoker go out when you are in the middle of examining a colony. All beekeepers have their favorite fuel, which they may use exclusively. The best fuel to use is the type that works best for you and is readily available in your locale.

Some commonly used fuels are:

- Straw or dried grass mixed with something else (wood chips or bark).
- Dry, rotted, or punk wood, which produces good smoke but can burn too hot: use it with other fuels.
- Sumac bobs.
- Brown evergreen (pine) needles and cones that are old and open.
- Cedar bark or other bark chips or bark mulch.
- Peanut shells.
- Rice hulls.
- Twigs.
- Burlap, untreated and dry.
- Wood shavings, sawdust mixture.
- Cotton stuffing.
- Rags, 100 percent cotton only, something like muslin.
- Corncobs.
- Dried horse or cow dung burns well, not much odor.

To cool the smoke, put a thin layer of green grass on top of your burning fuel. This layer catches the ash and keeps your bees from getting burned.

Synthetic materials, petroleum starters, or rags treated with pesticides should NEVER be used, because they may give off a toxic smoke when burned. Also, newspaper should not be used as the only fuel; the ash is too big and could burn the bees. Some beekeepers use binder twine (from hay bales), but this twine is usually treated with rot-retarding chemicals, which may be toxic to bees, so do not use it in your smoker.

To light a smoker:

Step 1. Drop a small amount of blazing fuel (even a small piece of newspaper) to the bottom of the smoker.

Step 2. Puff the smoker bellows and slowly add unburned material; puffing in air will help ignite the fuel as you pack the smoker.

Step 3. If your fuel is damp, add bits of beeswax, or start your fuel with a small, handheld blowtorch.

Step 4. Puff hard until the smoker stays lit.

Step 5. Once it is going, put a handful of grass or green leaves on top of the fuel to cool the smoke and catch hot ashes; make sure you don't put out your fire.

Step 6. Do not pack it too tightly, and keep filling it periodically with fresh fuel. A well-stocked large smoker should last 30–45 minutes.

After finishing work in the apiary:

Step 1. Place the hive tool(s) in the opened smoker and puff a blaze to sterilize the tools.

Step 2. Empty the remaining fuel and ashes onto dirt or pavement and drench them with water. Always have on hand some water to drench old, smoldering smoker fuel. Some beekeepers stuff a cork or green grass into the nozzle of the smoker to suffocate the fire if they have no water to put out the smolders.

Step 3. Make sure the fire is out and the smoker is cool before putting it away, and never leave a lighted smoker in a vehicle, even the back of a truck. It can be fanned into a blazing fire during just a short drive down the road.

Step 4. Sand clean the smoker and scrape out any clogging soot from inside the nozzle.

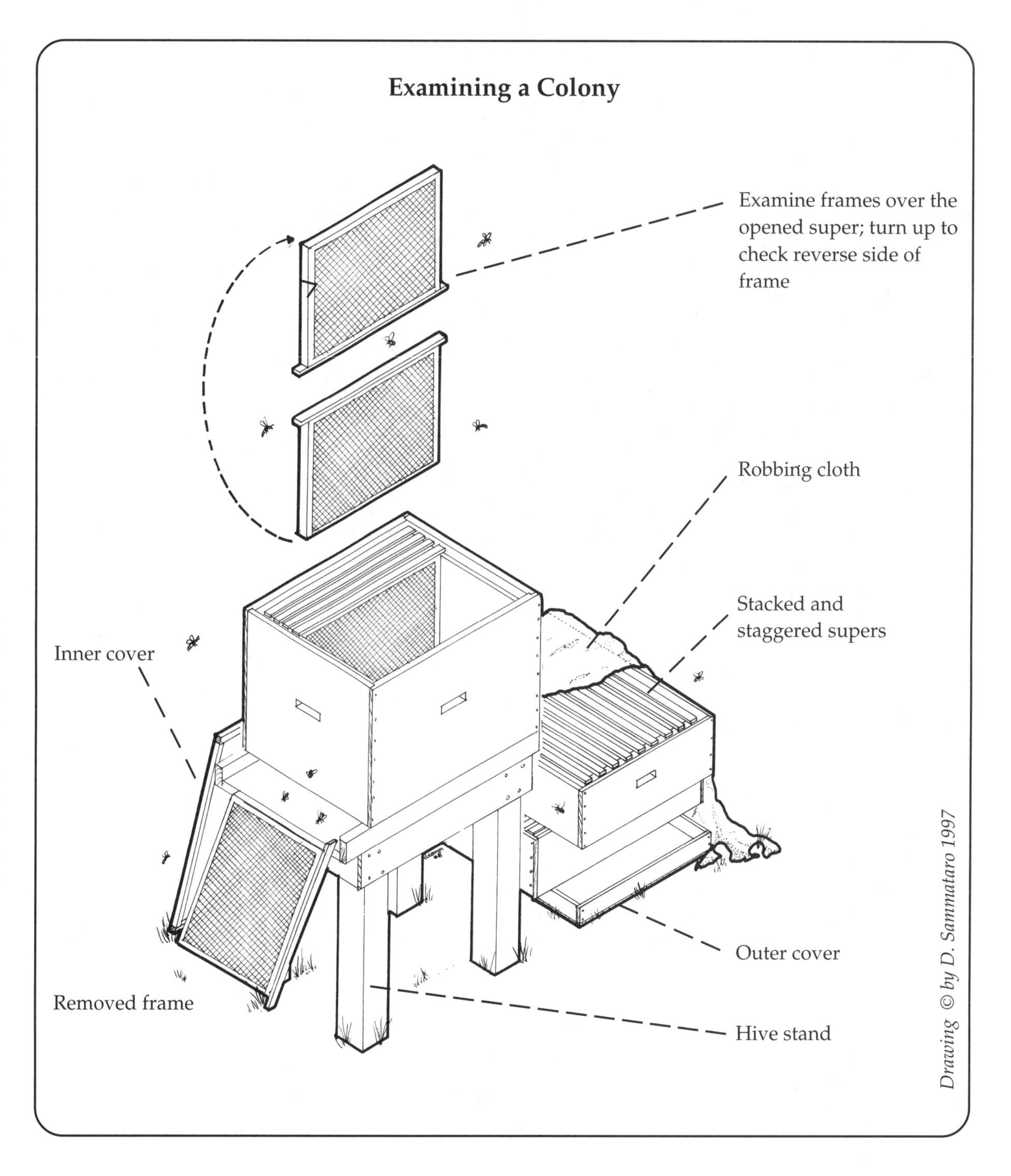

Examining a Colony

The general method used by most beekeepers to open and examine a hive is outlined below. The procedure may vary somewhat, depending on the number of supers on the hive and the purpose of the examination.

Step 1. Approach the hive from the side or back.

Step 2. Do not stand in front of the hive at any time, since the flight path of outgoing and incoming bees will be blocked.

Step 3. Puff some smoke into the entrance (be sure it gets inside), and wait 30 seconds so the bees can begin to gorge on honey.

Step 4. Gently pry or take off the outer cover and direct a few puffs of smoke through the oblong hole of the inner cover, and again wait 30 seconds for the bees to gorge honey. Then gently pry off the inner cover, puffing some smoke underneath. If an inner cover is not used on the hive, puff some smoke under the outer cover as you take it off, and wait 30 seconds.

Step 5. Lay the inner cover at the entrance so clinging bees can reenter the hive; do not block the entrance.

Step 6. After the covers have been removed, smoke the bees down from the top bars of the frames; smoke must be used judiciously—too much will cause the bees to run in every direction, making your work more difficult and decreasing the likelihood of finding the queen. In addition, too much smoke can damage brood cappings. Smoke bees just enough to make them move; experience will teach you what amount is right.

Step 7. Use the outer cover (underside up) or a spare hive body or stand as a base for stacking supers as they are removed from the hive (see the figure on examining a colony).

Step 8. As you begin to remove frames and set the supers aside, avoid quick body movements or jarring the equipment, since such actions tend to increase the defensive posture of bees. Slow, deliberate actions and the gentle handling of equipment tend to keep bees calm.

Step 9. Throughout the examination, smoke the bees as needed to keep them out of your way and to keep them from getting squashed.

Step 10. The purpose of the examination will dictate whether first to remove all supers above the bottom ones (to inspect the broodnest) or whether to work from the top down (to see if nectar or honey is being collected) during your inspection. Most of the time you will want to inspect the broodnest, which is where the queen, eggs, brood, the amount of drone larvae, and queen cells will be seen.

Step 11. Each time a super is pried off, puff a bit of smoke onto the super below and to the bottom of the one you are moving.

Step 12. If the hive is very populous, it is best to start by examining the bottommost hive body, after stacking all other supers on the upturned cover nearby (give them an occasional puff of smoke as you work). If you were to begin by working at the top, many of the bees smoked from successive operations on the upper supers would crowd to the lowest super, making it very full by the time you reach it and almost impossible to find the queen.

Examining Frames

Now you have made it to the hive body, where you want to start examining the bees.

Here's how you should proceed:

Step 1. Whenever you decide to begin your examination, smoke the bees off the top bars and down between the frames. Before removing frames, choose the one closest to the hive wall and push all other frames away from it with the hive tool; this will create sufficient space for easy removal of the frame. Avoid removing frames from the center of the hive first, as the queen may be crushed in the process of pulling out the frame.

Step 2. Once the first frame is removed, you have created more space to remove subsequent frames. Lean the removed frame against the bottom hive body or some other object, out of the sun and where it won't be kicked or jarred, or place it in an empty hive body.

Step 3. As each frame is examined, hold it vertically over the hive; in this way, if the queen falls from the frame, she may drop back into the hive.

Step 4. Continue to examine each adjacent frame until your objective has been met.

Reading the Frames

For each frame you inspect, quickly check for these items:

1. Sealed brood—It should be compact, in a concentric semi-circle at the bottom half of the frame, and have few open cells. If there are many open cells, it may mean some of the queen's eggs are not viable and had to be relaid at a later date.
2. Ratio of eggs to open larvae to capped pupae—A ratio of approximately 1:2:4 is ideal. This means there are twice as many larvae and four times as many capped pupae as there are eggs. It indicates the queen is laying constantly and the bees are of sufficient numbers to incubate those eggs.
3. No eggs found—If no eggs are found in the open cells, you can estimate how long ago the queen stopped laying by opening up some capped worker brood. Young pupae with white eyes will emerge in about seven days; if the eyes are purple, they will hatch in two to three days.
4. Queen cells—If you find supersedure cells, the queen is failing for some reason. If you find queen cups with larvae, lots of drones, sealed brood, but no eggs, the colony may swarm in about a week. If you find sealed swarm cells, some with holes in them, sealed brood, and few bees in the honey supers, the colony swarmed and a virgin queen is out. You should see eggs in about a week.
5. Other observations—Note changes in the behavior of a colony since your last visit, especially if the bees are more volatile; this could indicate lack of forage, pesticide use, pests, mites, queenlessness, or disease. Observe the amount of incoming honey and pollen in case bees are starving (no honey, dead brood on bottom board) or are becoming honeybound (honey filling all available space, even into the brood combs). Also note the physical condition of combs and frames, including any wax moth damage and uneven comb or foundation; fix any broken frames.

Step 5. Frames should be returned to their original positions and spacing unless you are adding frames of foundation, honey, drawn comb, brood, or eggs. If frames with brood and eggs are separated from the broodnest and placed elsewhere, those frames might become chilled, because the bees will have a hard time maintaining the proper temperatures in a scattered broodnest. This can result in the brood chilling, making them more susceptible to chalkbrood disease; if chilled too long, the whole frame of brood could die.

Step 6. If while working bees you see them fighting on uncovered frames, supers, or at the entrance, robbing may be in progress. This happens when there is no honeyflow. You must quickly cover the exposed equipment with a robbing cloth (such as wet burlap or other cloth) or, better yet, curtail hive examinations for the day (see "Robbing" in Chapter 11).

Step 7. When replacing supers, the bees in the super below will be milling on the top bars and rims; smoke the bees down so they will not get crushed as you replace the hive furniture.

Step 8. Whenever possible, scrape excess propolis and burr comb (comb not in the proper place) from the frames with a hive tool. These materials should be placed in a closed container; the wax can later be melted down in a solar extractor (see "Beeswax" in Chapter 12). Never discard propolis and wax around the apiary or anywhere else. Not only will this material attract such animals as skunks and bears, it could also promote robbing and can transmit diseases. Remember propolis and wax are also marketable products.

Working Strong Colonies

Working a colony more than two deep supers full of bees can be a challenge, even to experienced beekeepers. Some colonies occupy three or even four deep supers, making their examination a daunting task.

To handle such a colony, it may be best to break down the colony into a more manageable size. This can be done by taking off the top deep supers without looking at the frames; in other words, smoke the entrance as before, then take off the outer cover. Smoke the inner cover hole, and without removing the cover, crack the top deep and smoke and remove it. Place the super on the upturned outer cover, then proceed to remove the second deep.

This method will keep the house bees on the frames, not smoked to the lower supers. The end result is keeping the population of bees almost equal in all the supers. Starting at the bottommost super, begin your examination. If the colony is especially populous, you may find swarm cells, in which case you may want to divide the colony into one or two new colonies, each with queen cells, keeping the original queen with the parent hive. You don't have to use queen cells if you are concerned over the unknown quality of the queen that may result; instead, introduce a purebred or purchased queen to these splits (see "Relieving Congestion" in Chapter 11).

You may have to use more smoke on a populous colony, but the rules are the same. Work slowly, trying not to kill too many bees or knock or bump the equipment.

What to Look For

In the spring the colony must build up strength in order to achieve the peak population of 40,000 or more; such numbers are needed to secure a good honey crop.

You should be able to verify that:

- A queen and/or eggs or young larvae are present.
- There are adequate food stores (pollen, honey, or stored sugar syrup). If not present, you must provide these.
- The brood pattern is compact for both uncapped (larvae) and capped (pupae) brood; if not present, determine the cause.

Also check for and take measures to correct the following adverse conditions:

- Queenlessness: add a queen or queen cell or unite a queenless colony.
- Queen cups and/or queen cells (supersedure or swarm cells): manage appropriately.
- Presence of a failing queen or a drone-laying queen: replace queen.
- Presence of laying workers: unite with a queenright colony.
- Leaking feeders: replace.
- Crowded conditions: add extra honey supers or brood chambers.
- Overheated conditions: provide shade or additional ventilation.
- Diseases, mites, and other pests: treat accordingly.
- Robbing activities: reduce entrance and seal cracks and other openings.
- Bottom board filled with bees, debris, or propolis: clean off or replace.
- Wet, damp, or rotting bottom board: replace.
- Dwindling populations: if free of disease and mites, unite.
- Broken combs or frames: replace or repair.
- Cracked or broken equipment: replace.
- Obstructions in front of the entrance (grass, weeds): clear and mow.

Finding the Queen

The queen's presence and her reproductive state can be determined indirectly, without finding her. If you find brood frames with a concentrated pattern of capped worker cells, frames mostly filled with eggs or larvae (uncapped brood), or a combination of both, her presence and quality are indicated. The colony is healthy, headed by a fertile and productive queen.

Queen and Court

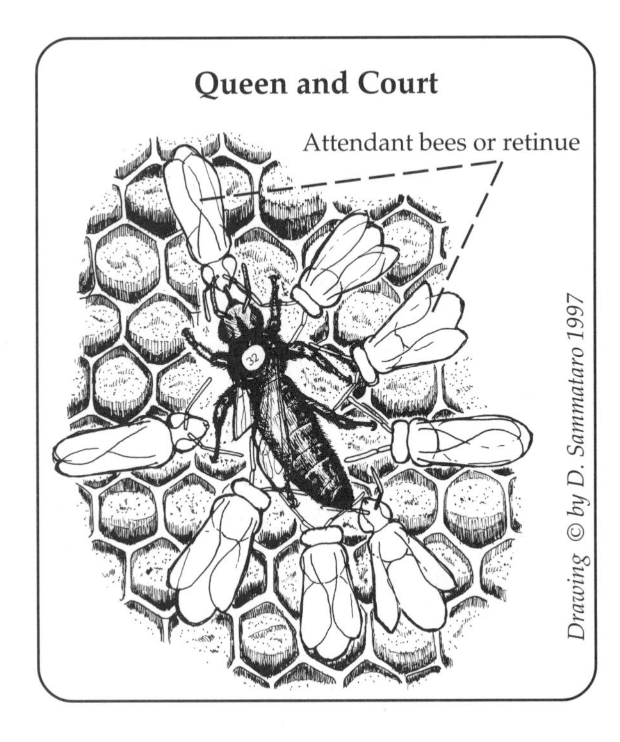

If it is necessary to find her, use as little smoke as possible, open the hive gently (as outlined in "Examining a Colony") and remove the outermost frame. She will seldom be found on frames with just honey and pollen or on frames with capped brood; she will most likely be found on or near frames containing eggs and uncapped larvae.

The queen can often be spotted in the midst of her encircling "attendants" or retinue. When a queen moves slowly along the frame from cell to cell, the other bees will begin to disperse, but the circle will be re-formed when she pauses (see the illustration of the queen and court).

If the queen must be found—whether before requeening, to kill her before uniting colonies, to mark her, or just to satisfy the need to see her—but cannot be located within 15 minutes or without disrupting the entire hive, it may be helpful to use the following method:

- Place a queen excluder between the two brood chambers (usually the two lower hive bodies).
- Five days later, the queen will be in the hive body whose frames contain eggs. Because all eggs hatch in three days, the brood chamber from which she was excluded will have no eggs.

If you do not see the queen or eggs or if the colony is not behaving normally, refer to sections of Chapter 11 for potential problems with your colony or have another beekeeper or a bee inspector check your colony.

Bee Temperament

Good Disposition

To minimize the likelihood of being stung and encounter fewer defensive foragers, work the hive on days when most field bees are foraging. Generally, it is best to work bees:

- In the spring, when populations are low and a honeyflow is on.
- During a good honeyflow.
- On warm, sunny, calm days.
- When populations are low, as with package bees.
- When bees are well gorged with food, as with a swarm or package bees that have been fed.
- Between late morning and early afternoon (roughly between 10:00 A.M. and 1:00 P.M., depending on season and time zone).

Irritable Disposition

Bees are more prone to sting when most of the foragers are in the hive. Conditions outside the hive (usually weather) are the reason for the foragers not being out. Other conditions may also occur that cause bees to become defensive.

Some things that can cause bees to become more defensive include:

- Queen temperament, genetically passed on to her offspring
- The effects of insecticides
- Disturbance by skunks or bears
- Poor honeyflow, when there is little food coming in
- Autumn, after the honeyflow has ceased
- Impending thunderstorm
- Cool, wet, cloudy days
- Hot, humid days
- Windy days
- Early morning, late afternoon, and evening
- Improper handling resulting in the killing of many bees
- Jarring of the hive or a hive part
- Disease/mite infestation
- Examining without using smoke
- Removal of honey or leaving supers or frames exposed during a dearth, thus stimulating robbing activities
- Reaction to pungent hair oils, lotions, deodorants, or perfumes
- Queenlessness
- Presence of laying workers

To minimize your getting stung, remember these rules:

- Work in slow motion; avoid rapid, jerky movements around the bees.
- Don't swat at flying bees; ignore them, and they will ignore you.
- Don't drop, bang, or bump hive parts, as vibrations upset bees.
- Don't stand in front of the flight entrance of the colony; stand to one side or at the back.

Unexpected Occurrences

When working with bees, situations may sometimes arise for which you are not prepared. Here are some of the more common unexpected happenings:

A bee gets in your veil. Kill it quickly, before it stings your head; or walk behind a tree or bush, trying not

Effects of Venom Compounds

Compound	Percentage in venom (dry weight)	Effects
Melittin	30–50	Bursts blood and mast cells; pain. Releases histamine and serotonin from mast cells; itching and swelling. Depresses blood pressure and respiration.
Phospholipase A_2	10–20	Destroys cell membranes; causes pain; synergistic with melittin.
Hyaluronidase	2	Hydrolyzes hyaluronic acid, which glues connective tissue together; allows greater penetration of other components.
Apamin	3	A neurotoxin that causes tremors.
Mast Cell Degranulating (MCD) peptides	2	Releases histamine from mast cells; pain; anti-inflammatory.
Histamine	<1	Burning itch; redness; local skin reaction.

Each sting contains about 50–150 μg of venom. Other compounds in venom that have an effect are protease inhibitor, antigen 1 (unique to honey bees), and peptide 401.

Source: Adapted from M. L. Winston, *The Biology of the Honey Bee* (Cambridge: Harvard Univ. Press, 1987), and J. O. Schmidt, *Bee Products,* ed. Mizrahi and Lensky (New York: Plenum Press, 1996).

to let other bees follow you, and remove the veil quickly to release the trapped bee.

Your smoker goes out. Cover exposed supers with extra outer cover(s) or cloth to prevent robbing, and relight the smoker.

You are chased by many bees. Blow smoke on yourself and walk casually behind bushes or trees. Be sure your smoker does not throw off flame, otherwise your clothing might ignite. Bees see movement (like fleeing bodies) very easily but are confused if many objects like branches or leaves are between them and their target.

The queen is balled. If the queen is balled when released directly into a colony of package bees or when introduced into a hive that is being requeened, don't panic. Balling may also happen when requeening a colony that already has a queen or when the hive is roughly handled. In these cases the bees consider the queen foreign and commence to surround or "ball" her. Balling will either suffocate or kill a queen by raising her body temperature too high, thus "cooking" her. Do one of the following:

- Cover the hive quickly and hope for the best (usually not the best choice).
- Break up the ball with smoke or water and cage the queen; reintroduce her using the indirect release method discussed in Chapter 6.
- Break up the ball and spray the queen with syrup, then place her on a frame of uncapped brood.
- Requeen.

The colony is exceptionally defensive. Close the hive as quickly as possible and wait for another day. Try to determine the reason for the bees' unusual behavior. Check other colonies in the same apiary to see if they exhibit the same behavior. A skunk or bear may be bothering your hives (see "Animal Pests" in Chapter 13). If only a particular colony is behaving defensively, it may be a genetic trait. If such is the case, requeen that colony (see "Requeening Defensive Colonies" in Chapter 10).

The queen flies away. During package installation, or other times that the queen is directly released or handled, she may fly off. Virgin queens are more prone to fly than mated, laying queens. In either case, do not panic. Shake a frame or two of bees at the front entrance of the hive she came from. Many of the bees will start to fan and scent at the hive entrance and will probably attract the queen to land. Watch for a cluster of bees to form at the hive entrance or on a nearby branch. If the latter occurs, the loose queen will likely be in that cluster, and you should collect it like a swarm. Lay it down in front of the hive or inside the hive (see "Swarming" in Chapter 11).

What to Do When Stung

If a worker pierces your skin with the barbed lancets of her sting, she cannot withdraw them. As the bee struggles to free herself, the poison sac attached to the lancets is ripped from her abdomen. This means that the bee will ultimately die; and having left most of her sting in your tissue, she obviously will not be able to sting again before she dies.

Other stinging insects have either smooth lancets

or lancets with ineffectual barbs; they can, therefore, withdraw that portion of the sting and repeatedly reinsert it. The queen honey bee, hornets, and yellow jackets have such a sting. Queens rarely sting beekeepers even when handled—they use the sting against rival queens.

Scrape off the sting with a fingernail or hive tool as soon as possible to minimize the amount of venom pumped into the wound. Start to scrape the skin with your nail about an inch away from the sting and continue scraping through the sting, which will pull out easily. *Speed of removal, not method, is what's important in reducing the amount of venom injected.*

Because an alarm pheromone is associated with the sting, other bees are likely to sting in the same vicinity. Apply smoke to the area of the sting to mask the alarm odor.

Treatment of Bee Stings

Local Reaction

Bee venom contains enzymes (hyaluronidase) and peptides (melittin) that cause pain (see table on venom compounds). For local reactions, there is very little an individual can do except to relieve the itching. Since the sting barbs are so tiny and the puncture so small, no treatment will be effective in reducing the amount of venom other than the prompt, proper removal of the sting structure.

Every beekeeper has a favorite treatment for bee stings. Although the treatment does not "cure" the sting, it does give a different sensation to the area and thus takes one's mind off the momentary pain. The following items are often used to relieve the pain and itching of bee stings:

- Bee sting treatment kits
- Ice packs or cold water
- Vinegar
- Raw onions rubbed on the area
- Paste made of aspirin tablets
- Honey
- Juice from the wild balsam, jewelweed, or touch-me-not (*Impatiens pallida*)
- Baking soda
- Ammonia
- Meat tenderizer, made into a paste
- Mud
- Hemorrhoid treatment cream

These treatments work best if applied immediately after being stung. But immediate application is usually impossible if you get stung away from home or through a bee suit. To give relief to the itching red welt that appears following a bee sting, apply calamine lotion, another insect bite or poison ivy preparation, or hot water.

Systemic Reaction

A good summary of bee stings can be found in J. O. Schmidt, "Allergy to Venomous Insects," in Chapter 27 of Graham's 1992 edition of *The Hive and the Honey Bee.* If you break out in a rash (hives) or have difficulty breathing after being stung by a honey bee, you are probably having an allergic reaction to bee venom. IN SUCH CASES, CALL FOR EMERGENCY MEDICAL HELP OR TAKE THE PERSON WHO HAS BEEN STUNG TO THE HOSPITAL IMMEDIATELY! TIME IS CRITICAL.

Medication for bee sting reaction can be obtained only by prescription. The drugs commonly prescribed are an antihistamine and epinephrine (adrenaline). Consult your doctor if you have questions. Here are some examples of medications:

- Injected: Anakit (Hollister Stier) and EpiPen (Center Laboratories) or other insect sting kits are available with a prescription and include a syringe filled with epinephrine (adrenaline), with instructions for it to be injected under the skin (subcutaneously). Read instructions carefully (some kits may require refrigeration) and become familiar with how to administer the medication.
- Oral: Some over-the-counter antihistamines are helpful; your doctor may prescribe more powerful ones.
- Aerosol: An aerosol bronchial applicator, as for asthma sufferers, will offer quick relief of breathlessness as a result of a bee sting. Epinephrine inhalers are also effective. The dosage of two puffs should be repeated after fifteen minutes.

Although this information provides an outline of what might be done for systemic or general allergic reactions to bee stings, exact and precise medical information should be strictly adhered to. No one should attempt to self-diagnose a response to bee stings or to prescribe medications—seek the advice of a physician.

If you develop an allergy to bee stings and wish to continue keeping bees, the only other alternative is to go through an immunotherapy session or venom allergy shots. It is expensive, complicated, and inconvenient but does usually eliminate future systemic reactions. Consult your doctor or allergist for more details (see "Bee Sting Reaction" in the References).

6 Package Bees

About Package Bees

A package of bees is a screened box containing several pounds of bees, a laying queen in a small separate cage placed inside the larger screened box, and a feeder can of sugar syrup (see the illustration of a package of bees). The package is prepared by package producers from southern states or California. In general, to make a package for sale they open a populous colony, isolate the queen, and shake the bees clinging to the frames into a funnel attached to the circular opening of the screened box. After the desired number of bees (measured in pounds) has been shaken into the package, a newly mated queen taken from a queen mating box is enclosed in a queen cage and placed in the package, usually suspended next to the circular opening for the feeder can. The feeder can containing sugar syrup is then inserted into the circular opening, a lid is placed over it, and the package is ready to ship.

The bees in the package now have a foreign queen (not their own), but because she is caged, they are unable to harm her. While in transit, the bees will come to accept her as their own.

There are many methods used to install bee packages, the basic differences being in the manner in which the queen is released from her cage. In both indirect-release methods, the queen remains caged, and the bees are allowed access to a candy plug, which they must remove to release her. This method simply delays the queen from being freed among the other bees for a few more hours or days and increases the likelihood of their accepting her.

In the direct-release method, the screen or cork is removed from the queen cage, allowing the queen to walk out onto the top bars of the hive among the other bees or into the entrance. When the queen is released directly, the bees may still not be fully acquainted with her. As a consequence, they may form a tight ball of bees around her and, by raising her body temperature, suffocate her. This process, called *balling the queen,* may result in the queen's death or permanent injury (see "Unexpected Occurrences" in Chapter 5). Combinations and variations of the methods for direct and indirect release are covered in "Installing Packages" in this chapter.

Ordering Packages

If possible, packages should be ordered directly from the producers or obtained through a local bee supply house or bee club. Sometimes a local bee supply dealer will drive south (or west) to pick up package orders in person; you could go along to help if you wanted to buy a large number of packages at a reduced price.

Order early in the year: November of the year before the bees are wanted is not too early, and January at the latest. Request that bees arrive about a month before the fruit trees bloom in your area. Advertisements by local dealers may be found in the publications of state beekeeping organizations and local bee clubs; such beekeeping journals as the *American Bee Journal, Bee Culture,* and the *Speedy Bee* include advertisements for almost all package bee producers (see "Bee Journals/Publications" in the References).

A week before the bees are expected, call the post office and leave telephone numbers where you can be reached (day and night phone numbers) so that the postal clerks can contact you when the bees arrive. If the bottom of the package has two inches of dead bees, contact the package producer for additional instructions. The package should be guaranteed against such a loss—check before buying.

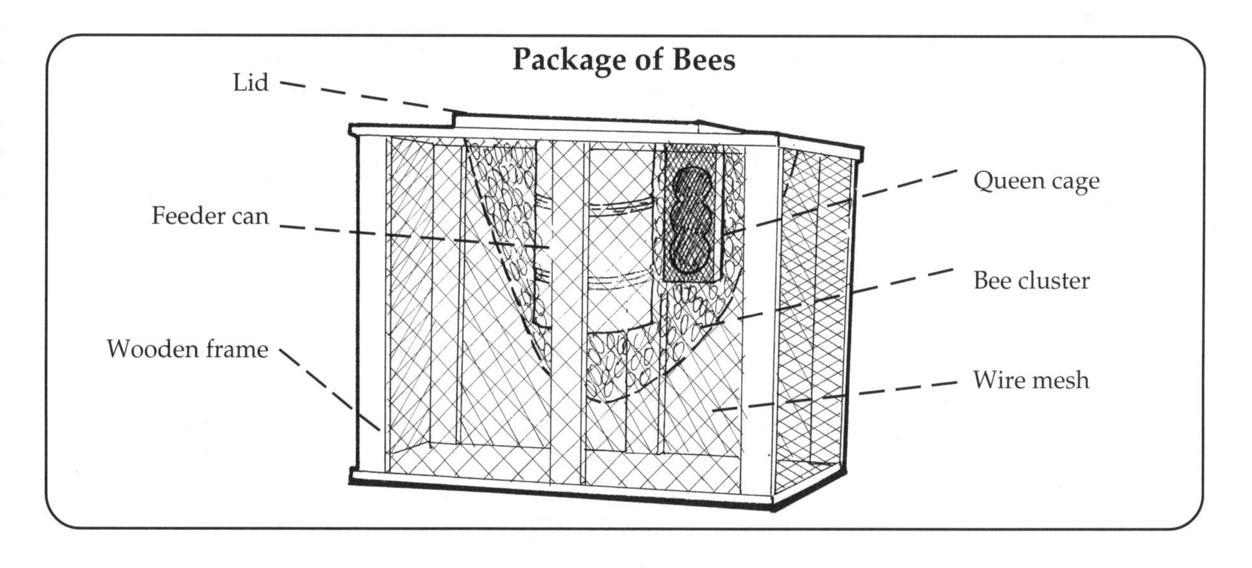

If so many bees are dead in the bottom (and even if there are few dead bees), check to make sure the queen is alive. If she did not survive, notify the producer immediately and ask for a replacement and when it can be expected. Any replacements must not be delayed or some workers will undergo ovary maturation and begin to lay eggs. Because laying workers can produce only drone eggs, the colony would be doomed (see "Laying Workers" in Chapter 11). If the queenless package bees can be provided with one or two frames of eggs and uncapped larvae from an established colony, laying workers will not develop, and the bees will raise a new queen. But the colony would lag behind in growth because it takes three weeks for a new queen to develop, be mated, and lay eggs. Because few drones are available in the spring, the queen may not be mated, further delaying colony development. A better solution would be to add a queenless package to an existing but weak colony, which would bolster its population and morale and could make it a productive colony.

When the Packages Arrive

When the package finally arrives and everything seems to be in order, check for the certificate of inspection of each package. This certificate means that the dealer's bees were inspected for pests and diseases prior to their sale (see "Mites" in Chapter 13).

The bees may be buzzing loudly and wandering all over the package. They are not "mad" or ferocious. As soon as possible, the package should be placed in a cool (not cold), draft-free, quiet, and darkened area and fed heavily with sugar syrup: the bees will soon become calm. Feed the bees liberally with sugar syrup from a spray bottle (but do not soak them), or sprinkle the syrup on the screened sides of the package. Some beekeepers brush the syrup on the screen, but doing so can injure the bees, many of whom will have their tongues and feet protruding through the screen.

Prepare the syrup, which should consist of a mixture of from one to two parts white sugar to one part warm water, before the bees arrive (see "Sugar Syrup" in Chapter 7). The syrup should be medicated with Fumidil-B and antibiotics (see "Foulbrood Disease Chemotherapy" in Chapter 13).

Install the package in the late afternoon. If the weather is unusually cold, wait for the weather to improve (but do not wait more than a few days) and continue the feeding.

All equipment should be readied and in place well before the bees' arrival; equipment should include a deep hive body with ten frames of foundation or drawn comb, a bottom board, inner cover, outer cover, and entrance cleat (see "Basic Hive Parts" in Chapter 3). Hive entrances of the empty equipment should be closed until the bees are installed to prevent mice from entering the hive and damaging the combs.

Installing Packages

Indirect-Release Method I

The bees should be fed with syrup almost continuously for the last half hour before the package is installed so that they will remain calm (see "Sugar Syrup" in Chapter 7). You can start packages on frames of foundation, empty drawn comb, or a combination of both. If using a combination, place the drawn combs in the center, because the queen can start laying eggs in the cells almost immediately. Make sure that any used frames are from hives that had no diseases; foulbrood and nosema spores last a long time. If there is any question, disinfect used combs (see Chapter 13). In addition, it may help new packages, especially if started on foundation, to have additional pollen patties or supplements. If the weather is bad for a few weeks, the bees may not be able to forage for pollen, which is required for feeding larvae; see Chapter 7 for more information.

To install a package, follow these steps:

Step 1. Take the package to the preassembled hive.

Step 2. Shake or jar the package so the bees drop to the bottom of the package. This will require you to bump the package firmly on the ground, so all the bees fall to the bottom.

Step 3. With your sprayer, lightly mist the bees with syrup or water to coat their wings, but do not soak them.

Step 4. Remove the lid, exposing the top of the feeder can and the queen cage.

Step 5. If the queen cage is attached to a metal tab adjacent to the feeder, remove the cage and replace the lid.

Step 6. If the queen cage is hung from a wire or piece of screen next to the feeder, grasp the wire tab to keep the cage from falling into the package. Remove the feeder can and then the queen cage. Replace the lid to contain the bees.

Step 7. If the queen is alive, remove the cork from the end of the queen cage that contains the white candy; poke a hole in the candy plug with a nail, being careful not to impale the queen. The candy will delay the queen's release, helping to ensure her acceptance by the other bees. Various types of queen cages are now in use; see "Types of Queen Cages" in Chapter 10.

Step 8. If no candy is present, after removing the cork, plug the hole with a midget marshmallow (have some on hand, just in case).

Step 9. Suspend the queen cage between the fifth and sixth frames of the hive, with the screen face oriented so the screen is not against the comb surface and the candy end is up (see the illustration of the first indirect-release method). Do *not* place the cage directly under the oblong hole of the inner cover, because syrup dripping on the cage could kill the queen.

Step 10. Remove the package lid and shake approx-

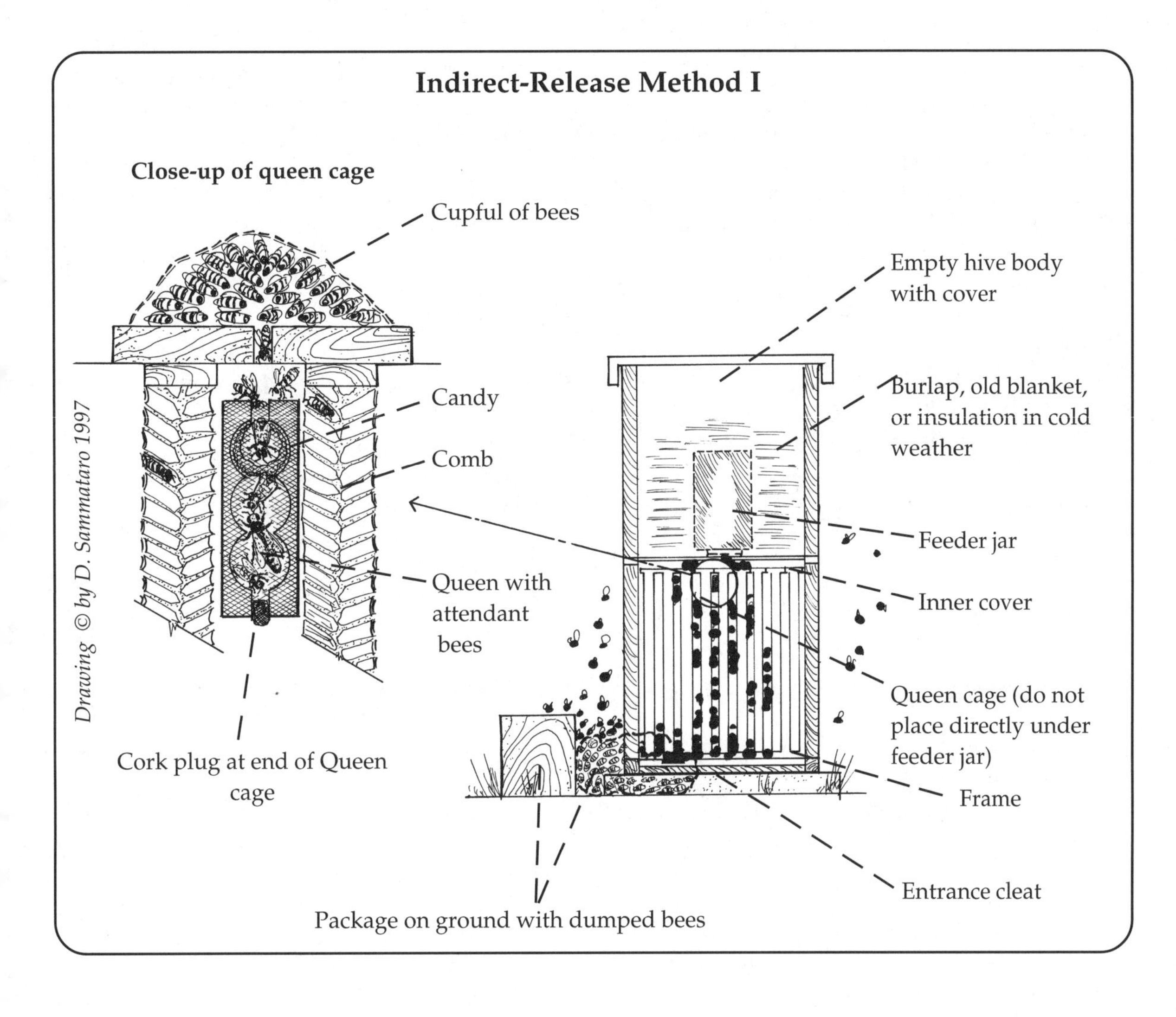

imately a cupful of bees onto the queen cage suspended in the hive; replace the package lid.

Step 11. Place the inner cover on the hive, rim side down, to allow extra room above the top bars. Invert the feeder can or jar over the oblong hole of the inner cover; invert the jar so that initial drippings will fall on the ground away from the hive, otherwise syrup dripped on the hive or inner cover may invite robbing bees. If the feeder leaks, get another (see "Screw Top Jars or Feeder Pails" in Chapter 7).

Step 12. Place any additional patties (pollen or grease) on the top bars of the frames. Then place an empty hive body over the inner cover and feeder can and cover with the outer cover.

Step 13. Again lightly spray the remaining bees in the package with syrup.

Step 14. Remove the package lid and shake a third of the bees out in front of the hive, allowing them to walk into the entrance. If it is cold, however, use the direct-release method instead.

Step 15. The freed group of bees will soon begin to *scent* (their heads will face the entrance, abdomens will be raised, with wings fanning), releasing a pheromone from the Nasanoff gland, to attract loose bees to the hive.

Step 16. When the bees begin to enter the hive, slowly shake the rest of the bees from the package directly in front of the hive. By shaking bees outside, all the dead bees are left outside the hive. Drifting may be a problem if the weather is warm, so do this late in the afternoon and when the weather is cooler.

Step 17. After most of the bees have entered, partially block the entrance of the hive with a reducer cleat or with grass; leave the entrance partially blocked for two months (replacing grass when needed) to discourage robbing.

Step 18. Leave the package near the hive entrance overnight, open end up, to let any remaining bees escape into the hive.

It is not necessary for you to put Apistan strips in a package colony, for the bees should have been exposed to a strip in the package. For the proper timing of using strips to control varroa mites, see "Varroa Mite" in Chapter 13.

After one week, check to see if the queen is released. If she is, then remove the cage. Light your smoker and gently blow smoke into the empty hive body on top, then at the entrance. Smoke the bees down just enough to retrieve the cage and close up the hive. If she is still in the cage, pull off the screen and let her walk onto the frames.

For the next 14 days, do not disturb the colony except to replace syrup in the feeder can. When replacing the syrup, have a lit smoker ready, and first blow smoke into the empty hive body at the top. Smoke around the feeder; then tilt up the empty feeder, direct smoke into the oblong hole of the inner cover to move the bees away, place a full feeder on top, and close the hive.

The temptation to look at a new colony is more than most people can stand. To satisfy that urge, observe the entrance activities of the colony instead. For example, look for:

- Undertaker bees removing dead bees.
- Incoming workers with pollen (observe colors of pollen).
- Guard bees at the entrance, challenging incoming foragers.
- Bees fanning at the entrance.
- Orientation flights, during which the bees learn to locate their hive.

If you do this every day, you can learn a lot about the foraging activity, colony organization and structure, and pollen plants blooming in your area and if the colony is progressing normally. If you do not see these activities, there may be something wrong that requires your attention, so you should look into the colony.

Finally, open the hive on day 15 after installing the package (or before, if the colony is not behaving in a normal fashion and the weather is warm); use smoke as needed. If one or more of the frames shows a fairly compact brood pattern (capped cells and open cells full of eggs and larvae), all is well. Close the hive and leave it undisturbed for another week.

During the next visit to the hive, remove the queen cage and refill the feeder and patties as needed; continue feeding the colony until the first major honeyflow or until they stop eating what you provide. Two months after installing the package, you should be able to add a second hive body with frames. If the first is not full of drawn comb and brood, call a bee inspector or other experienced beekeeper to determine the problem. If there is no major honeyflow and the new hive body still contains frames with foundation only, continue to feed the bees with syrup. If the bees are not fed, they will chew the foundation.

Advantages:

- Excellent chance of queen being accepted.
- Syrup located in vicinity of bees and queen, so likelihood of starvation is slight.
- Easy way to feed medicated syrup.
- Bees will not leave hive if queen is caged and unable to fly.
- Dead bees in bottom of the package will fall on the ground and not inside the hive.

Disadvantages:

- Some drifting occurs.
- An extra hive body is needed.
- May take a little more time than other methods.
- Queen cage must be removed at a later date.
- Egg laying delayed because queen is not released immediately.

Indirect-Release Method II

Follow the same procedures as in the first method as far as removing the queen cage, and then follow this sequence:

Step 1. If the weather is cool, place the queen cage in your pocket, screen side away from your body; if the weather is warm, place the queen cage in the shade.

Step 2. Remove four or five frames from one side of the hive body.

Step 3. Suspend the queen cage between two frames, after poking a hole in the candy plug (as in method I).

Step 4. If the weather is cool, shake some bees onto the queen cage to keep her from becoming chilled.

Step 5. Place the entire package in the vacant space in the hive (where the frames have been removed), being sure that the open end of the package is up, to allow the bees to escape (see the illustration of indirect-release method II).

Step 6. Replace the inner cover, rim side down, and feed the colony, as described in the previous section, with sugar syrup from a feeder can or jar.

Step 7. After placing an empty hive body on top and covering it with the outer cover, reduce the entrance with a cleat or with grass (as in method I).

Step 8. Inspect the hive to check the queen and to refill the feeder can as outlined for the first method.

Advantages:

- Excellent chance of queen being accepted.
- Bees disturbed less than with method I.
- Less drifting.
- Easiest for beginners.
- Dead bees are left in the package and not on the hive floor.

Disadvantages:

- Additional trips must be made to remove the queen cage, remove package, and replace frames.
- Egg laying delayed because the queen's release is delayed.
- Bees may build comb in the package, not in the hive proper, and the queen might start to lay eggs in these combs.
- Extra hive body needed.

Direct-Release Method

Follow the procedure outlined for the first indirect-release method up to removing the queen cage. Then follow this sequence:

Step 1. If the weather is cool, place the queen cage in your pocket, screen side away from your body; if it is warm, put the cage in the shade.

Step 2. Remove four frames from the middle of the hive.

Step 3. Remove the lid from the package, and shake all the bees onto the bottom board in the vacant space (or onto the ground at the hive entrance).

Step 4. Spray the bees with syrup to reduce their flying ability, but do not soak them.

Step 5. Dip the queen cage in syrup or spray it so the queen will not fly off when released; do not dip it if the weather is cold.

Step 6. Carefully remove the screen from the queen

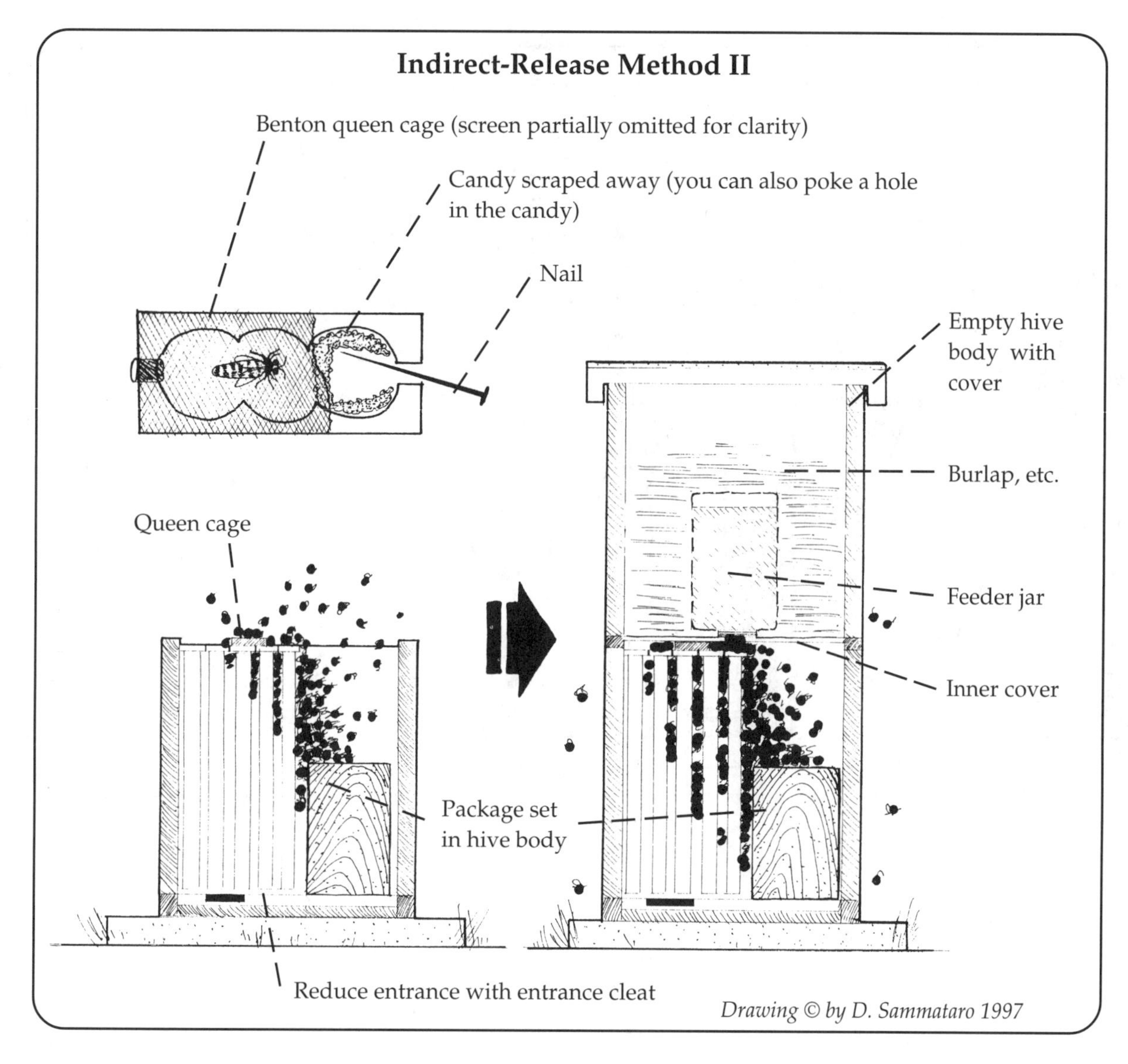

cage (see the illustration of the direct method).

Step 7. Allow the queen to walk or drop gently on top of the bees.

Step 8. As soon as the pile of bees disperses, carefully replace the frames, taking care to avoid crushing the bees (unless you released the bees on the ground).

Step 9. Replace the inner cover; follow the remaining steps in indirect-release method I.

Advantages:

- Queen released and can start to lay sooner.
- Easiest, fastest method; complete in one operation.
- No return trips to apiary needed except for feeding.

Disadvantages:

- Extra hive body needed.
- Queen could be killed or balled.
- Bees could leave (abscond) with free-flying queen.
- Queen could fly away during installation (taking bees with her) or be otherwise lost.
- Queen could be superseded.
- Some bees, the queen, or both could be injured or killed when frames are replaced.
- Drifting occurs.
- Any dead bees in the package are also shaken into the hive, making extra work for bees to remove them.

Combination Method

The combination method follows the same sequence as the direct-release method except that the queen remains caged and is released by an indirect-release method. In other words, the bees are shaken directly into the hive, but the queen is kept caged so the bees will have to free her. The advantages of this method are the same as for the first indirect-release method; the disadvantages are the same as for the direct-release method, except for references to the queen.

Reasons for Package Failures

Beehives started from packages sometimes fail after they are installed. Reasons for failure include, but are not limited to, the following:

- Queen has been superseded (due to nosema disease or other reasons, such as improper mating of the queen or her injury in shipment).
- Queen is balled as a result of too many hive disturbances by the beekeeper (especially during the first 10 days after installation).
- Weather has been too cold for bees to forage or obtain syrup because of the feeder's location or the type of feeder.

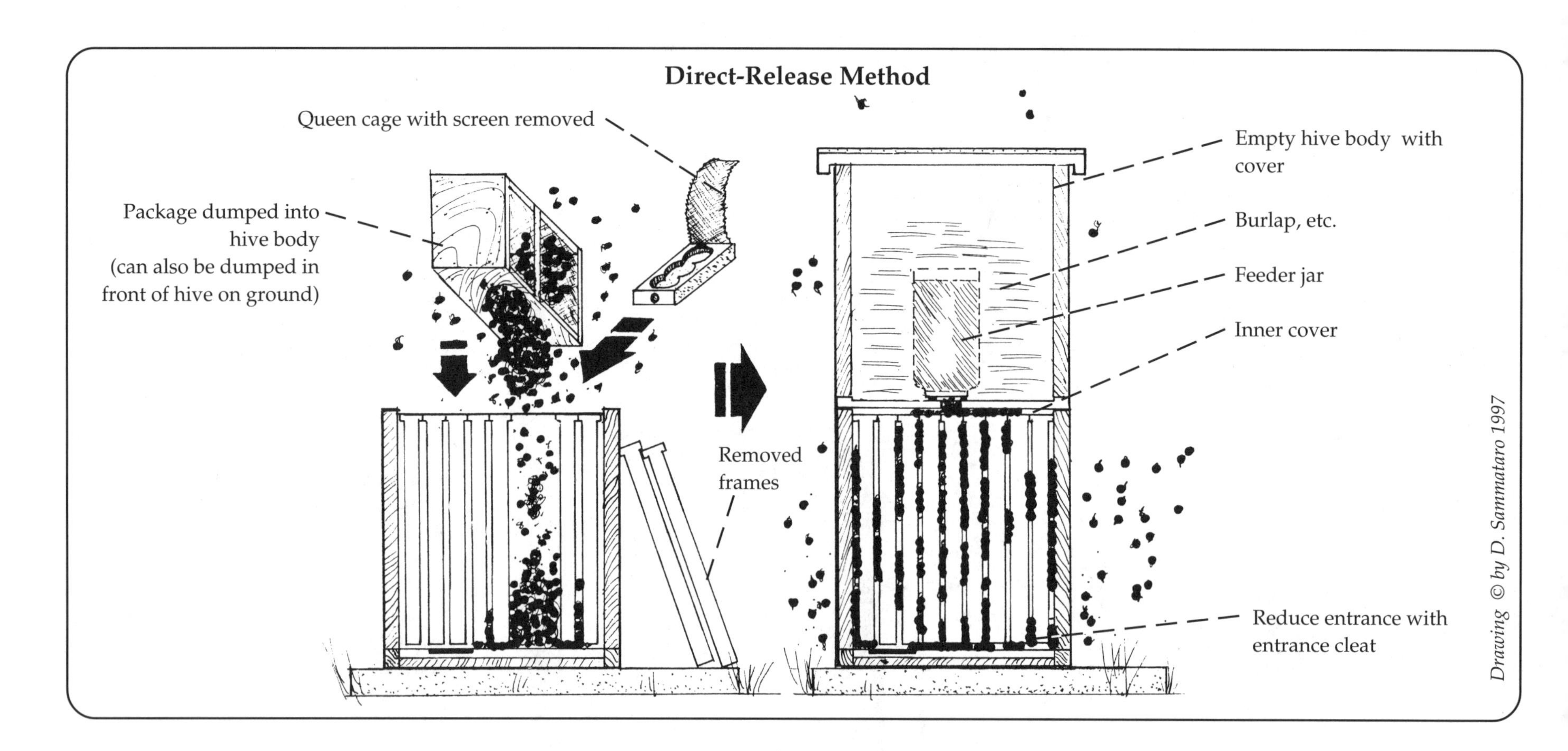

- Bees have starved.
- Bees died because of disease or mites.
- Bees have left the hive.

7 Feeding Bees

At certain times, it is necessary to feed bees supplemental food, either honey or pollen or substitutes. Recognizing when bees are near starvation is an important step in learning beekeeping and could save the colony. Bees can exhaust their own food stores or for other reasons be unable to build up existing stores and eventually deplete them. In either case, the colony will be hard pressed to stay alive and must be fed. If this situation occurs during the flowering period, the colony may continue to exist on a day-to-day basis, but it will be weakened and, should an interval of inclement weather set in, may perish.

Bees should be fed under the following conditions:

- When no natural honey or pollen is available (in late winter or early spring), in order to stimulate brood rearing.
- When the colony is in any danger of starving.
- When it is necessary to supply medication (chemotherapeutic agents).
- When installing a package or hiving a swarm, as well as to stimulate wax glands when these bees are given foundation to draw.
- When requeening.
- When rearing queens and no natural honeyflow is on.

If stores are exhausted in the fall, winter, or early spring, the colony will die. Stores may have been reduced in these ways:

- The beekeeper removes too much honey, particularly in the fall.
- The bees eat up the last of the winter food in late spring.
- The number of field bees becomes reduced because of spring dwindling (see "Spring Dwindling" in Chapter 8).
- The bees' food consumption increases when egg laying resumes in midwinter, to provide heat and food for brood.
- An expected honeyflow fails to materialize, inclement weather sets in at the time of the honeyflow to prevent bees from collecting fresh food, or a plant fails to yield expected food.

When colonies are in a condition such that starvation is imminent, the bees must be fed to ensure their survival. The various methods of feeding bees with sugar syrup, dry sugar, honey and pollen, and substitutes are discussed in this chapter (see also "Feeding Bees" in the References).

Sugars

Carbohydrates are organic compounds composed of carbon (C), hydrogen (H), and oxygen (O) atoms with a general formula of $C_nH_{2m}O_m$; in other words, there are twice as many hydrogen atoms as oxygen atoms, and the number of carbon atoms can vary. Because carbohydrates are composed of many C—H bonds, which liberate a great deal of energy when they are broken, these molecules are excellent candidates for energy storage. Sugars, starches, and cellulose are examples of carbohydrates.

Three different types of sugars that bees feed on are glucose, fructose, and sucrose. *Glucose* (also called dextrose; $C_6H_{12}O_6$) is found in all living cells. It is produced by photosynthesis in green plants and is the primary energy storage unit. Glucose is a simple sugar (a monosaccharide) that comes in two forms: a straight chain molecule and, when mixed with water, a six-sided ring structure or hexagon (see the figure on sugars).

Fructose (levulose or fruit sugar) is another monosaccharide and has the same formula as glucose. It is the sweetest of the simple sugars because of the placement of the double-bonded oxygen, which is attached internally in the molecule, rather than on the end, as in glucose. Two different ring structures are possible in fructose molecules: a hexagon and a pentagon (five sides).

Sucrose (cane sugar; $C_{12}H_{22}O_{12}$) is a disaccharide, or two sugars, made up of glucose and the pentagon form of fructose molecules linked together. The major component of nectar, sucrose is broken down into the two monosaccharides by the action of the enzyme *invertase* (see Chapter 12). An enzyme is a protein that serves as a catalyst in chemical reactions, building up, breaking down, or rearranging the atoms.

Sugar Syrup

One gallon of sugar syrup (2:1 ratio of sugar to water) will increase the food reserves of a colony by

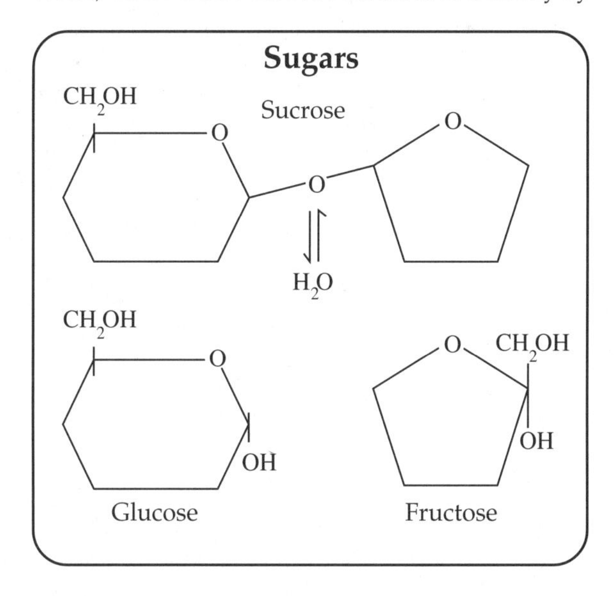

about 7 pounds. The following proportions (by volume) of sugar to water should be fed, depending on the season and the purpose for the feeding:

- 1:1 ratio, sugar : water, for spring feeding.
- 2:1 ratio, sugar : water, for fall feeding.
- 1:2 ratio, sugar : water, to stimulate brood rearing. Make only two holes in the lids of the gravity feeders so the bees will be able to obtain just small amounts over an extended period of time; this effect will be similar to a light nectar flow.

Use white, granulated cane or beet sugar only. Never use brown or raw sugar, molasses, or sorghum, because these contain impurities and can cause dysentery in bees.

Mix the desired proportions of sugar and hot water and stir adequately until all the sugar is dissolved. Hot water from the faucet is hot enough to dissolve the sugar; or mix the sugar with hot—not boiling—water that has been heated over a stove. Never let the sugar-water solution boil over direct heat; syrup that is burned, or caramelized, will cause high bee mortality. Heating the mixture over steam or in a double boiler will prevent caramelization.

To prevent fall syrup from crystallizing, some beekeepers add cream of tartar (or tartaric acid) to the solution of sugar and warm water. Tartaric acid breaks down the sugars, but there is some concern that it may be detrimental to bees, thus lately, its addition has not been recommended. To keep the syrup from molding too fast, some beekeepers add a teaspoon of apple cider vinegar to each gallon.

Bees should be fed early enough in the fall so that the sugar has time to cure—that is, the bees have time to reduce the water content near to that of honey (about 18 percent). If the syrup does not cure, it could ferment or freeze.

SCREW TOP JARS OR FEEDER PAILS

One of the best ways to feed bees at almost any time of year is with a 5- or 10-pound glass jar or a 1-gallon plastic pail with screened hole, turned upside down over the top bars of the hive or over the oblong hole of the inner cover. The jar lid can be perforated with a few small holes so the bees can insert their tongues into the holes to withdraw the sugary solution.

Plastic Feeder Pail

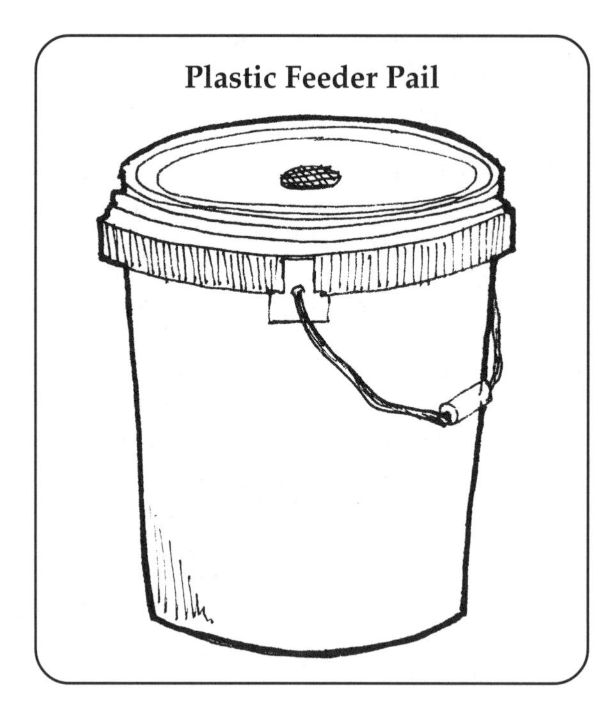

Don't use plastic jugs, because they often collapse after being filled and inverted. Tin pails become rusty, are difficult to clean, and are expensive to purchase; these are not recommended either. Plastic buckets work well (see the illustration of the feeder pail), as do glass jars, which often can be obtained free of charge and, although breakage does occur, are easier to clean and inexpensive to replace—you can readily see if they need to be refilled.

Use a shingle nail or a three- or four-penny nail (1/16-inch diameter [1.6 mm]) to punch from six to ten holes in the jar lid, after removing the cardboard washer from a screw top lid. Plastic pail tops can have a 1-inch hole burned out of the center and a fine metal screen melted in place to cover the hole.

At the hive, first invert the jar or pail so drippings will fall into some container or on the inner cover, rather than on the hive or ground, so as not to encourage robbing. As soon as the dripping stops, place the feeder over the top bars near the cluster if the colony is weak; otherwise, place it over the oblong hole of the inner cover. Put an empty hive body around the feeder, and replace the outer cover.

The syrup will not leak as long as the holes are not too large and the feeder is level. If there is empty drawn comb in the hive, the bees will remove the syrup from the feeder and store it in the comb below.

DIVISION BOARD FEEDER

The division board feeder is a frame-sized container that can be inserted in place of a frame within a deep hive body. Feeders of this type have some kind of flotation device, a strip of screen, or roughened sides that allow the bees to reach the syrup without drowning. Older-model division board feeders are made with Masonite, wood, or plastic and usually have Styrofoam, screen, or wooden floats. Most plastic division board feeders now have roughened inside walls to provide footing for bees.

Some beekeepers keep one division board feeder in each deep hive body. On cold days, however, bees in weak colonies will be less able to obtain syrup when this type of feeder is used, unless it is located near the cluster. If the cold weather continues for a long duration, the bees may be unable to move to the feeder to take up the syrup and could starve.

MILLER FEEDER OR TOP FEEDER

Another type of feeder, called the Miller feeder (also called a Miller super or top feeder) is composed of two aluminum or plastic pans fastened together and hung within or otherwise attached to an empty

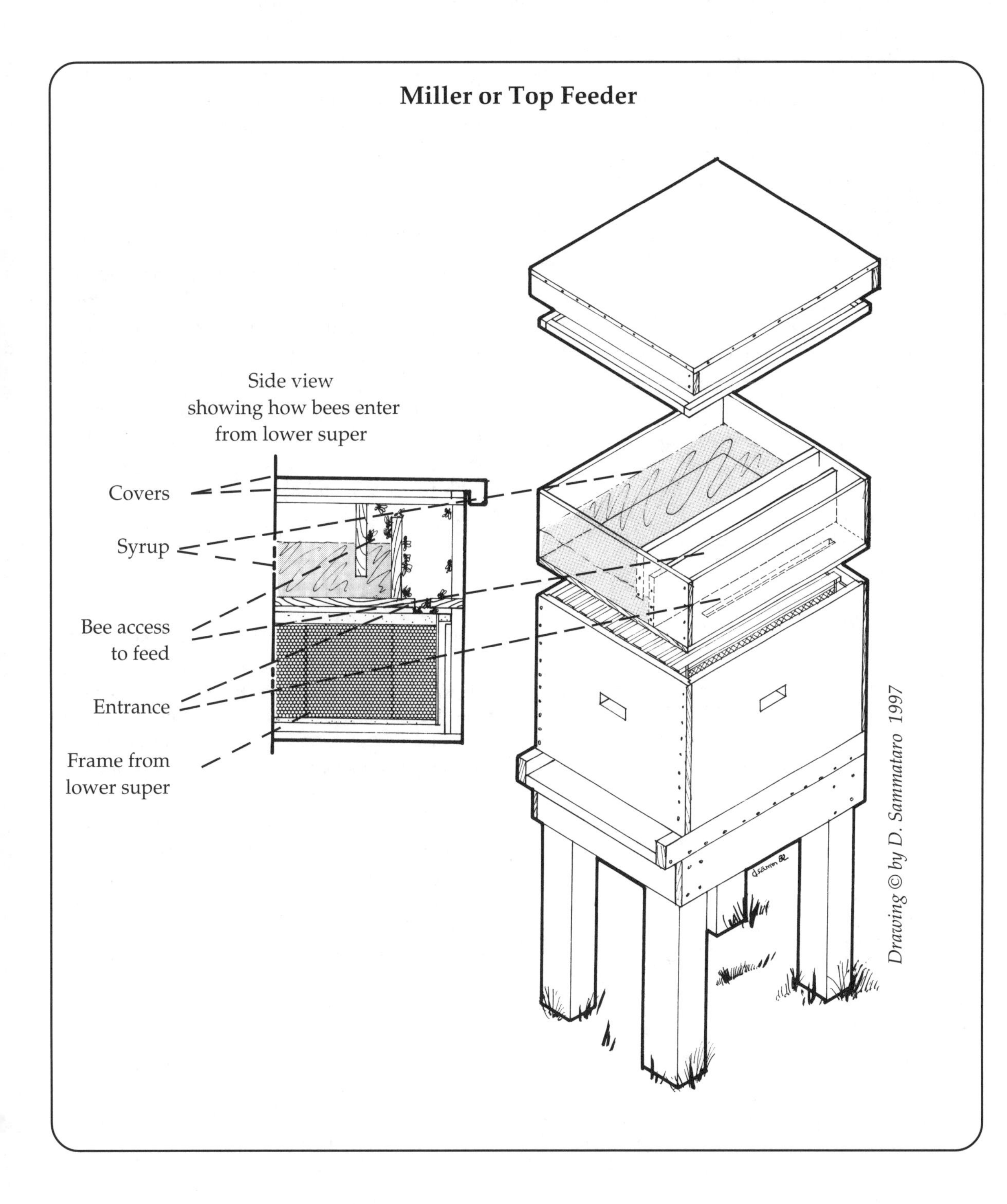

shallow super. Some Miller feeders are made entirely of wood or plastic. A bee space between the two pans allows the bees to crawl up and over the sides to feed. A control strip of wire mesh enables the bees to cling to its sides without drowning in the syrup (see the illustration of the Miller feeder). Some top feeders have only one feeder pan and an entrance space along one end wall.

The Miller feeder is placed on top of a hive, beneath the inner and outer covers; it can hold up to 4 gallons of syrup and can be rapidly filled or refilled. Variations on the same principle are also available. If this type of feeder is used, the Miller super should be bee-tight on the outside or else robbing may result.

EMPTY DRAWN COMB

Frames of empty drawn comb can be filled with syrup and placed in the hive. Slowly dip the frames into a tub of syrup or sprinkle them with a sprinkling can or other device. A steady stream of syrup poured directly from a container will not fill the cells, because the air in the cells will act as a barrier to the liquid (see the illustration on filling drawn comb). This feeding method can be used for emergency feeding, especially if the combs are located near or adjacent to the broodnest. As with other feeding methods, you have to look into the hive (and remove these frames) to determine whether they need refilling.

This method is often used when installing package bees in a hive with drawn comb. Newly hived swarms, however, should not be fed by this method. Bees in a swarm have full honey sacs and, if put on drawn comb, may regurgitate the contents of their sacs into these cells; if this honey contains disease spores and is later fed to larvae, such brood diseases as American foulbrood could result (see "Collecting and Hiving a Swarm" in Chapter 11). It is very important to ensure that any combs used for feeding syrup have no history of diseases.

Filling Drawn Comb

BOARDMAN FEEDER

A Boardman feeder is a wooden or plastic holder for a quart-size or larger mason jar (see the illustration of the Boardman feeder). The front portion of the holder's base is an entrance platform and is inserted into the hive entrance. The bees can obtain syrup by crawling into the holder's entrance.

The Boardman feeder is not recommended for feeding syrup, because bees from other colonies can rob from it, which tends to encourage further robbing activity. If the weather is cold, the colony being fed will not break the cluster to reach the feeder and may starve. The feeder holds only a quart of liquid (if you use a quart jar) and may require frequent refilling when the bees are actively feeding. Larger jars may not be emptied fast enough, and the syrup could start to ferment. Furthermore, in this highly exposed condition, the liquid could freeze or the sun may decompose chemicals in the syrup that were added to medicate the bees. Unless your hives are on stands, Boardman feeders can be easily knocked off by hungry raccoons, which will scatter the jars throughout your yard.

The best use of the Boardman feeder is for watering bees during the summer months. If the weather is quite hot, bees can go through a quart of water in one or two days. This method of feeding water also helps to keep bees from bothering your neighbor's swimming pool.

Boardman Feeder

Side view
showing how the feeder fits
in the entrance

Top view

Drawing © by D. Sammataro 1997

PLASTIC BAG FEEDERS

Half-gallon plastic, zippered bags, the kind used to store frozen food, are good emergency feeders, if jars or pails are not available. Fill bags one-half to three-quarters full, expel the air, and seal. Take them out to your bees, lay one over the top bars or on the inner cover, and make a slit 1–2 inches long near the top. The bees will come up and feed as the syrup oozes out. Do not fill the bags too full, as you won't be able to cover them. Take out the plastic bags when empty.

High-Fructose Corn Syrup

High-fructose corn syrup (HFCS) is a sugar syrup produced by breaking down cornstarch, which is composed of two major chains of glucose (glucose molecules linked to one another in long chains), into single units of glucose, fructose, and water. In 1969, researchers used two enzymes, one that split the long chains of glucose into smaller ones and another that reduced the smaller chains into individual glucose molecules. Then, with the use of a third enzyme, a percentage of the glucose was altered to fructose. The resulting syrup contained 52 percent glucose, 42 percent fructose, as well as other substances. Although syrups with higher fructose levels are available, the one with 42 percent fructose is commonly fed to bees. The methods for feeding HFCS are similar to those described for other syrups. Unfortunately, HFCS is not available to hobbyist beekeepers, because most companies sell it by the tanker load. Check with your local bee club or bee dealer. Make sure it is fresh.

Dry Sugar

Dry, white, granulated sugar can be used as an emergency food in late spring when outside temperatures are high enough to permit the bees to obtain water for dissolving the sugar; occasionally, water that has condensed inside the hive may be used by the bees for this purpose. If the bees are unable to store honey in the early spring, feeding dry sugar in late spring, before a honeyflow, may help prevent starvation.

The sugar should be located as close to the bees as possible. It can be spread around the oblong hole of the inner cover (rim side up), on the back portion of the bottom board, or on the top bars of the frames near the bee cluster. Or the sugar can be spread over a single sheet of newspaper placed over the top bars of the hive body where the bees are located; the bees will chew through the newspaper to obtain the sugar.

Only strong colonies will benefit from the feeding of dry sugar. Weaker colonies may not have sufficient numbers of bees to obtain the needed water.

"Mock" Candy

A quick candy can be made simply by mixing clean honey (honey not from diseased colonies), a thick sugar syrup, or HFCS with confectioners' sugar or Drivert sugar. Knead it like bread dough to form a stiff paste. Store it in the refrigerator or freezer and use it as emergency feed, for queen cages, or for feeding small, queen-rearing nucs. As an emergency food, place a thawed piece flattened on the top bars or near the inner cover hole.

Fondant or Sugar Candy

Fondant candy can be made and fed to bees in small molds or with a special rim feeder or candy board (see the illustration of the candy board). This method is less sloppy than feeding syrup or dry sugar, but the preparations take much longer. The basic fondant or sugar candy recipe (to feed one colony) is:

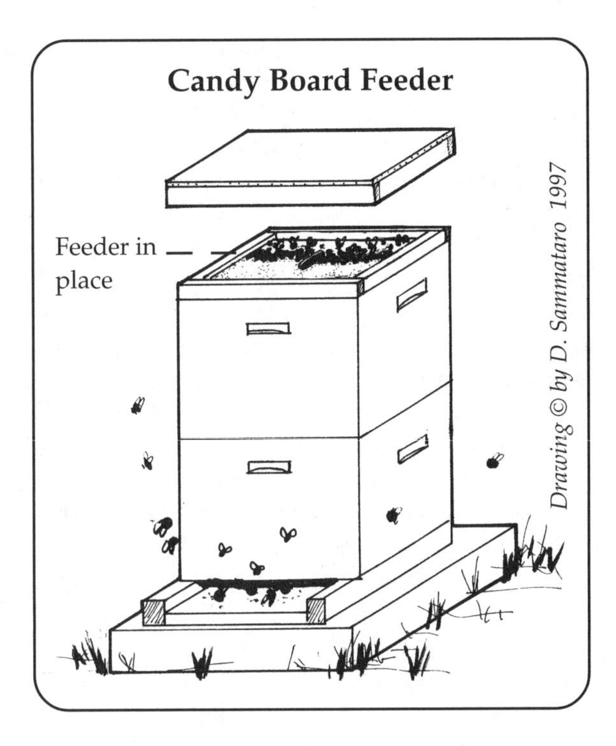

Candy Board Feeder

- 2 cups white sugar
- 2 tablespoons corn syrup (light) or 1/8 teaspoon cream of tartar (tartaric acid)
- 1½ cups boiling water

Combine and heat ingredients, stirring until the sugar dissolves. Heat without stirring to 238°F (114°C) or until medium ball on a candy thermometer). Pour out onto a cold platter, and cool until warm to the touch. Beat until light, and pour into molds or shallow dishes.

The molds can be inverted over the top bars near the cluster.

Another variation is to make up a rim feeder, which is a hardwood or Masonite board the size of an inner cover but with a 1-inch or deeper rim. It is filled with sugar candy and then inverted over the cluster.

Another recipe for candy (from C. Collison, *Fundamentals of Beekeeping*) is:

- 15 pounds sugar
- 3 pounds glucose (look for this in health food stores)
- 4 cups water
- ½ teaspoon cream of tartar

Heat to dissolve sugars until the temperature reaches 242°F (117°C); cool to 180°F (82°C) and then beat with a rotary beater until thick. Pour into molds to solidify. Feed bees by placing a cake of candy above the cluster of bees.

Honey

The best food of all—when properly ripened, capped, and free of disease—is, of course, a super of honey placed above the broodnest or several frames of honey placed next to the broodnest in weak hives. Honey extracted from old, irregular combs and cappings, as well as crystallized honey, can be diluted and fed to the bees by way of any method used for feeding syrup. Crush the honey-filled combs in warm water, and feed the syrup to bees. Supers with frames that are "wet" or sticky after having been through an extractor can be placed above or below the broodnest or over the inner cover for the bees to clean.

Be careful when feeding bees with diluted honey or wet combs; the odor of honey will stimulate robbing. Therefore, feed honey or place wet combs on the hives in the early evening so that the bees will have sufficient time to remove and store it before morning. If this food is given to weak hives, reduce their entrances as a further precaution against robbers (see "Robbing" in Chapter 11). Supers with wet combs should not be put on colonies in the late fall or in winter. The entire colony of bees might move up into them and not return to their full stores below, which could result in their starvation.

Store-bought honey should not be fed to bees. Honey from hives with foulbrood is sometimes extracted, bottled, and sold. There is an excellent chance that such honey contains foulbrood spores, which remain viable in honey for up to 50 years, and that it would likely cause an outbreak of American foulbrood disease in your hives. (Fortunately for us, American foulbrood spores contained in honey are not harmful to humans.) Be certain that any honey fed to the bees is free from spores that cause brood diseases.

Honey mixed with cappings, scrapings, or debris can be fed to bees if it is placed in a container above the inner cover. The remaining wax can then be recovered and melted.

Pollen

Pollen, the male germ plasm of plants, is the principal source of protein for honey bees. The protein content of pollen ranges from 8 to 40 percent. An average-size colony (approximately 20,000 bees) collects 125 pounds (57 kg) per year; a single bee can bring back a pollen load of 0.0035–0.0042 ounce (100–120 mg), about one-half her own body weight. Other estimates indicate a range of 40–110 pounds (18–50 kg) a year for average-size colonies (see Appendix E for metric conversions, and Appendix F, "Fun Facts").

When a bee collects pollen in her leg baskets, or *corbiculae,* these pellets are stored in the comb back at the hive, surrounding uncapped larval cells. House bees pack the many pellets dropped off by foragers into a cell; when full (about 18 loads), they mix the contents with some glandular secretions (to prevent the pollen grains from germinating) and cap it with honey and wax. The pollen thus sealed is called *bee bread.* The honey top serves as a physical barrier, but its high acidity also deters bacterial growth and preserves the bee bread. Pollen is a necessary part of the diet of larvae and young adult bees, which must have this protein during the first two weeks of life to develop normally.

In addition to proteins, which can be broken down by bees into simpler components called *amino acids,* pollen is practically the only source of fatty substances, minerals, and vitamins for larvae and adult bees. Foraging bees are apparently unable to discriminate between highly nutritive pollen and that which has low nutritional value.

Bees increase their consumption of pollen in the fall. This factor, coupled with a decrease in foraging and brood-rearing activities and increased numbers of fat bodies, seems to extend the longevity of worker bees beyond their usual summer life expectancy of six weeks. Many winter workers live over three months.

During the brood-rearing period, the consumption of large quantities of pollen by young adult workers stimulates the head glands to secrete a milky-white, protein-rich food. This substance, called *royal jelly,* is fed in abundance to all larvae less than four days old and to queen larvae during their entire larval stage and throughout their adult lives. Worker and drone larvae more than four days old are fed modified jellies and, later, a mixture of brood food, diluted honey, nectar, and pollen. Without pollen, bees could not manufacture royal jelly. About 0.0042–0.0051 ounce (120–145 mg) of pollen is needed to rear one bee from egg to adult.

When flowers are available, bees usually have sufficient supplies of pollen in the hive. The demand for pollen increases in the winter when brood rearing resumes and the remaining pollen stores can be consumed quickly. A good rule of thumb is for bees to have 500 to 600 square inches of stored pollen reserve going into northern winters. Because bees cannot forage for pollen late in the year, there must be ample bee bread stores to enable bees to feed the larvae that will replace them in the spring. If these stores are used up or not available, bees must be fed some supplemental protein.

Bees are fed pollen, pollen supplements, and pollen substitutes to initiate, sustain, and increase brood rearing in bee colonies. By stimulating brood rearing, beekeepers profit by an increase in bee populations. Such colonies can send many foragers out to harvest nectar and pollen. Beekeepers utilize strong colonies to make splits and divisions and to supply bees for pollination or to sell in packages.

Our knowledge of all the factors that return a honey bee colony from a broodless condition to one with brood is incomplete. Pollen seems to be one of the factors governing the initiation and maintenance of brood rearing; however, length of day, pheromones, physical and chemical stimuli, and the commencement of nectar and pollen collecting may all play an important role in the brood-rearing activities of honey bee colonies.

The sequence of events that initiates brood rearing may begin with an increase in the amount of sugar intake by the queen, after which she begins to lay eggs. The presence of brood activates the special glands of nurse bees. These glands stimulate bees to feed on pollen, which provides the glands with the nutrients necessary for their production of larval food. Certain ingredients in the pollen may then initiate the elaboration by these glands of the larval food.

If one could choose the best protein source for bees, pollen would be it; in second place is pollen supplement, which is better than a pollen substitute. When feeding these substances, additional stimulus should be provided by feeding sucrose or high-fructose corn syrup. Syrup feeding also stimulates the cleaning or hygienic behavior of bees, which in turn stimulates bees to forage.

Trapping Pollen

Various pollen traps are available to trap and collect pollen. If the trap's design or the hive's position does not allow for the placement of the trap at the existing entrance, a new entrance should first be established. Once the bees are familiar with their new entrance, close up the old entrance and set the trap at the new one. Some traps come with removable grids (the part the bees pass through); others do not (see "Equipment" in the References and the illustration of the pollen trap). In any case, bees should be permitted frequent access to the free entrance of the hive so that they can replenish their own pollen stores.

Collected pollen has a variety of uses: its characteristics can be studied, its structure can help identify which flowers the bees are visiting, and it can be used both for human consumption and to feed bees during periods when fresh pollen is not available. Such feeding during the late winter and early spring may stimulate brood-rearing activities, which could result in a stronger colony.

An especially important reason for trapping pollen is your suspicion of its being contaminated with insecticides. For example, sweet corn pollen is avidly collected by bees but could be poisonous if the corn has been sprayed. To reduce bee kills, trap and dispose of the collected pellets (see "The Pesticide Problem" in Chapter 13).

There is always the danger that collected pollen may contain chalkbrood, American foulbrood, or European foulbrood spores. Collect or purchase only pollen that has been obtained from disease-free colonies. When in doubt, collect your own.

When the trap is in position on a hive, bees entering or leaving the hive must pass through the grid (a mesh of 5 wires per inch). The grid's dimensions will not permit a bee laden with pollen to pass through until its load (the pollen pellets) has been removed.

Ideally, a pollen trap is put on the hive during pollen flows and is kept on for only short periods of time. Some beekeepers keep the trap on throughout the summer but actually collect the pollen (set the trap) on alternate weeks or every three or four days.

Auger-Hole Pollen Trap

Side view showing how the trap fits in the side of the hive

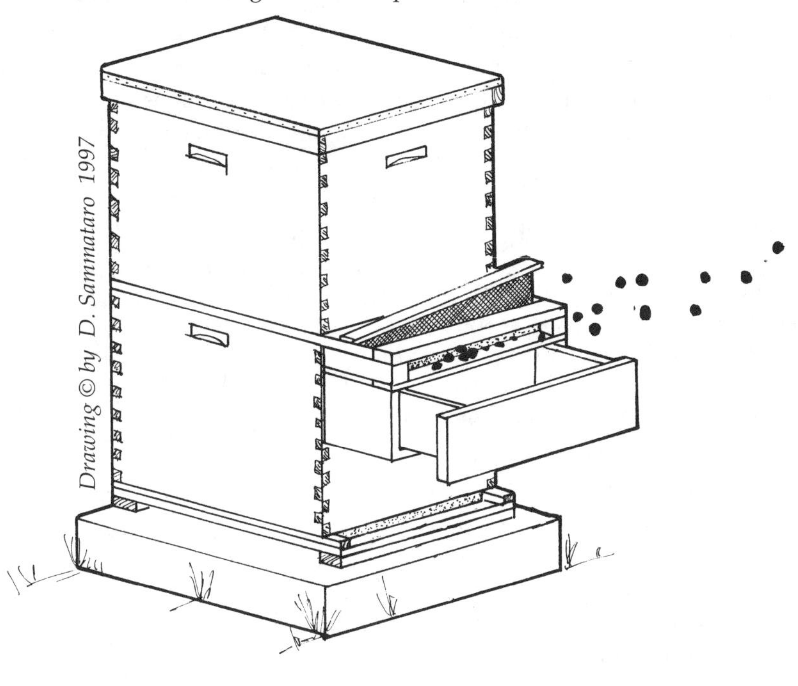

Source: E. R. Harp, *A Simplified Pollen Trap for Use on Colonies of Honey Bees,* Pub. no. 33-111 (Beltsville, Md.: USDA-ARS, Entomology Research Division, 1966).

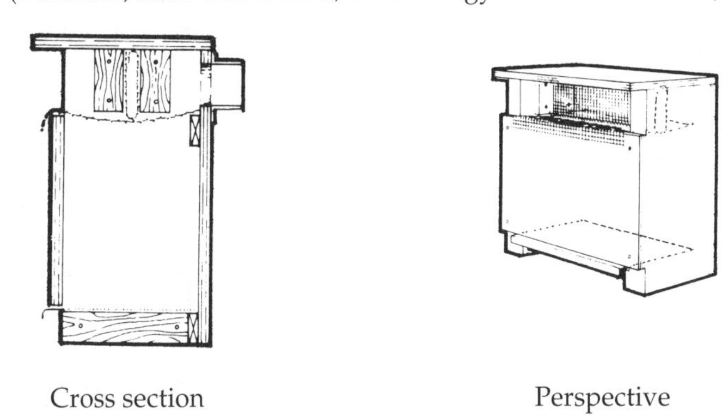

Cross section

Perspective

A colony deprived of pollen for an extended time may quickly decline. Trapped pollen must be preserved and stored carefully to prevent it from spoiling; fresh pollen molds quickly, especially in hot, humid weather. To preserve its quality, collect it every other day. Pollen loses its nutrient qualities quickly on hot days and after two years of storage. Make sure pollen is free of debris and insects before storing it.

Storing Pollen Pellets

Drying Fresh Pollen Pellets. Fresh pellets collected from a pollen trap can be dried in the sun for a few days, in a warm oven, or with a lamp or food dryer. Heat to 120°F (49°C) for the first hour to kill yeast spores, then dry for 24 hours at 95–97°F (35–36°C). The pellets are ready when they will not crush if rolled between the fingers and will not stick to one another when squeezed together. Store them in closed containers at room temperature.

Dry pollen may be fed directly to the bees or mixed with other dry materials. If the dry pollen is to be added to wet mixes, it should first be soaked in water for an hour.

Advantage:

- Inexpensive way of preserving pollen.

Disadvantage:

- Less attractive to bees.

Freezing Pellets. Place fresh pollen pellets in containers and store them directly in a deep freezer (at 0°F [−18°C]) until ready to use; they will be moist when defrosted. Use immediately, or dry them, as mentioned above.

Advantages:

- Attractive to bees.
- Can be used separately or added to mixes.

Disadvantage:

- More costly to preserve.

Sugar Storage. Cleaned pollen pellets can be preserved with sugar. Fill a container alternately with layers of pollen and white sugar, topping the layers with several inches of sugar. Close the container tightly and store it in a cool place. Pollen should be mixed with twice its weight of sugar (1 part pollen to 2 parts sugar). Careful labeling of the container as to its amount of sugar and pollen will ensure that proper proportions are maintained when preparing mixes with brewer's yeast and soy flour.

Advantage:

- Attractive to bees.

Disadvantage:

- Difficult to separate pollen and sugar if you want to feed straight pollen.

Methods of Feeding Pollen

One method of feeding pollen is to place dried pellets on the top bars of frames beneath which most of the bees are located or to pour pollen around the oblong hole of the inner cover, if it is close to the active broodnest. Dry pollen is not always attractive to bees and is not usually recommended. A better method is to pour the pellets into frames of empty drawn comb according to the following procedure:

Step 1. Fill the comb on one side of a frame with pollen pellets. Insert the comb into the hive.

Step 2. If both sides of the comb are to be loaded, spray a thick syrup onto the loaded side before filling the reverse side, so the loose pellets won't fall out.

Another method is to feed pollen to bees by putting it in any cardboard box placed anywhere in the apiary. The container should be covered (that is, have an outer cover) to prevent spoilage of the pollen by rain or moisture. Naturally, it is necessary that the bees have access to the containers. Because only strong colonies will benefit from open feeding and inclement weather will prevent bees from foraging to the container, we recommend internal feeding over open feeding whenever possible.

POLLEN PATTIES

Pollen can be made into a dough by kneading it with clean honey, corn, or sugar syrup and water (that has been boiled before use). The dough should be stiff, not runny. Sandwich the patties between two pieces of waxed paper, not plastic wrap, to keep them moist. The bees will eat holes in the waxed paper to get to the patties, discarding the paper. An easy formula is 4 parts hot water to 1 part pollen to 8 parts sugar. Patties can also be made with high-fructose corn syrup instead of honey.

POLLEN SUPPLEMENTS AND SUBSTITUTES

If you find it necessary to feed weak colonies, packages, or nucs, feed sugar syrup over honey, and feed trapped pollen over anything else. Only if pollen is not available or comes from a questionable source should you consider supplements or substitutes.

The terms *pollen supplement* and *pollen substitute* have been used in connection with feeding bees protein. These two terms have caused some confusion and are used here according to the following definitions: A pollen supplement feed consists of pollen and other substances of nutritional value to bees. A pollen substitute feed contains no pollen but consists of substances nutritional to bees.

Previously, soy flour made by the low-fat "expeller" or "screw press" method was a common ingredient, but now that the fat in soybeans is being expelled chemically, this flour is hard to obtain. Experimentation is currently under way to test other types of low-fat (5–7 percent) bean flour. Dairy products have also been used, but remember that the milk sugars lactose and galactose are toxic to bees. The substances in the following list have been developed for honey bee feeds. These ingredients can be purchased from bee dealers listed in bee journals and can also be mixed with pollen.

- Brewer's yeast
- Torula yeast, sold as Torutein 10, a spray-dried brewer's yeast
- Whey and yeast
- Skim milk powder (not instant)
- Wheast
- Bee Feast
- Pollenex
- F-200 defatted soy flour
- Toasted Nutrisoy (T-6)
- ProMix
- Beltsville Bee Diet

FEEDING POLLEN SUPPLEMENTS AND SUBSTITUTES

These materials can all be mixed with pollen, honey, or sugar (for greater palatability) or with medication and fed either dry or as patties. Numerous formulas are already published for feeding bees (see "Feeding Bees" in the References). Also, many bee suppliers have their own formulation of bee food, some with added vitamins and minerals, so try several different types to see what works best. Whatever is fed, it is important to start feeding in the early spring for 10-day intervals or as fast as the bees consume it. Do not stop feeding once a regimen has begun, but continue to feed until the bees will no longer take it. Once natural pollen is available, bees will not take the artificial feed.

DO NOT BE TEMPTED TO MIX VEGETABLE OIL INTO THE POLLEN PATTIES. THE ADDED LIPIDS MAY BE HARMFUL TO BEES.

Place moist patties directly over active brood frames. Make sure patties are not runny and do not drip down between the frames. If a patty is not consumed within a few days or if it starts to mold, remove it.

Some beekeepers have added anise oil, fennel oil, or chamomile oil to make these feeds more attractive to bees. Here are some common formulas and suggestions for feeding supplements and substitutes to bees:

- One pound (0.45 kg) of the dry material can be mixed with 4 cups of syrup consisting of a 2:1 ratio of sugar to water to make several 1-pound patties. Make sure this preparation is thick enough not to drip between the frames.
- A mixture of 4 pounds pollen, 12 pounds soybean (defatted) flour, 11 pounds water, and 21 pounds sugar will make 32 cakes. Some beekeepers add poultry vitamins to pollen substitutes.
- When making patties, sandwich the mixture between two pieces of waxed paper (not plastic wrap); that way the patty will remain moist.
- Tearing a few holes in the waxed paper on the underside of the patty will get the bees to start feeding.
- Dry material can also be fed like pollen pellets.

Pollen substitutes available in bee supply houses may contain eggs or larvae of grain pests; these can usually be killed by freezing the material for several days at 0°F (−18°C); the material should then be placed in sealed containers to prevent subsequent contamination.

8 Winter/Spring Management

Timing is as critical to beekeeping as it is to most endeavors. To time your beekeeping activities properly, it is necessary to keep accurate records of past seasons, to observe carefully the flowering periods, and to recognize the needs of your colonies. Even then, there are capricious fluctuations from year to year or season to season that will make beekeeping a continual challenge. Be prepared—no two seasons are alike.

This section is organized to take you through the year from winter—if your bees are in a location where they will experience several months of winter—to spring. If you live in southern climates where the winter is much shorter, read the section "Wintering in Warm Climates." The next chapter, "Summer/Fall Management," is designed to help you anticipate the major tasks that must be attended to during the warmer months.

Wintering

General Rules

Colonies can survive very well without elaborate wintering techniques as long as the bees are protected from winter winds. Following the minimum procedures for wintering hives, however, can be the difference that makes for overwintering success. These are the most common wintering practices:

- Invert the inner cover, to rim side down, to allow warm, moisture-laden air to escape through the rim hole (increased space). If your inner cover does not have a rim hole, you may want to invert it and raise one side with a small piece of wood to allow for the escape of the warm, moist air.
- Reduce the main entrance with a wooden cleat or a piece of quarter-inch wire mesh hardware cloth to keep mice from invading hives. Make sure that the opening of the cleat is turned up against the hive body, not against the bottom board; this will help prevent the opening from becoming blocked by a layer of dead bees, which may accumulate on the bottom board during the winter.
- Provide top ventilation by propping up the inner cover slightly. You can make another entrance by boring an auger hole, not more than 1 inch (2.5 cm) in diameter, in an upper corner of the top super; this opening lets moist air out of the hive and serves as another entrance.
- Unite weak but healthy hives. Kill all diseased (or heavily mite-infested) ones; take your losses in the fall.
- Place weights on top of hives so the covers will not blow off. Such weights are very important and could save your colonies from dying of exposure.
- Remove queen excluders and escape boards.
- Remove bee escapes.
- Leave ample honey (or cured sugar syrup) and pollen stores. Feed syrup early, while the days are still warm, to allow bees to cure such stores.

The Winter Cluster

In the late fall and winter, bees form a winter cluster. Honey bees do not "hibernate" in the winter but cluster in a ball when the air temperature is below 57°F (14°C). At this temperature, the cluster is a recognizable, defined, and compact ball. On days or weeks when the air temperatures are 43–46°F (6–8°C), most of the bees have joined the cluster. The winter cluster expands and contracts as the outside temperatures rise and fall. Bees remain relatively active in the cluster, eating, moving, rearing brood, and generating heat by shivering (contracting their wing muscles). The by-product of these activities is metabolic water vapor, which must be allowed to escape. Ventilation during the cold months is as important as during the warmer months.

The amount of heat generated by the cluster depends, among other things, on whether brood is present. In the late fall, the colony is without

Winter Cluster

brood and therefore, the cluster will be producing only enough heat to keep the colony from freezing, maintaining a temperature range of around 57–85°F (14–29°C). When the queen resumes her egg laying in midwinter, the cluster temperature in the vicinity of the eggs and brood will be maintained at around 93°F (34°C) (see the figure of the winter cluster).

Connective clusters of bees join the main cluster to the food stores. If these connectives are cut off or if the winter is unusually long and very cold, the bees could starve even though there is honey elsewhere in the hive. The cluster must be able to move to the food periodically throughout the winter. In general, bee clusters will move upward throughout the winter, which explains why the bees are found in the upper hive bodies by spring.

Thus it is important to have stored honey above the winter cluster in the fall: when bees can move up, the food stores are accessible. If you find during a late winter or early spring inspection that bees are not moving up into honey, hit the side of a hive to stir the bees into moving around.

Bees retain their feces during periods of confinement, a common situation during the winter. Periodically, air temperatures may reach over 57°F (14°C) in the winter, and on such days bees are able to break their confinement to take cleansing flights. If the bees are confined to the hive over long periods, the hive floor and frames can become littered with fecal material, and dysentery can weaken the bees further.

Winter survival depends on:

- A young, vigorous queen.
- Large population of bees (20,000–30,000, or 8 to 10 frames, covered both sides with bees); see the information on estimating colony strength on page 71 and Appendix F.
- Adequate supply of honey (or cured sugar syrup) and pollen (usually one shallow or medium super in addition to the one or two deep hive bodies used normally in your region).
- Disease-/mite-free condition (medicate if this condition does not exist).
- An upper entrance for winter/spring cleansing flights (auger hole in top super).
- Top ventilation to release moist air (prop up or reverse inner cover).
- Protection from prevailing winds.
- Reduced front entrance.
- Periodic inspections.
- Maximum sunlight.
- Proper protection from cold winds.
- Cold winters interspersed with warm, sunny days (to permit cleansing flights and cluster movement).
- Dry, cold winters (so bees do not eat up their stores too fast).
- Early springs with moderating temperatures.

The average loss of bees over winter, before mites were a concern in the United States, was around 10 to 25 percent. Now, with the mites, winter losses can be much higher. Even without mites, reduced winter survival can be a result of:

- Wet, cool winters.
- Long, cold winters with few sunny, warm days (reducing or eliminating opportunities for cleansing flights).
- Unmated, injured, or diseased queens or drone layers (no queen).
- Weak colonies in the fall, with improperly cured or poor quality stores.
- Nosema disease, dysentery, mite predation, and diseases.

Wintering in Extreme Climates or at High Altitudes

In regions where average temperatures during the coldest months are below 20°F (−7°C), you should leave about 90–120 pounds (41–54 kg) of honey on each colony (or feed an equal amount of sugar syrup). This is about one deep super full of honey. If sugar syrup is to be fed to bring the food stores up to these numbers, it must be fed to bees while the weather is still warm so it can be properly cured. Bees will not feed once a defined cluster is formed. Make the syrup in a 2:1 ratio of sugar to water. You may also try cellar wintering if you live in extremely cold northern climates, where winter lasts more than five months; see the discussion in this chapter.

The essential elements for a colony to overwinter successfully, as covered above, are similar in extreme conditions, with the addition of the need to provide adequate windbreaks, wrap colonies, and insulate hives.

Windbreaks

Apiaries should be located where they will be sheltered from prevailing winds to reduce the amount of cold drafts in winter and spring. Situate hives where barriers such as fences, evergreens, thick deciduous growth, walls, or buildings will take the brunt of the prevailing winds. When no windbreaks are present, construct temporary ones to lessen the velocity of the wind as it approaches the hives, while still permitting air drainage to take place.

A suitable windbreak is a 6-foot high snow fence or slotted board fence set up on the north side of the apiary. The boards should be about 1 inch apart both to allow air to filter through and to slow the wind velocity. The first row of hives should be about 5 feet (1.5 m) from the windbreak.

Wrapping or Using Hive Covers

Hives can be wrapped with tar paper to protect the bees from chilling winds; the dark color will absorb the sun's heat. There are now several new hive covers made from plastic corrugated paper that are easier to install. Check what other beekeepers do in your area, and see "Management of Bee Colonies" in the References.

There are several procedures for wrapping hives, and most of them incorporate these features:

- Top and bottom entrances.
- Top ventilation.
- Absorbent material enclosed in a super over an inner cover to draw off moisture (straw, shavings, porous pads, corrugated paper, fiberglass, Insulite board, or other building insulation).
- Dead air space underneath hive.
- Use of mouse poison, such as treated grain, on the bottom board or the inner cover, or other protection from mice.

Advantages to wrapping:

- Protects colonies from piercing winds.
- Allows colony to warm up when sun is out.
- Bees can move and recluster on honey if inside hive temperature is warm enough.

Disadvantages:

- Time consuming.
- Vapor barrier may form between hive and tar paper, resulting in excess moisture accumulating in the colony, which, if it freezes, will encase bees in an "icebox."
- Hive may warm up too much and bees may begin premature cleansing flights before air temperatures are high enough to protect them from being chilled.

Insulation

Insulation will provide colonies with extra protection against cold winter temperatures. Before the turn of the century, beekeepers would double wall their hives with another wider hive and fill remaining spaces with sawdust or straw (called *chaff* hives). Today, more modern insulating materials can be used, with caution. If experimenting, go slowly: use materials on only a few colonies at a time.

Any one or a combination of these insulating materials, methods, and devices has been found to be of some aid (see the figure on insulation):

- Provide dead air space underneath the hive.
- Place follower boards against the inside walls. A follower is a solid piece of board or other durable insulation material the size of a deep frame, of variable thickness, that hangs like a frame. It can be used to reduce the interior size of a deep super by substituting it for one or more frames.
- Insert an empty division board feeder (can substitute for follower board).
- Insulate the top with moisture-absorbing materials (newspaper, straw, leaves, old blankets, or shavings, or a combination of these) placed between the outer cover and the inner cover (rim side down); make sure the oblong hole is open.
- One-inch Insulite board on top of inner cover.
- An empty hive body with screened or burlapped bottom on top of inner cover and filled with absorbent material.

Some beekeepers paint hive bodies with insulating paint or put a super of dry, drawn comb on top of the inner cover, although these combs can collect excess moisture and be damaged. Supers of honey are also good insulators, so leaving extra honey on never hurts.

Hives can also be double walled and the spaces between the walls can be filled with insulating material. Such hives are more often used in other countries, so check out "International Beekeeping" in the References for more information.

Preparing for the Next Season

The following tasks should be attended to during the winter months in preparation for the next spring season:

- Order queens/packages to arrive by the time fruit trees bloom in your location.

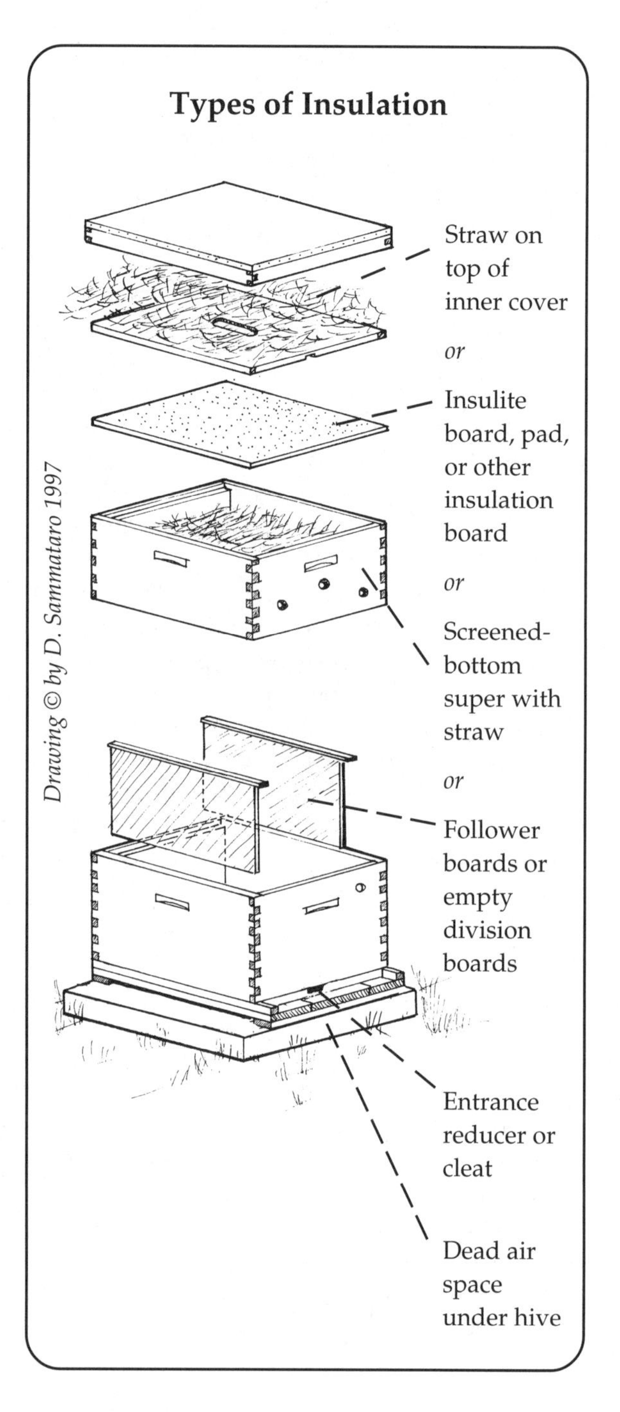

- Clean supers and frames of burr comb and propolis.
- Paint and repair equipment; replace with new equipment when necessary.
- Sort and cut out old, sagging, diseased, damaged, and drone combs and replace with foundation.
- Store wax foundation and scrapings in mothproof containers or in the freezer.
- Build new equipment.
- Order new equipment.
- Check apiary periodically for damage by wind, vandals, or predators.
- Attend bee meetings; read club newsletters, journals, and books for the latest information.

Late-Winter Maintenance

There are important tasks that need to be done in late winter and early spring, weather permitting, before the fruit trees begin to bloom. Some of this work must be done again in late spring and thus may be repeated in the next section on spring maintenance. These tasks are:

- Check for dead colonies; remove or close up any dead hives to prevent them from being robbed.
- Determine if the dead colonies succumbed to disease (nosema, foulbrood), starvation, or mites. If they died from mites, any honey remaining may be fed to colonies that need it. But *never feed honey from diseased colonies to any other colony!*
- If the bees died with their heads inside the cell, they probably starved. If there is a sour or a foul smell, check for scales or dead brood tongues sticking perpendicular to the cell wall, which could be foulbrood scale, in which case the colony equipment and bees should be burned (see "American Foulbrood Disease" in Chapter 13). Before burning any equipment, call your local bee inspector for positive identification.
- A large amount of fecal matter on the frames, covers, or bottom could indicate nosema disease or dysentery. Clean off the feces and fumigate or otherwise clean the dirty equipment. If you feed nosema-infected frames to hungry bees, you are not helping them unless you also medicate for nosema (see "Nosema Disease" in Chapter 13).
- If a colony is queenless or weak but healthy (i.e., no mites or diseases), it can be united with another colony (see "Spring Dwindling" in this chapter and "Uniting Weak Colonies" in Chapter 11).
- Unite weak colonies—those with fewer than five frames of bees—and kill the inferior queen. If the weak colonies have high mite loads, kill the colony and close up the hive. Combining a mite-infested colony with another will only propagate mites in your apiary.
- Look for signs of nosema or dysentery, especially numerous yellowish to dark brown spots on the outside of the hive bodies. If dysentery, the honey stores might be to blame; if nosema was the cause, mark these colonies to be treated with fumagillin.
- If the weather is very cold, determine the amount of stores by lifting or tilting the hive; if the hive seems light, feed the colony. Emergency feeding in cold weather includes fondant candy, frames of honey, or dry sugar sprinkled on top of the frames or inner cover (see Chapter 7).
- If warm enough, begin feeding sugar syrup and pollen patties to stimulate brood rearing. Feed the patties made from syrup-moistened dried mixtures of natural pollen, supplements, or substitutes. Feed frames of sealed honey and pollen if the weather is too cold.
- If the air temperature gets above 75°F (24°C), the colony may be checked for the queen's condition by examining the brood pattern: a compact pattern of worker brood indicates that a healthy queen is present. This examination should be brief, otherwise the brood can become chilled.
- If a healthy colony is starving, provide clean frames of honey on either side of the cluster as well as above it. Otherwise, feed it depending on the present and the future weeks' weather conditions.
- Diseases, mites, or animal pest damage should be attended to at this time. Collect adult bees and freeze or store them in alcohol to have them checked for mites or nosema; tracheal mite populations are highest at this time of year (see Chapter 13).

A few related tasks out of the beeyard should also be undertaken in the spring:

- Update the hive diary—look for floral sources that bloom this early (mostly elms, alders, maples, and willows).
- Prepare for the arrival of package bees or nucs.
- Make new equipment for the coming year, such as frames, hive bodies, tops, and bottoms.

If you are preparing hives to move for pollination services, you want to get them stronger fast. First, select strains of bees (notably the Italian races) that overwinter with lots of bees, and encourage brood rearing. Feed these hives with a good amount of stores to keep them alive all winter. In the early spring, feed again with pollen patties and sugar syrup (2 parts sugar to 1 part warm water). Follow the overwintering procedures discussed above, but mark those hives destined to be moved for pollination so you can give them special treatment. You may need to treat the colonies with Apistan strips and extender patties at this time to keep mites and diseases under control.

Spring Maintenance

The following apiary tasks should be started when the dandelions and fruit trees are in bloom in your region and when all danger of frost is past:

- Clean the apiary of any winter debris.

- Unwrap or take down winter protection.
- Remove temporary winter windbreaks.
- Remove insulating or moisture-absorbing materials above the inner cover.
- Remove entrance reducers from strong colonies. Keep them in weaker colonies, but switch the entrance hole to the larger opening.
- Clean off bottom boards. Note which colonies have few dead bees, or clean boards, for the bees may have a strong hygienic instinct.
- Feed colonies that require additional food.
- When temperatures are about 75–80°F (24–27°C), inspect the hives for diseases, mites, brood pattern, and amount of remaining stores.
- Mark colonies to be requeened to replace poor performers (see Chapter 10).
- Reverse brood chambers if the cluster has broken and bees are flying. If the lower brood box is essentially empty, the upper one should contain the queen, brood, and most of the bees. Put this upper super of bees, brood, and honey on the bottom, and let the bees move up (see "Relieving Congestion" in Chapter 11).
- Replace dark, old, sagging, and uneven combs or broken frames. It is a good time to start recycling old, dark frames for newer ones or for frames of foundation. Provide foundation frames when the spring bloom of fruit and dandelions is under way. Recycle all your frames every three to four years to reduce the buildup of toxins and disease spores that are in the wax.
- Clean bottom boards, and scrape off excess propolis and burr comb. Note which colonies are drier, cleaner, and healthier, and breed queens from these colonies.
- Investigate clean water sources, or provide fresh water.
- Provide additional space (hive bodies or supers) as needed.
- Make increases in space only when the weather is warm enough so the brood will not become chilled.
- Medicate or treat the colonies against mites, brood diseases, and nosema. Do not give medication once the bees have begun to store nectar, or the nectar and its end product, honey, could be contaminated. Read labels carefully for proper dosage and timing.
- Look for signs of swarming, and if necessary, initiate swarm prevention/control techniques (see "Swarming" in Chapter 11).
- Register hives and apiaries with your state agriculture department.

Spring Dwindling

In some colonies, older bees may begin to die faster than young bees emerge, such that the number of bees is reduced to a point at which the process cannot reverse itself and the colony dwindles to nothing. This is called *spring dwindling*, because it usually happens in the spring. This condition may result from pathogens (spiroplasmas, bacteria, or viruses), mite predation, poor queen genetics, race of bees, condition of colony going into the winter, poor winter stores, or current weather conditions.

Spring dwindling may be prevented or checked by:

- Strong populations of young bees and a young queen in the fall.
- Wintering only strong colonies with ample stores of honey and pollen; combining or destroying weaker hives in the fall, if necessary.
- Providing high-quality winter stores, such as capped honey and pollen supplements.
- In the fall, having a young queen of a race that is known to overwinter well in your area.
- Using windbreaks to protect your hives from winter and spring drafts and dampness.
- Having ample colony strength in spring; if only three or four frames of bees exist, add them to a stronger colony, or destroy the weakened hive.
- Medicating the bees against nosema in the fall or spring or feeding extender patties for disease control in the fall (see Chapter 13).
- Preventing poisoning (from pesticides) during the previous summer. Because some toxins could be stored in older comb, replace dark, old comb with new foundation.
- Preventing drifting, to avoid making some colonies weaker by fall.

Other Wintering Options

Cellar or Indoor Wintering of Hives

Beekeepers who live where winters are extreme are always interested in learning about successful methods of overwintering their colonies as an alternative to purchasing spring packages every year. Overwintering bees indoors is one such technique. Lynn Birny Westcott, a Canadian beekeeper, shares some of the secrets of this technique. Cellar wintering, once extensively used in the northeastern states, is now practiced in Canada and other cold regions. This type of wintering requires that only the strongest hives be placed in a draft-free location where temperatures are kept between 40°F and 50°F (4–10°C); temperature and humidity control is very important.

Wintering bees in cellars was a common practice in the early 1900s and was described in detail in beekeeping publications into the 1920s. Problems with the timing for bringing hives indoors and placing them back outside, coupled with difficulties in controlling cellar temperature and humidity, were responsible for the decline in cellar wintering as a practice.

Today, with the availability of ventilation, equipment for the control of temperature and humidity, and inexpensive insulation material, indoor wintering is both financially and biologically feasible. It requires an initial outlay of money and space to build the cellar or building, but if you are in an area where this is a workable alternative to outdoor wintering, the method is described below.

INDOOR BUILDING/CHAMBER REQUIREMENTS

New buildings may be constructed specifically for wintering beehives. Existing honey extraction rooms or other buildings will also work well, provided the room or building is well insulated, is ventilated, and can accommodate the hives you want to winter inside.

Suggested space requirements range from 14 to 30 cubic feet (0.40 to 0.85 m^3) per hive for rooms without an air conditioning unit. The minimum space requirement with an air conditioning unit is 8.3 cubic feet (0.24 m^3) per single brood box.

Insulation is one of the most important details in construction of an overwintering chamber. Insulation thickness should be 5 inches (13 cm) on walls and 6 inches (15 cm) on the ceiling. A concrete floor is optimal and should be insulated with 2 inches (5 cm) of polystyrene insulation. A vapor barrier on inside walls is a necessity. Doors should also be insulated and be large enough to enable you to move hives in and out of the building with ease. Several lower-cost wintering buildings have been designed: see the references on "Wintering." One clever design includes an outdoor flight channel for each hive. The channels are closed in late fall when hives are placed in the building and reopened in late winter to allow free bee flight in warm weather.

The chamber must be kept completely dark, because light promotes bee activity and makes their successful storage more difficult. Outside light should be prevented from entering the chamber at all fan openings and air intakes. You can do this by making light traps, constructed from plywood with a rimmed edge, and placing them around fans and ductwork without obstructing the air movement. Exclude light around doors by covering the spaces with heavy black plastic. Red light (red lightbulbs) is recommended when inspecting hives and other work inside the building, such as sweeping up the dead bees. In addition, equipment noise and vibration in the building should be kept to a minimum, as honey bees are very sensitive to such disturbances.

VENTILATION, TEMPERATURE, AND HUMIDITY CONTROL

Ventilation is important because it controls carbon dioxide (a by-product of respiration), temperature, and humidity levels in the wintering room. Carbon dioxide concentrations in the facility should be less than 3,000 parts per million or else bees could suffocate. The process of ventilation entails drawing air in from outside, distributing the air throughout the building, and expelling an equal amount of air containing the impurities to the outside. Many overwintering facilities use exhaust fans in conjunction with a recirculation fan attached to a duct distribution system. This type of ventilation system is frequently used in greenhouses and poultry buildings and ensures uniform conditions throughout the room.

The necessary recirculated air flow is 21.2 cubic feet per minute (cfm) per pound of bees (or 2 liters per second per kilogram [kg] of bees). You may have to increase this to 42.4 cfm per pound (5 liters/second/kg) of bees in fall and spring when outside temperatures are higher. An average hive containing 5.5 pounds (2.5 kg) of bees would require air flow of over 116.6 cfm (5 liters per second) under normal winter conditions.

The optimal temperature inside the building is 37–41°F (3–5°C). A range of 36–48°F (2–9°C) is acceptable. Temperatures up to 59°F (15°C) are tolerable on a warm spring day, but adequate ventilation and air flow is a must. It is best to maintain steady, consistent temperatures in the building rather than aim for a specific temperature value. Below-freezing temperatures, however, should be avoided indoors.

A wide range of relative humidity is tolerated by bee colonies. Fifty to 75 percent is the acceptable; adequate ventilation, as outlined earlier, will keep the humidity within this range.

PREPARATION OF HIVES FOR INDOOR OVERWINTERING

Colonies should be requeened and equalized by late fall; complete this operation as soon as honey is removed if these hives are to be overwintered. Many beekeepers will take off all the honey, select the colonies that are to be taken indoors, and then redistribute the food stores once inside.

Single and double brood chambers have been overwintered with equal success. Single hives are prepared by placing two combs of capped brood, two combs of pollen, and five combs of honey in the brood chamber. Using a single brood chamber plus a second chamber full of stores, added after the hives are in the building, has proven to be a very successful overwintering method.

FOOD REQUIREMENTS

Single brood chambers need 53 pounds (24 kg) of stores for a total hive weight of 92 pounds (42 kg). For double brood chambers, 68 pounds (31 kg) of stores for a total hive weight of 133 pounds (60 kg) is necessary.

Syrup, composed of 2 parts sugar to 1 part water (a 2:1 ratio), fed to colonies in late summer or early fall is the best winter feed for bees unless noncrystallizing honey can be guaranteed. Crystallized honey is very difficult for bees to reliquefy, because they need a source of water, which is not available indoors. Furthermore, crystallized honey and other inappropriate forms of feed can cause dysentery in bees. Honey from canola (rapeseed) should never be used as winter feed, as it crystallizes rapidly, becomes very hard, and thus is impossible for bees to use.

Hive-top insulation added in the fall before hives are moved indoors has proven beneficial. The insulation may be a sheet of Styrofoam 2 inches (5 cm) thick or fiberglass 2.5 inches (6.4 cm) thick wrapped in plastic.

MOVING HIVES INDOORS

Syrup feeding should be completed by late fall. Hives are moved inside in late October to mid-November, depending on weather conditions and geographical area.

Stack hives 6–7 inches (15–18 cm) apart and the same distance from the building's walls to allow good air circulation. Hives are often arranged back-to-back in two rows with 3 feet (about 1 m) of spacing between the rows. Hive entrances are left open, and outer covers are removed to expose the oval hole of the inner cover, which may be used for feeding if necessary.

MOVING HIVES BACK OUTDOORS

Hives are moved outside when the first spring plants are beginning to bloom. In northern areas, dandelions, hazelnut, cherry and ornamental cherry, early blooming fruit trees or shrubs, willow, and salmonberry are among the first plant species to bloom.

Wintering in Warm Climates

In contrast to wintering bees in the middle and higher latitudes, wintering bees in southern areas does not require the same attention or effort. Here, the day is longer and nights are shorter in contrast to northern regions. In the south, winter temperatures fluctuate between 45 and 68°F (7–20°C), and although temperatures can fall below freezing for short periods, it is not necessary to provide the same winter protection (such as wrapping and insulation) for your colonies.

In addition, plants that yield nectar and pollen are available almost year-round, so brood rearing is interrupted for only a brief time. It is in these climes that the package bee industry is located and where many commercial beekeepers overwinter and rear new queens for their thousands of colonies.

Most of the published information concerning wintering honey bees is based on apiculture in temperate climates. As a consequence, these sources frequently concentrate on wintering because that is usually the time of most colony loss. It is often recommended that you overwinter colonies with two full-depth supers as a standard colony configuration, a brood chamber below (with a large population of bees), and a food chamber above, filled with honey and pollen stores.

In more southern climes, different conditions prevail. The cold weather is not as severe, and therefore, it is unnecessary to pack colonies for protection. Brood rearing goes on much longer during the year; there may, in fact, be no broodless period in some parts of southern Florida and Texas. The two full-depth supers are often reduced to one brood chamber with a medium-depth honey super as the standard hive configuration.

Dr. M. Tom Sanford, the Extension specialist in apiculture at Florida State University, offers some observations on wintering in the South. The length of day increases as one approaches the equator, and warmer zones have a greater variety of plants, which produce relatively less nectar per individual species. These two things together appear to be the most important reasons that honey production is not as pronounced or intense as in more temperate regions.

The relatively long season for brood rearing, the warmer winters, and the more sparse nectar production are the cornerstones of southern beekeeping. Thus colony management must be spread out across the year more evenly, and timing of specific tasks is radically different from that in the North. As examples, two management techniques that are extremely important in successful wintering stand out: requeening and managing varroa mite populations.

Requeening in the South is done in the fall and with queen cells rather than mated queens, as is common in the North. Mite treatments are scheduled differently during the longer growing season in the South, because varroa and tracheal mites can be present in the colony year-round. Because of the later honey flows, a summer/fall varroa treatment must be timed carefully to keep bees mite-free but not contaminate honey. In addition, another varroa treatment may be necessary in the early spring before the colonies build up.

Here are some points to remember when wintering in southern areas:

- Winter in one brood box and a minimum of a 6⅝-inch super.
- Install entrance reducers.
- Keep bee populations between 15,000 and 20,000 (5–8 frames covered with bees).
- Medicate/treat for diseases and mites.
- Provide an upper entrance.
- Protect colonies from cold winter winds.
- Periodically check to see how food stores are holding up.
- Feed as needed.

9 Summer/Fall Management

Summer Management

Each colony should be examined closely about once a week before the major honeyflow begins in your area. Check the colony strength to determine whether it is populous enough; a colony should reach a population of over 40,000 by the time of the first major honeyflow. Weak colonies should be united or destroyed.

Here are two methods for estimating colony size:

- Try to count bees coming and going at the entrance; if they can be easily counted, the colony is weak. Between 30 and 90 bees per minute indicates a strong colony.
- One deep frame completely covered with adult bees equals about 1 pound of bees (3,500 bees).

For additional methods, see the information on estimating colony strength.

Other tasks during this time should include the following:

- Requeen as necessary (i.e., when the colony is weak or otherwise not up to standard).
- Monitor colony strength, uniting weaker ones with healthy, stronger colonies.
- Check for diseases/mites.
- Check the colony's food stores.

Estimating Colony Strength

- A shallow frame fully covered with bees will hold 0.25 pound of bees.
- A deep frame fully covered holds 0.5 pound of bees or about 1750 individuals.
- There are about 3500 bees per pound.
- A sheet of wax and plastic foundation 16.75 × 8 inches (42.6 × 20.3 cm) equals 3350 cells per side.
- One square inch holds 25 cells (5 cells/linear worker cells, 4 for drone).
- Deep frame: 16.75 inches × 5 = 83.75 cells (linear); 8 inches × 5 = 40 cells (vertical). So 83.75 × 40 = 3350 cells/side, or 6700 cells double sided.
- 18 cells per 10 cm (linear) by 21 cells (linear) deep equals 378 on each side (756 cells double sided). If you do it by measurements, 18 cm × 18 cm = 324 cells.

For more information, see Appendix F.

- Reverse the brood chambers again if necessary. If the colony is unusually strong and may swarm, practice swarm prevention procedures, or split and use the colony to rear queens for fall requeening.
- Add honey supers as needed; when a super is two-thirds full (six or seven full frames), add another super (see "Rules for Supering" in this chapter).
- Add frames of foundation in supers only if a good honeyflow is on; otherwise, bees will chew holes in the wax, or the foundation will warp and bend.
- Keep burr comb and propolis scraped off frames and hive walls; collect and process.

Cooling the Hives

When the temperatures are above 90°F (about 32°C) for an extended period of time, you may need to help the bees keep from overheating. Here's how:

- Shade the hive from the noonday sun with fencing, boards, or shrubs, or break some branches and place them over the hive cover.
- Stagger supers slightly to increase the airflow throughout the hive. Some beekeepers raise the inner cover or the front of the bottom super with a small block; others bore a ¾-inch (18.8 mm) auger hole in an upper corner of the top deep super. Do NOT bore a hole above or in the handholds! Trying to lift these supers off is a stinging adventure you want to avoid.
- Make sure fresh water is available; this can be done with a water tank filled with water and sand (to keep the bees from drowning), with dripping hoses, or by giving each colony a Boardman feeder filled with water.
- Use slatted racks on the hive bottom to give additional room for fanning bees to work.
- Paint metal covers or wooden tops with white paint to reflect the maximum amount of sun.

Signs of Honeyflow

Prepare for the honeyflow in the winter or spring months; do not wait until the last moment to make extra frames or supers or you could lose a honey crop! Start by repairing frames or by preparing frames with foundation for the honey supers. Keep fresh wax foundation sheets in plastic bags to protect them against wax moth infestation and to keep them from drying, because dry foundation becomes brittle and breaks easily.

Honeyflows are periods during the year when bees are able to collect ample supplies of nectar. A honeyflow may be of only a few days' duration, or it may last a few weeks. Major honeyflows provide bees with more nectar than is needed to sustain the colony over short periods. Bees are natural hoarders, and if you provide them with more storage space than they need, the presence of empty combs (supers) stimulates hoarding behavior, inducing more bees to collect nectar. This behavior of storing more food than is immediately needed is a sound evolutionary trait, for the amount of food any one colony will consume over the year is unpredictable. By understanding this aspect of bee behavior, we increase the opportunity to obtain a surplus of honey.

This surplus is stored by the bees as honey in supers (located above the brood chamber) and is what we, as beekeepers, can harvest, providing there are sufficient stores in the second brood chamber. We only harvest the surplus, not the needed food reserves.

A honeyflow is indicated by one or a combination of the following signs:

- Fresh white wax evident on ends of drawn comb and on top bars.
- Dramatic weight gains in the hive over several days or weeks.
- Wax foundation drawn out quickly.
- Large amounts of nectar ripening in cells.
- Bees fanning at the hive entrance.
- Much foraging activity.
- Odor of nectar (ripening honey) often pervading the apiary.
- Bees in good temper and easy to work with.

During the main honeyflow, you should avoid opening the hive to look at the broodnest unless you are doing some major management operations (requeening, treating for diseases, and so on), nor should pollen traps be placed on hives. When necessary, check the colonies before a major honeyflow, because if you disrupt the bees by tearing down the hive during a flow, it will disrupt the bees' gathering activities and may even reduce the amount of honey being brought in for several days. You will also kill many more bees (the population is much higher at this time) and perhaps even the queen if you work bees intensively during this time. Thus, restrict your curiosity to smaller colonies, or look at your observation hive.

If you must look at your colony (e.g., for disease treatment or requeening), try to minimize your time inside the hive to reduce the disruption.

Apiary Tasks in Summer

In general, the tasks at the apiary just before and during the major honeyflow should include the following:

- Super hives as needed, placing honey supers above the broodnest (see the next section).
- Provide adequate ventilation so bees can cool the hive and cure the honey adequately.
- Keep supers on until honey is capped, as unripened honey will ferment when extracted.
- Avoid adding too many supers at once, because bees may partially fill all of them instead of filling one completely (called the *chimney effect*). Put on only the number that the bees can fill in a few weeks; this is a judgment call, one gained by experience.
- Never super a weak colony.
- NEVER MEDICATE colonies at this time. It is illegal, and you can contaminate the honey. Honey collected from diseased colonies that required medication must not be used for human consumption.
- DO NOT FEED syrup during this period, because it will probably end up in the honey supers. The bees will cure sugar syrup instead of honey.
- Requeen weak colonies or those heavily infested with parasitic mites so as to break the brood cycle (see Chapter 10). This is a perfect time to raise queens: take advantage of a natural honeyflow to raise high-quality queens.

Super Sizes

When a honeyflow is in progress, the bees will deposit nectar in supers placed above the brood cham-

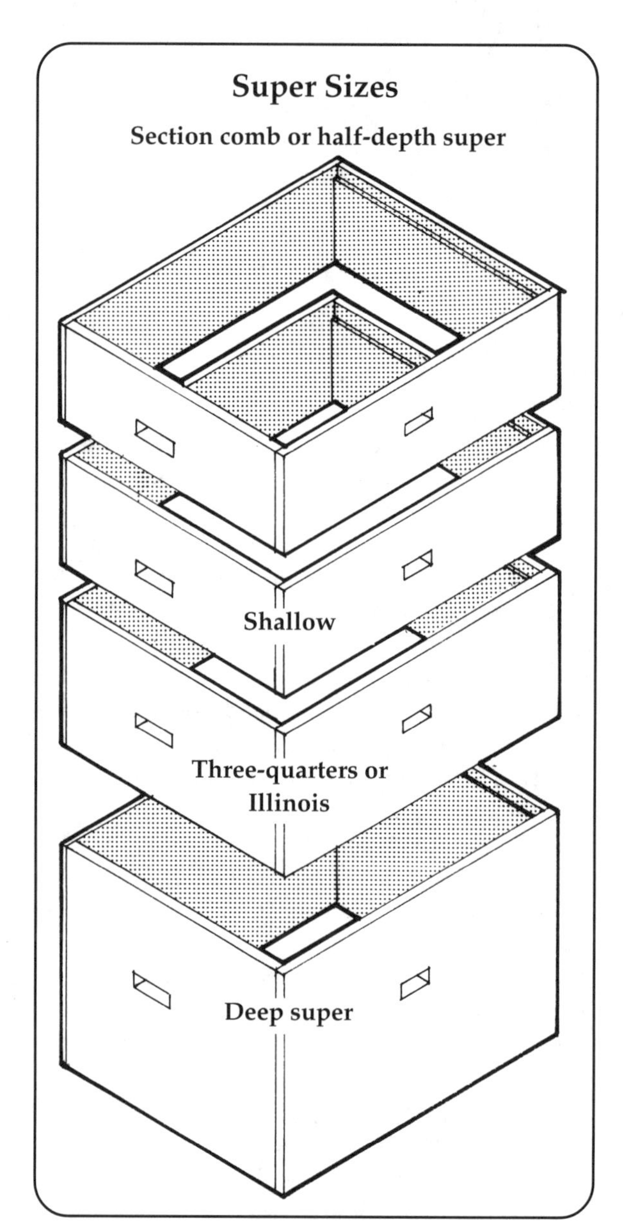

ber. These supers may vary in size from full-depth supers (deep hive bodies) to the shallow or section comb supers (see the figure on super sizes). There is no hard-and-fast rule about which super size to use; personal preference, one's physical strength, and the quantity of the expected surplus should be your guide. Some beekeepers keep all hive furniture the same size so they don't have to mix and match different sizes of supers and frames or foundation. There are pros and cons to mixing or to keeping the super sizes the same, so chat with some beekeepers about which super sizes are best in your region.

Rules for Supering

The most important rule for supering is to keep the queen out of the honey supers. The presence of brood would make the honey (when extracting) unsanitary because of larvae and pupae floating in the tank, and brood would be wasted. In addition, it is difficult to cut away the cappings of brood combs that have been darkened with propolis. Cull any dark and old frames and save them for other uses, such as bait for swarm traps; otherwise, melt them down for the wax.

Use one of the following methods to restrict the queen from laying in the honey supers:

- Place a queen excluder above the broodnest (see the figure on beehive components).
- Place a super of light-colored comb or foundation above the broodnest; as long as the queen is not crowded for space, she will prefer to lay her eggs in the darker comb.
- Keep a hive body filled with honey directly above the broodnest. Such a honey barrier often keeps the queen from moving upward.
- Place a section comb honey super above the broodnest; queens generally will not lay in the section boxes.

Some general guidelines for supering bees during a honeyflow are:

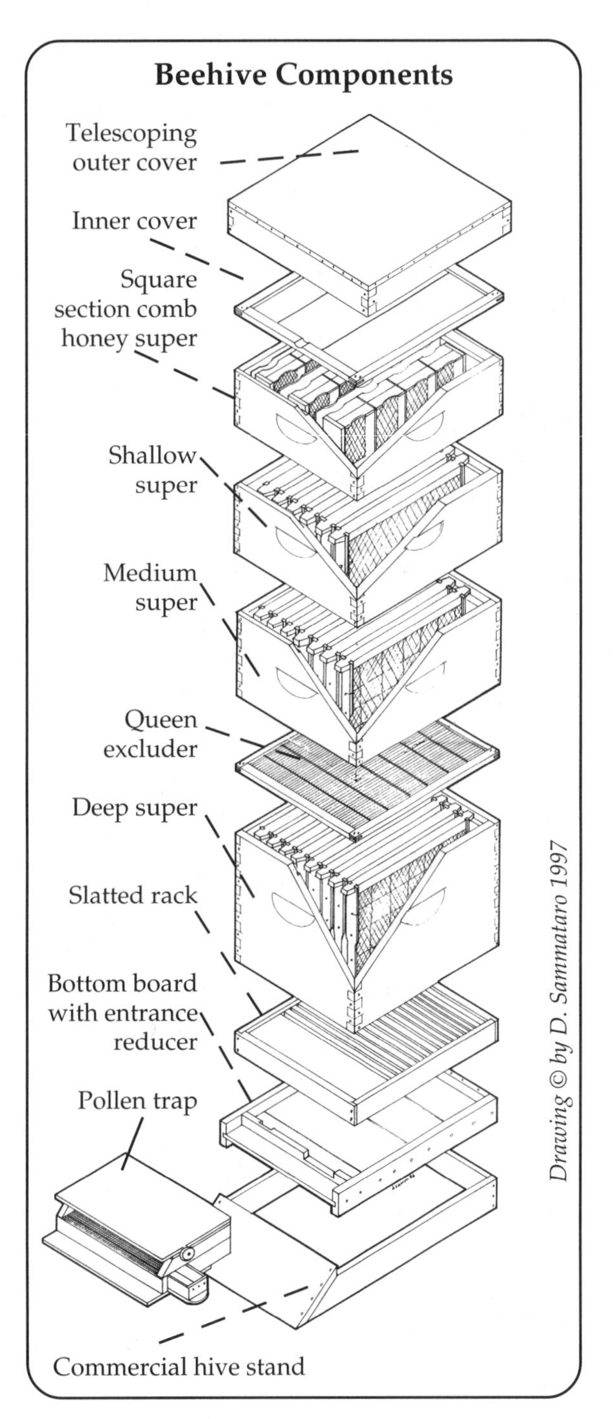

- Stagger the honey supers to hasten the ripening of honey, especially in hot, humid areas.
- Use only eight or nine frames in the supers destined to be extracted, so the bees will draw the foundation out wider than normal; this makes it easier to cut the cappings off when extracting honey.
- Bait an empty honey super with one or two frames of capped or uncapped honey if the bees seem reluctant to move up; this will attract bees to move into the super.
- Some beekeepers use drone comb foundation in their honey supers: the cells are larger, and honey seems to extract readily from them. Drone foundation can be obtained from bee supply houses.
- Rearrange frames in supers, periodically, so the full ones are at the ends and the empty ones are in the middle (bees fill the middle ones first). Doing this will buy you time if you run out of supers and the bees are filling only the center frames.

Methods of Supering

There are two basic ways to super for honey: reverse supering and top supering (see the figure on supering).

Reverse or Bottom Supering. This method generally requires a queen excluder to keep the queen from laying in the honey supers (see the figure on p. 74) and can also be used for comb honey production. A super with foundation or dry combs (S2) is always placed below a super at least one-half full of honey (S1). Because the emptier supers are on top of the broodnest, the queen excluder is necessary. As the supers are filled, they are taken off, or full supers can be stored above emptier ones.

Top Supering. This method does not require a queen excluder, because the queen rarely will go into a super full of honey. Put supers with dry comb or foundation (S2) *above* honey supers that are at

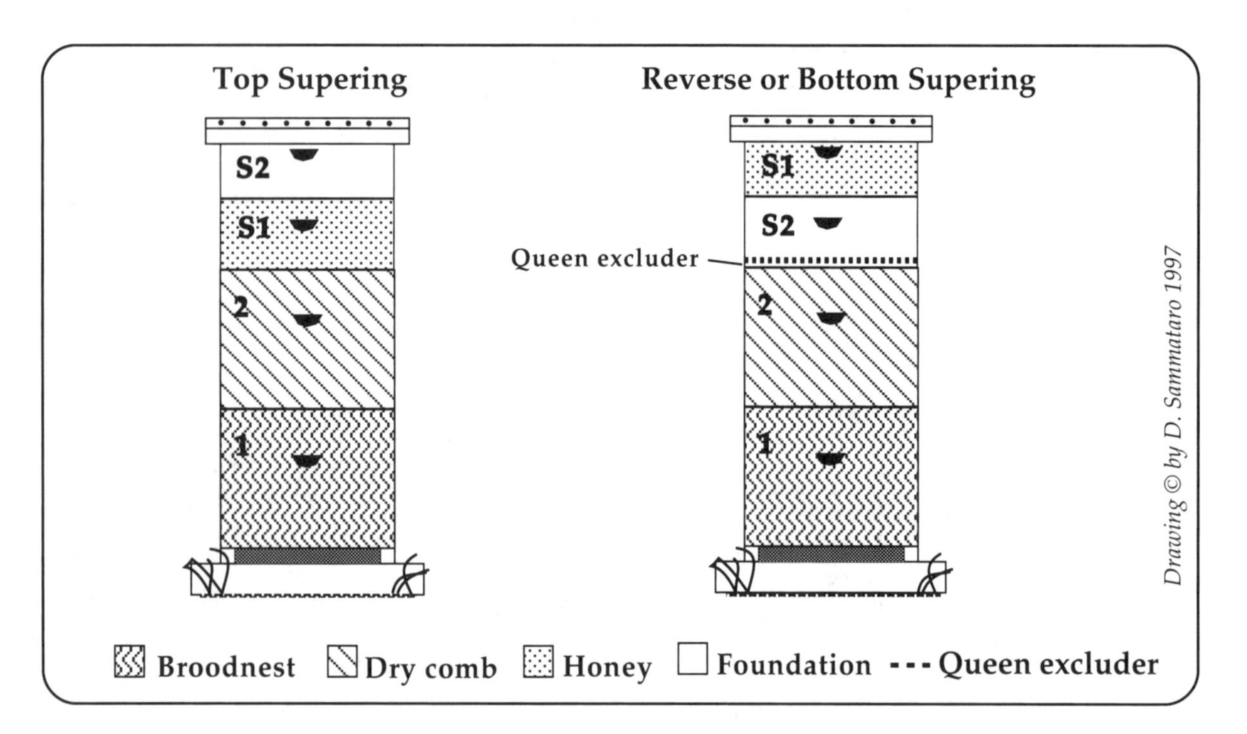

least half filled with honey (S1). Keep adding supers as the bees fill the ones below, until you take the honey off in the fall.

There are many methods of supering using these two themes; talking with local beekeepers may be helpful in determining how to super in your particular area. Success of either method often depends on the type of honeyflow in your location—fast and quick or slow and long. Fast and quick flows will enable you to make comb honey; slower ones should be reserved for extracted honey. See also figures on pp. 75 and 76.

Comb Honey

Harvested honey can be left in the comb or extracted from it. Honey in the comb is referred to by various names. Normally found at fairs and honey shows, *bulk comb honey* is an entire frame of capped honey that is packaged without cutting. If the honey-filled comb is cut and packaged, it is referred to as *cut comb honey.* Cut comb placed in a bottle that is then filled with extracted honey is called *chunk comb honey.* Comb honey contained in small wooden frames (section box), plastic rings (Ross Rounds), or plastic boxes (Half-Comb Cassettes) that is not cut out of the frames is referred to a *section comb honey* (see Chapter 12).

Foundation for the bulk, cut, chunk, and section comb honey should be the thin, unwired type. As soon as the combs are sealed, they should be removed from the hive to prevent the white cappings from becoming darkened with propolis, soiled by travel stains, or damaged by wax moths or the bee louse, which lays its eggs in honey caps.

The supers containing frames for comb honey production should be placed only on the strongest colonies, either those consisting of two brood chambers or colonies reduced to one brood chamber (as described for the production of section comb honey in the chart on p. 77). Place an excluder above the broodnest and super the hive using the same rotation as that illustrated for reverse supering. It is a good idea not to mix the comb honey supers and the extracting honey supers on any one hive, although if the honeyflow is very fast and strong, the bees will fill any available space quickly. Some bee races fill section boxes rapidly and with very white cappings—such colonies should be reserved for section comb honey only.

SUPERING FOR SECTION COMB HONEY

Comb honey, especially section comb honey, is difficult to produce because success depends on a heavy honeyflow, exceptionally strong colonies, and time-consuming hive manipulations at the correct intervals. The Miller method of supering is one that is used for section comb honey; this is described below and is illustrated as method A in the chart on page 77.

A colony used for section comb honey production is generally wintered in two deep hive bodies (1 and 2). In the spring, this colony must be built up to full strength before the major honeyflow, and the brood chambers should be reversed to provide ample room for the queen to lay. This may need to be done several times to maintain enough room for the queen.

As soon as the honeyflow begins, reduce the strong two-story colony to one deep (2). Set up this colony so it contains two empty brood frames (in the middle) and as many frames of capped brood as possible on either side, with accompanying queen and worker bees.

Follow this procedure (method A):

Step 1. Reduce hive to one deep (2); frames of honey and any remaining brood frames should be given to other colonies.

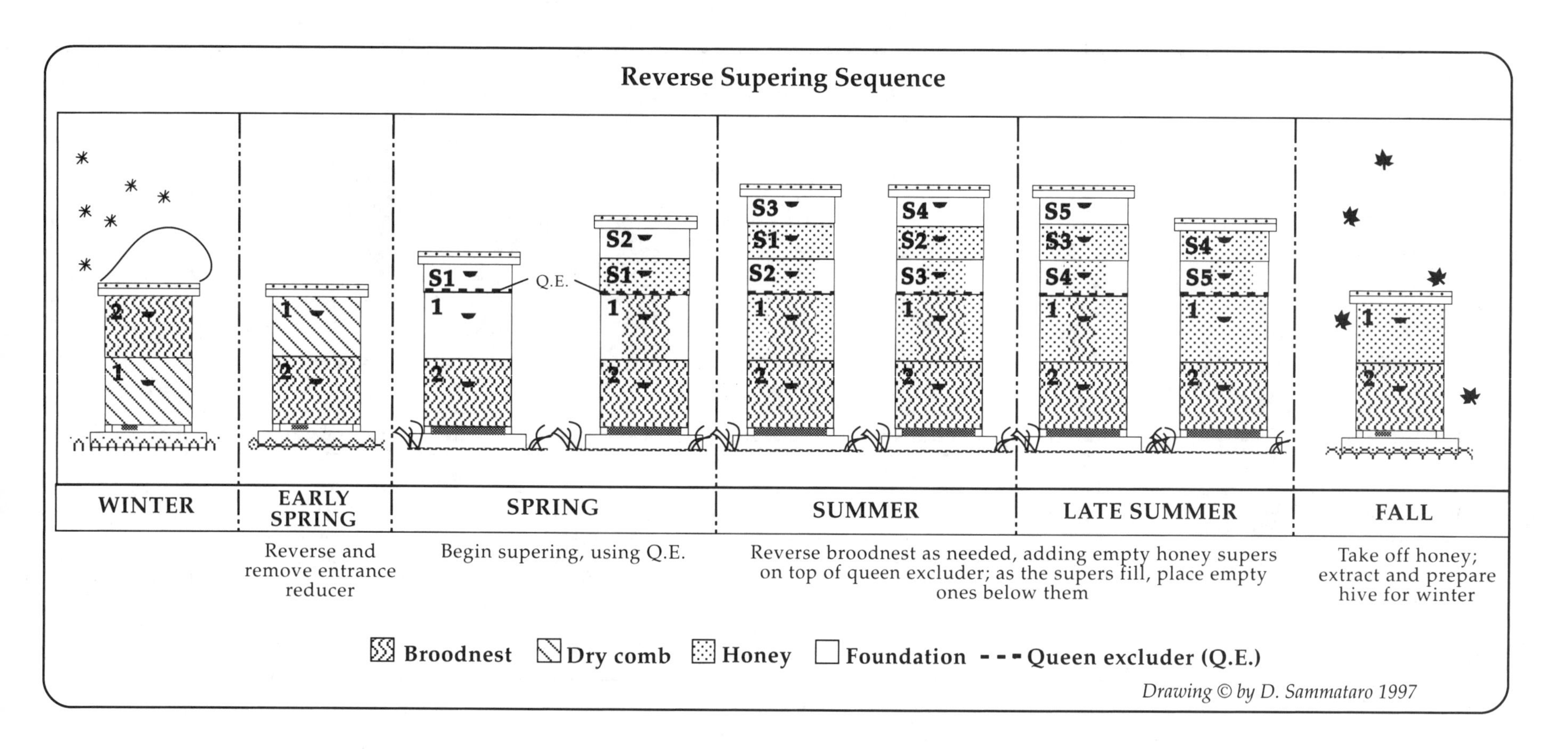

Step 2. Over the reduced hive (2), place the first section super (ss1), with thin foundation in the section boxes or rounds (or half-comb cassette supers that require no foundation).

Step 3. When ss1 is one-half full of honey, place a second section super (ss2) below it.

Step 4. When ss1 is almost filled, reverse it with ss2 (so the full super is above the empty one).

Step 5. If the honeyflow is strong, add a third (ss3) (and, later, subsequent supers) above the broodnest until ss2 is half filled, then reverse again so the full supers are above the emptier ones. Before placing another empty super on top, make sure the section supers are full from end to end, or the bees may funnel up the center, ignoring the end frames. You can correct this (the chimney effect) by removing the full frames or rearranging the super so the full frames are on the ends and the emptier ones in the center.

Step 6. Remove the completely filled section supers either as they are filled or all at once. Use bee escapes to clear the bees out of the supers. Fume boards are not recommended because the honey might be adversely flavored.

Method B is slightly different: use a queen excluder and a single deep brood chamber plus a shallow super. The idea is always to have a full honey super above the section boxes so as to encourage the bees to move upward.

Comb honey should be marketed as soon as possible to reduce the danger of its granulating or being damaged by wax moths or the bee louse (*Braula coeca*). Storing comb honey in the freezer will help eliminate these problems (see Chapter 12). After the honeyflow is over and the section comb honey production ceases, take off all section supers and unite the reduced colony with another hive or otherwise allow it to build up enough stores to overwinter in two deep hive bodies.

Harvesting the Honey

In some regions, two crops of surplus honey can be expected, one in the summer and another in the fall. Some beekeepers harvest the summer and fall

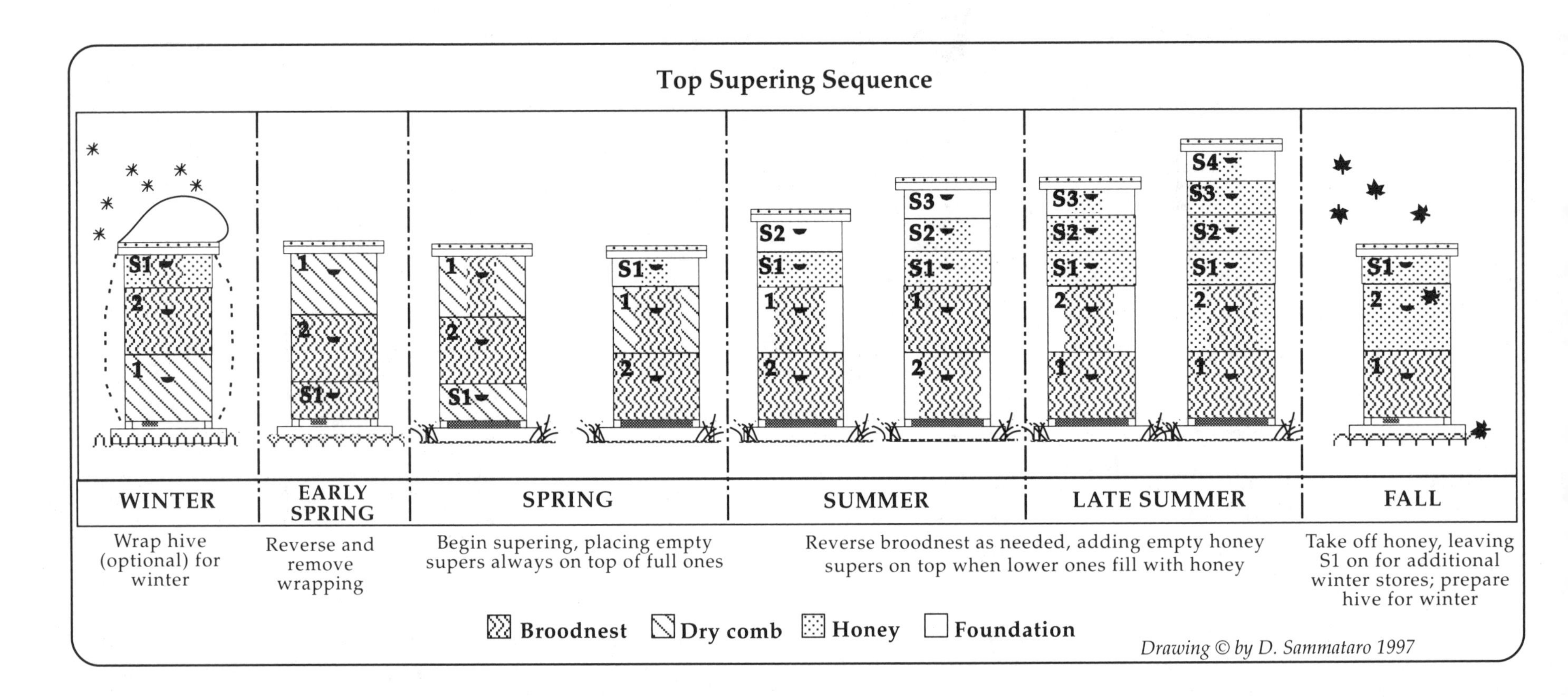

crops separately; others harvest both at the end of the fall honeyflow. Recently, it has become necessary to treat bees again in the fall with Apistan strips to control varroa mites. It is advised to take off honey supers as soon as possible, rather than leaving them on the bees until late fall.

Average yields of surplus honey depend on the amount of open land filled with honey plants. Yields vary from as low as 25 pounds of surplus per colony to over 100 pounds. For hives located in temperate climates, 90 pounds or more of honey should be left on for overwintering each colony (see "Wintering" in Chapter 8).

Today, the populations of varroa mites are at their height when beekeepers take off the late-summer honey harvest. Consider taking off your honey a little earlier in order to treat your hives with Apistan strips, thus allowing enough time for the strips to work for two bee generations and still be removed before winter. For more information, see "Varroa Mite" in Chapter 13.

REMOVING BEES FROM HONEY SUPERS

The five methods given below describe ways of removing bees from honey supers. Honey supers are often free of bees when it gets very cold (in the early fall), because the bees leave the supers to join the warm cluster below. But remember, once you commit yourself to taking off honey, be prepared to extract or otherwise process your honey in a few days. You cannot store supers of capped honey for very long, unless you freeze them, because of the danger of wax moth or braula infestations.

A good tip to make honey removal less messy is to go to your hives the day before you are removing honey and break apart the supers (or clean off the honey-filled burr comb). By at least breaking the seal, the bees can clean up any dripping honey before you remove the supers. But this works only if you have fewer than 30 colonies; otherwise, it is too time consuming.

Shaking or Brushing. Remove a frame with sealed honey from the super and shake the bees off in front of the hive entrance, or gently brush off the bees with a soft, flexible bee brush or a handful of grass. Allow the bees to fall at the hive entrance. Then place the frames, free of bees, into an empty su-

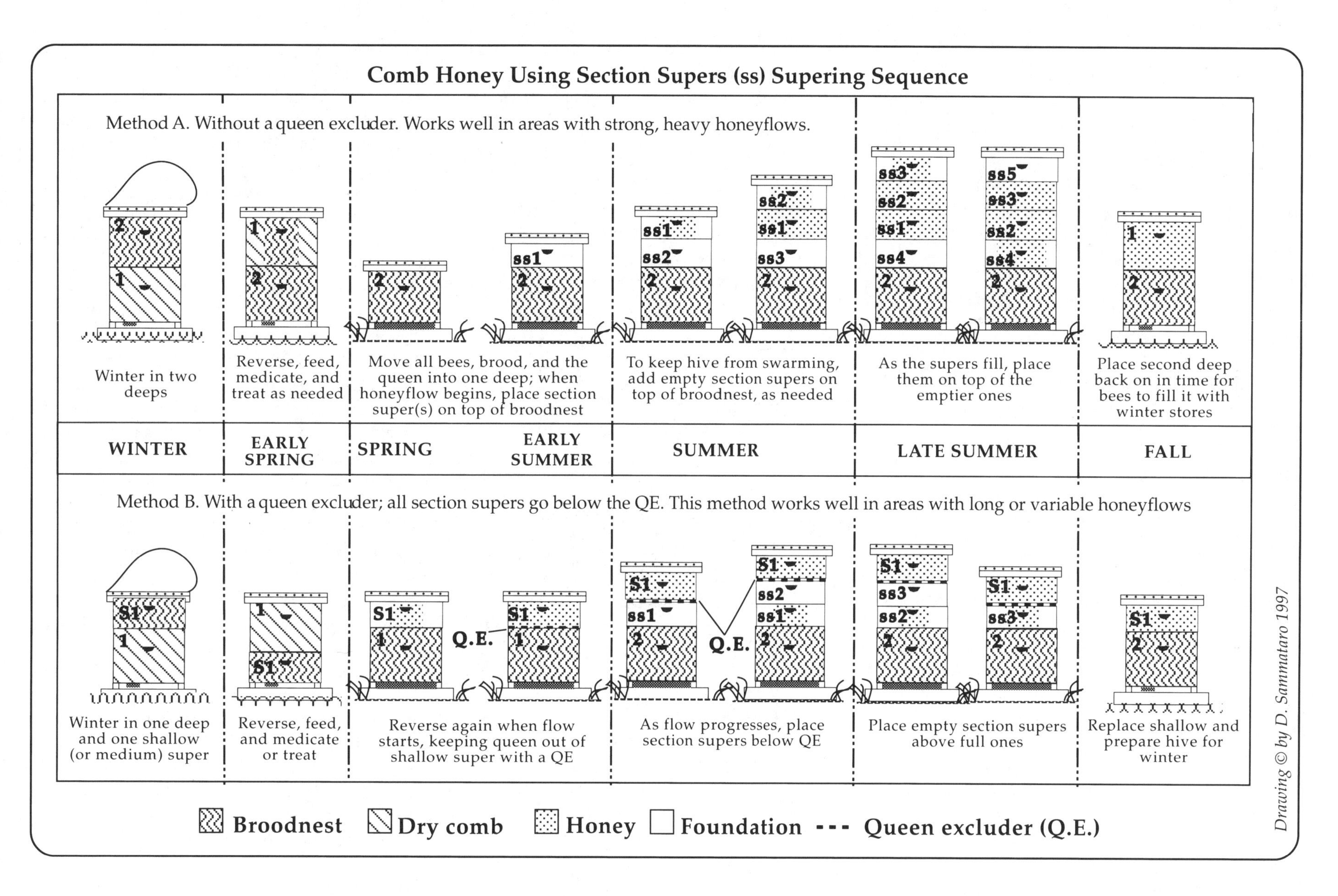

per and cover it with burlap or a thick, wet cotton sack (robbing cloth) to keep out robbers. If robbing is particularly intense, an additional cloth might be needed to cover the super you are working. If robbing becomes unmanageable, put the honey frames into a vehicle and close all doors and windows; stop taking honey from colonies in that apiary for the day.

Advantages:

- Able to select frames containing capped honey (honey covered by thin layer of wax).
- Relatively easy if bees remain calm and only two or three colonies are involved.
- Inexpensive.

Disadvantages:

- Timing is critical to avoid robbing.

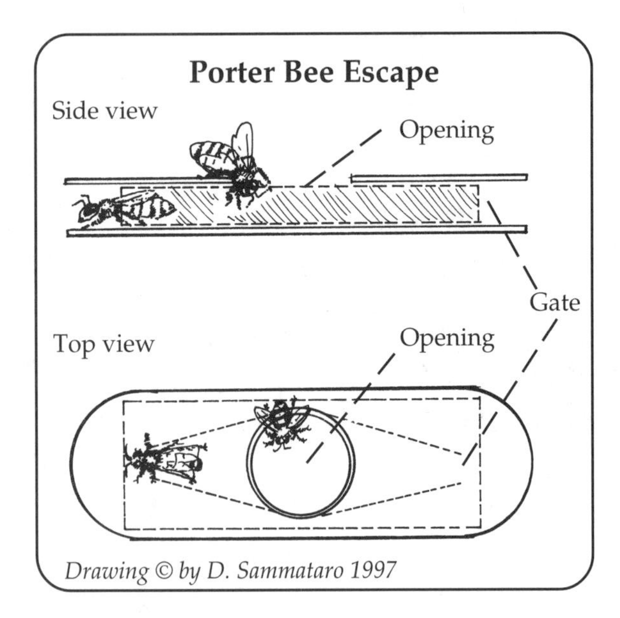

- Method is time consuming.
- Brushing may excite bees to sting.

Bee Escapes and Escape Boards. These devices are primarily used to remove bees from hive bodies containing honey so that the honey can be harvested free of bees. The inner cover can be made into an escape board by placing a Porter Bee Escape in the inner cover hole (see the illustration). A Porter Bee Escape is an inexpensive metal or plastic gadget that allows bees to pass through it in only one direction. The escape fits into the oblong hole of the inner cover—or any cover that has been modified to hold four or five bee escapes—to facilitate the passage of bees.

All types of bee escapes are available besides the Porter, and all provide passage in one direction; check with bee dealers and look at bee supply catalogues. When an inner cover or modified board contains one or more bee escapes, it is referred to as an *escape board.* Escape boards made with several bee escapes are more efficient and will more quickly clear supers of bees. Some of these boards are made with cones, screens, and other escape devices.

Escape boards can also be used to move a hive that consists of more than two hive bodies. First, place the escape board below the honey supers. After the bees exit, the extra bodies are removed, and the remaining hive can easily be lifted and moved (see "Moving an Established Colony" in Chapter 11).

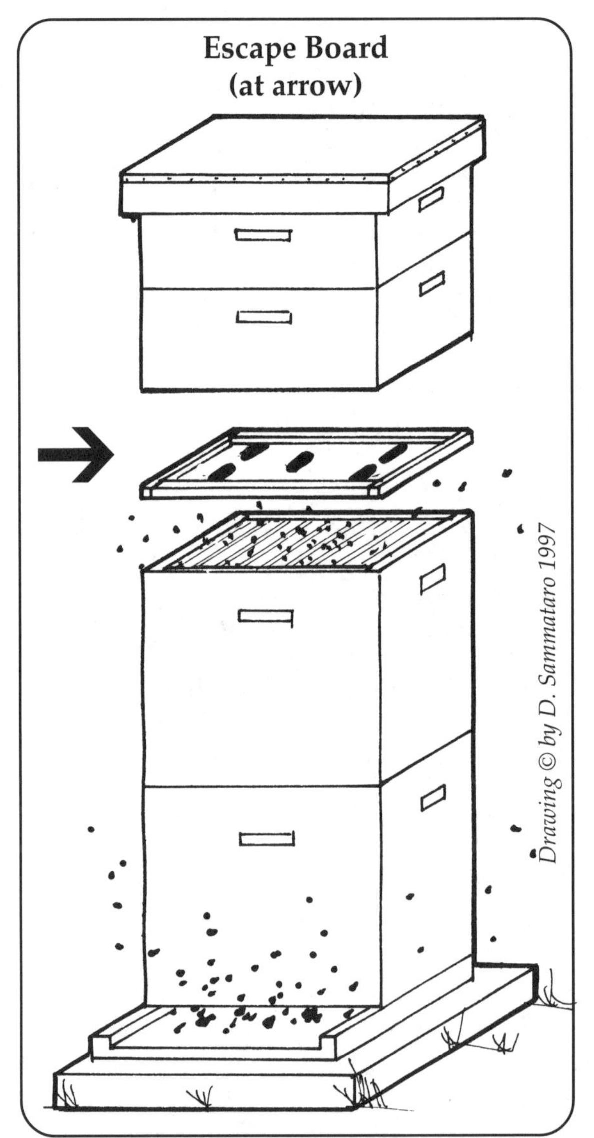

The escape board is placed directly below the honey supers you wish to remove (see the illustration of the escape board). Usually within 48 hours after the escape board is in place, the bees will move down to seek the warmth of the broodnest or the bee cluster, and because many of these bees are field bees, they will leave the honey supers to resume foraging activities. Do not leave the board on for more than 48 hours, for the bees will figure out how to reenter the supers, especially through the cone escapes.

There must be no cracks or holes in supers placed above an escape board, because bees from the same hive, robbing bees, and other insects (yellow jackets) will invade and remove the honey. Tape or otherwise close off these inadvertent entrances to the unprotected honey supers. If the outer cover is warped and you are using the inner cover as an escape board, put an extra inner cover above the topmost super to close off the top and to keep out all robbers.

In extremely warm weather, install the escape board during the late afternoon, and remove the supers as soon as they are free of bees—the next day, if possible. A word of caution: in warm weather, comb may melt if bees are not available to fan it.

If the supers contain brood, the bees will be less likely to abandon them. If this is the case, pull out the frames of brood and place them in an empty super (the same size as the frames you removed); once filled, add this new "brood" super to a weak hive to allow the young bees to emerge. Meanwhile, you can select honey-only frames to place above the escape board and proceed normally.

If you find brood in your honey supers, try another supering technique. If you used a queen excluder, check for the leak that allowed the queen to move through. Whether your excluder leaked or another supering technique failed, when you place another queen excluder below these brood-filled honey supers, in twenty-five days the supers will be cleaned of brood as they emerge. Repair, if possible, or discard "leaky" excluders.

Advantages:
- Does not excite bees.
- Easy.
- Inexpensive.
- Usually effective.

Disadvantages:
- Honey could be removed by bees from the same hive or by robbers if supers are not bee-tight.
- Not always effective.
- Drones or dead bees may block escape, keeping bees in supers to be vacated.
- Involves extra trips to the apiary to insert board, remove supers, and so forth.

Repellent (Acid) Board or Fume Board. Fume boards are used by many commercial beekeepers to drive bees out of the honey supers with a chemical odor offensive to bees. An absorbent pad or cloth is placed inside a spare outer cover or other holder. You can also buy premade fume boards that consist of a wooden frame with a metal cover and flannel lining to hold the repellent. A chemical that repels bees (sold as Bee Go or under other names) is sprinkled on the pad and allowed to evaporate inside the hive. Some fume boards have black metal or wooden tops that will absorb solar heat and make the chemicals work better.

To use a fume board:

Step 1. Sprinkle about 1 teaspoon of the repellent chemical (follow label directions) on the pad.

Step 2. Remove the outer and inner covers, using smoke as needed.

Step 3. Scrape any burr comb off the top bars.

Step 4. Use smoke to drive the bees downward between the frames.

Step 5. Place the fume board over the frames (see the illustration of a fume board).

Step 6. After no more than five minutes, the bees will have left the super.

Step 7. Remove the first super and repeat the process for subsequent supers below.

Step 8. Air the supers thoroughly in the honey house, to prevent robbing.

Use the repellent board only long enough to get the bees out of the supers. Do not leave it on the hive for more than a few minutes. Do NOT use this board on comb honey supers, as it will impart an adverse flavor to the cappings or the honey.

Two chemicals legally permitted to be used as repellents are:
- Butyric propionic anhydride (works best between 80 and 95°F [27–35°C])
- Benzaldehyde (works best between 60 and 70°F [15–21°C])

Federal and state laws may restrict the use of some chemicals as bee repellents; comply with all regulations in this regard. These chemicals are hazardous and must be treated with respect; you are responsible for following label instructions for their use, storage, and disposal. The use of these chemicals is further complicated by the fact that their efficiency is governed by the air temperature; therefore, the desired result is not always certain. Never store or transport the fume boards or the supers treated with these chemicals in a closed vehicle.

Advantages:
- One trip to remove honey.
- Easy.
- Inexpensive.

Disadvantages:
- Excites bees.
- Dependent on temperature.
- Could adversely flavor honey.
- Rapidly crowds bees to lower hive bodies and may force them outside.
- Queen excluder needs to be removed to facilitate movement of exiting bees.

The Bee Blower. A bee blower is a portable gas or electrically powered device (that approaches 200 mph air speeds) to blow the bees off the frames and out of the supers. Set the super on end, not on its side, and blow air up from the bottom bars to the top bars—move the frames from side to side as you blow the air up through them, to clear out all the bees. Some beekeepers have a special stand that holds each super as it is blown free of bees. In this case the stand is placed near the hive entrance and the bees are blown down from the top. If, on the other hand, the weather is cool, blow down through a super that is still on the hive before removing it.

Sometimes leaf blowers can be modified to become a bee blower: they are cheap, commonly available, and adequate for most jobs. Add more hose (15–20 feet of 2½-inch hose) to give you room to maneuver. Take the supers away from the hives. If supers remain too close to the hives, the bees can get into the blower's air intakes and clog them up (or screen the intakes and the motors). A blower is very useful if you misjudge how quickly bees will leave the supers when using a fume board (i.e., supers are still full of bees when you are ready to load) or if there is some brood and the bees have not left the super. Blowers are a good backup system, especially if you are harvesting more than 50 colonies.

Advantages:

- Fast.
- Effective.

Disadvantages:

- Expensive.
- During cold weather, bees blown out may be unable to return to hive.
- Queen could be blown out and lost.
- Requires at least two people—one to load supers, the other to operate the blower—to work efficiently.

Abandonment (Tipping). This is a good method, but it should be used only by experienced beekeepers because it requires considerable expertise and know-how. W. Allen Dick of Alberta, Canada, outlined his method of tipping. To use tipping successfully, you must be able to recognize the difference between bees leaving a hive (or super) and robbing bees. Be sure you understand the conditions, both seasonal and weather related, that influence how well the method works; remember, conditions could change quickly.

Here's how tipping works:

Step 1. Choose a day when outside temperatures are warm enough for free bee flight and a good honeyflow has been on for several days.

Step 2. Remove full or partially filled supers; try to get supers that are brood-free, preferably from above a queen excluder.

Step 3. Place each super on end, either on the ground to one side of the hive's entrance or on top of a hive nearby that has its lid on normally. (Perhaps that latter hive has just had its honey removed and a super added.) Do not block flight paths.

Step 4. In a short while the bees in the tipped supers should finish their tasks, clean up any drips from burr comb, and fly back to the hive from which they came. This may take minutes or it may take hours, depending on the temperature and the intensity of bee flight activity. Or do tipping in the late afternoon and pick up the supers the next morning.

Step 5. Pick up the supers and take them away. You are done.

Advantages:

- Fast and easy.
- Extra equipment or chemicals are not used.

Disadvantages:

- Weather can change fast, as can the temperament of the bees. Bees that were happy at one moment may turn to heavy robbing, resulting in (total) loss of the honey.
- Queens, if excluders are not used, may be in the super(s) of some hives. Careful blowing, brushing, or shaking toward the correct hive is then required, as the bees may not leave by themselves. Brood in supers will have the same effect; that is, bees will stay with the brood.
- Requires a second trip to the apiary to collect supers.
- Requires an experienced beekeeper to know the temper of bees in the hives and appropriate weather conditions.

GETTING ALONG WITH YOUR BACK

Lifting off supers full of honey might be the reward of a productive year, but it can also be a literal pain in the back. Unless you are careful in lifting these heavy boxes, you can do serious damage to your back. Proper lifting and strengthening exercises might be needed if chronic back pain is a problem. In any case, medical advice should be sought. For some general information on back care, see "Honey and Honey Products" in the References.

Extracting the Honey

Now that you have removed the honey, place the supers in your honey processing area (a basement, garage, trailer, or special honey house); you are now ready to process it. Decide what type of honey you are going to process—extracted, cut comb, section comb, or a combination of all these. Try to do all your extracting at one time to avoid cleaning up the floor, the extractor, and other equipment more than once or twice. It is a messy job and will try the patience of all involved, but the reward is worth the effort!

The usual process for getting honey out of the wax cells is to remove the cappings with a hot knife called an *uncapping knife* (see the illustration on cutting wax cappings) or plane and to put the uncapped frames into a huge centrifuge, called an *extractor.* As the extractor spins, the honey is forced out of the cells and against the cylindrical wall of the extractor, leaving the frames of wax combs empty of most of the honey. A small gate at the bottom of the extractor can be opened to let the honey flow out into other containers. Keep your eye on the honey draining into these containers—overflows are costly and messy.

Honey left in supers before extraction can be ruined if it is stored in a humid or wet area, since even capped honey can absorb water vapor. To prevent absorption, stack the supers in a staggered arrangement to allow ventilation, and either use a dehumidifier or blow warm, dry air over the combs

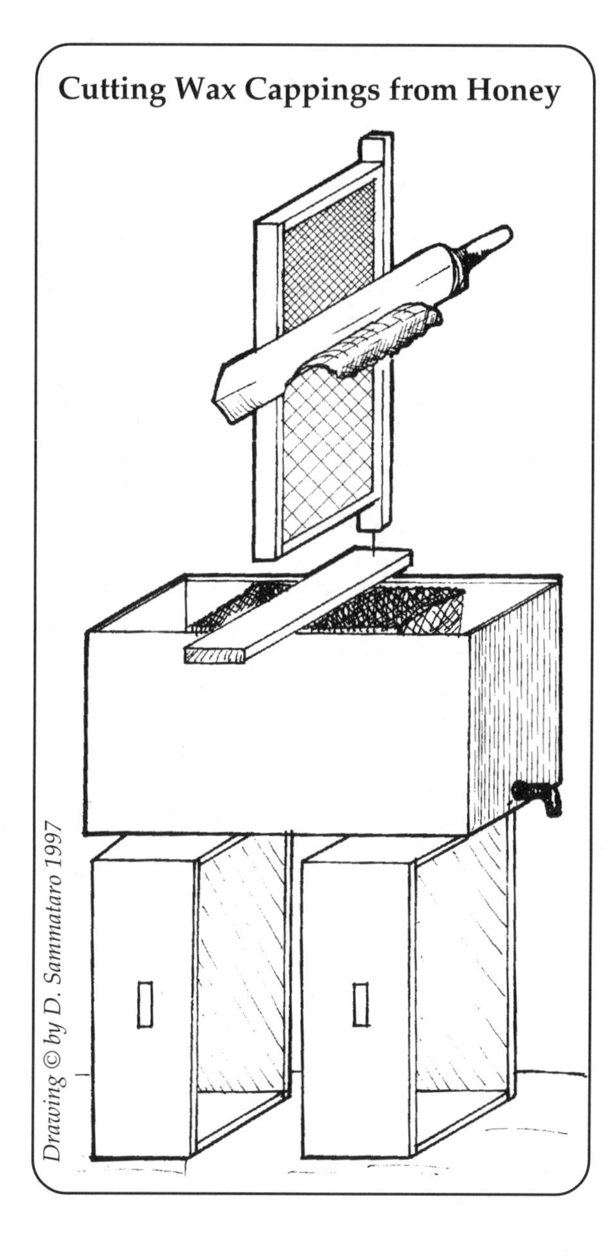

before extracting. You can also place extracting supers in a warm, well-vented room for a day to "dry down" the honey and make extracting easier. This action will further reduce or maintain the existing low moisture content of the honey. Warm your supers at least one day before extracting, if the combs are cold; warm honey will move through the extractor faster and will filter in less time.

Frames to be extracted should be completely or almost completely capped. Uncapped cells will contain honey with a higher moisture content. Extracting too many partially capped frames will increase the moisture of the extracted honey and invite spoilage by fermenting yeasts (see "Extracting Honey" in Chapter 12).

Remember: HONEY IS A FOOD PRODUCT—SOMEONE WILL BE EATING IT. KEEP THE HONEY AS CLEAN AS POSSIBLE TO HAVE A HIGH-QUALITY PRODUCT.

Laws require certain minimal sanitary practices, especially if you are selling your product. Honey is a quality hive product, and it is up to all beekeepers to keep it pure and clean. Check with local or state authorities for food-processing rules and legislation.

In addition to nectar collected from floral or extrafloral nectaries (e.g., cotton), bees collect another sugary liquid called *honeydew*. Honeydew is excreted from the intestinal tract of aphids and scale insects (order Homoptera) as they feed on plant sap. Bees will sometimes collect honeydew in large quantities and store it as honey. Such honeydew is usually dark in color, contains less of the two principal sugars found in nectar honey, has more protein, and is less acidic (has a higher pH) than honey made from nectar. It is considered to be a low-quality honey in this country but is quite popular in Europe and other areas. If you find you have lots of it, seek out special markets for this particular honey rather than blending it with your regular crop.

Harvesting Comb Honey

When harvesting comb honey, it is easier to go into the supers the day before harvest and scrape off any burr comb the bees may have built up between supers. They will clean up any honey drippings overnight, and the combs can then be packaged without any leaking honey or drips.

Take off these section supers carefully, trying not to crack or bend the bodies or the frames, because the section can leak if broken. Do not jar or drop such supers, keep them covered to exclude dirt and debris, and once in the honey house, process them in the next day or two.

Take out the sections from the frame, and cover and store them in the freezer (to kill wax moth eggs or larvae) until ready for sale. Once you are ready to sell, take the sections out of the freezer and allow them to come to room temperature before putting on the labels (see "Selling Your Hive Products" in Chapter 12).

Fall Management

After the fall crop has been removed and the supers have been cleaned and stored, each colony should be checked and attended to as follows:

- If possible, pick a day when there may still be a light honeyflow and forager bees are out. Bees are more prone to sting in the fall when a hive is being manipulated (most of the foragers will be inside the colony).
- Check for brood diseases and mites.
- Do not attempt to overwinter a colony found to have American foulbrood disease or excessive mite infestation.
- Medicate for nosema and American and European foulbrood as a preventive measure (see Chapter 13).
- Remove the queen excluder.
- If requeening, check after seven days to see if the queen has been accepted.
- Check winter stores; about 90 pounds (40 kg) or more of surplus honey should be left for each colony in areas where winters are not severe (see

"Wintering in Extreme Climates or at High Altitudes" in Chapter 8).

- For those colonies whose stores are low, feed early enough in the fall so the syrup can be properly cured by the bees.
- Treat for mites and add extender or grease patties (see Chapter 13).

Fall Management of Southern Colonies

Many beekeepers live in the southern states where beekeeping practices are a little different. Dean Breaux of HybriBees (Florida) suggests that the best way to overwinter colonies in the South is to start with young, vigorous queens. Requeening in the fall is necessary because honeyflows occur so early in the year (usually in February) that you could miss your first honey crop unless your bees are strong. The bees do not have a chance to utilize a queen raised in March, so begin fall management in October by requeening your colonies on the flows in late fall. Requeen with ripe queen cells; use mated queens for the colonies that were not successfully requeened with the cell.

The fall requeening procedure for southern states is:

- First remove all surplus honey and the old queen from the colony.
- Install Apistan strips, queen cells (or caged queens), and medicated grease patties; do all this in one trip.
- Return in two to three weeks to check and see that there are mated queens in the hives; if queenless, install a young mated queen in a cage. Give grease patties to all the hives that need them.
- By the middle of November, requeening should be completed.

Winter preparations for southern colonies can now proceed. Here are some tips to remember:

- Overwinter in a deep brood box with a minimum of one 6⅝-inch super of stores above a queen excluder.
- Install entrance reducers.
- Provide upper entrances, if you have several months of cold weather.
- Make sure the bee population is between 15,000 and 20,000.
- Test for mite levels to ensure mite- and disease-free colonies; treat if necessary.
- Protect colonies from cold, wintry winds in marginal regions.
- Periodically check to see if stores are holding up; feed if necessary.

Once most of the cold weather is over, prepare for the first spring honeyflows by doing the following:

- From the first of December to the end of January, you can move hives from fall locations to spring yards (primarily citrus groves) to prepare for the spring flows.
- Try to locate spring yards in areas that have red maple (*Acer rubrum*) trees; they bloom in late December through January and provide pollen for the bees to start building brood, in preparation for the early citrus flow in the beginning of March.
- Begin stimulatory feeding with light syrup at the beginning of February to ensure that the queens are laying well. This feeding will ensure a strong population of young bees in March for the early honeyflows.
- Add supers to colonies in March in preparation for the flows.

10 Queens and Queen Rearing

Although queens may live for four years or more, the most productive queens are usually between one and two years old. Many beekeepers replace an older, existing queen with a younger queen annually or every other year (see "Marking or Clipping the Queen" in this chapter). Studies have shown that a colony with older queens is more likely to swarm than a colony with young queens. Others replace only queens who perform poorly. Recently, requeening colonies annually has been found to help keep parasitic bee mites from overwhelming colonies. Read current research articles in bee magazines for the latest information on this topic.

If the bees are preparing to swarm or supersede their queen, they are in effect requeening the colony themselves. This natural process of requeening is not beneficial to the beekeeper. Queen replacement as a consequence of swarming, for example, results in a loss of a portion of the colony along with the old queen and reduced honey production, unless the swarm is captured and reunited with the colony. Furthermore, the new queen in the original colony may mate with drones of undesirable genetic traits (see "Swarming" in Chapter 11). If excessive swarming is experienced, requeening with other stock is recommended.

Queen supersedure, on the other hand, takes place only after the colony has been declining because its queen is failing (see "Queen Supersedure" in Chapter 11). Her replacement may be inferior, especially if the colony numbers and stores are not adequate for rearing good-quality queens. She may also mate with drones whose genetic stock is undesirable.

You should think of requeening colonies that show the following:

- Low bee populations for no apparent reason other than a failing queen.
- Bees that are particularly susceptible to diseases or mites.
- Queen laying more drone than worker eggs.
- Unmated or injured queen that is laying drone eggs, or drone and worker larvae scattered over the comb.
- Diseased queen, brood, or workers.
- Defensiveness.
- Excessive propolizing.
- Too much debris on hive floor (nonhygienic).
- Poor wintering success (very weak in spring).
- High honey consumption.
- Poor honey production.
- High tendency to swarm.

Types of Queens to Purchase

Queens can be obtained by purchasing them, raising one's own, or obtaining them from colonies preparing to swarm or supersede their queen.

There are five categories of purchased queens:

1. Virgin queens are unmated and require extra work to introduce them successfully into a colony and get them mated. If purchased when too old, they will not fly to mate and will become drone layers.
2. Untested queens have been observed to lay; most queens sold to beekeepers today are untested.
3. Tested queens are not shipped until the first brood emerges in order to determine purity of mating. Other variables may not be observed.
4. Select-tested queens are tested not only for purity of mating but also for other characteristics such as tolerance to disease or mites, gentleness, and productivity.
5. Breeder queens are tested for one to two years, evaluated, and used to raise daughter queens to sell. Commercial beekeepers and people who sell queens often buy breeder queens.

Whenever queens are purchased, make sure they come with Apistan Queen Tabs to protect them (and your bees) from varroa mite infestations. It is also recommended that all new queens be marked, as they could be superseded or replaced by the bees. If unmarked, you would never be able to tell whether the queen was the one you purchased.

Occasionally commercial queens display unusual behaviors. If you have a colony with some of these characteristics, requeen it with queens from a new source.

- Queen does not attract workers.
- Bees are unusually loud, as if queenless (decibel level above 65).
- Excessive supersedure or emergency cells evident.
- Colony suddenly queenless.
- Queen retinue does not form.
- Eggs and larvae are scattered about the broodnest, not in compact, concentric circles.
- Pollen is stored in the honey or otherwise not near brood rearing areas.
- Honey is stored below the broodnest or scattered throughout the colony.
- Colonies with lots of honey abandoned.
- Many larval and pupal deaths are evident.
- Hybrid brood cappings—more domed than worker but flatter than drone caps, like a cross between the two—are seen.
- Supersedure rates are high.
- Bees fail to cluster.
- Bees are seen scent fanning in large numbers.
- Bees abscond from hives at unusual times (during winter).

Although diseases or mites may cause some of these behaviors, such is not the case in general.

Purchasing and Ordering Queens

Queens can be purchased alone or in packages, both from reputable queen breeders throughout the United States. If the purpose of ordering queens is to requeen an existing apiary, order queens in the early summer, when there is a good honeyflow on in your area. You can requeen either in the summer or in the fall, for mite control. If the latter, you can still have your queens delivered in the summer, but first install them in nucs to get them laying and then requeen the other colonies later in the season.

Order queens as you would packages, as early in the year as possible, because most breeders start rearing queens when snow is still on the ground in northern states. Some breeders will ship queens with or without attendants; others will place the cages in a package or other shipping container with loose worker bees and some kind of feeder.

Treat your queens carefully, and follow the installation instructions given in this chapter. Do not treat new queens with Apistan strips until they have been laying for about 30 days.

Many journal advertisements are claiming "mite-resistant" queens. Be careful. To which mite are they resistant, and what is the basis for such claims? Ask for research articles. To date, the only known race of bees that has some mite-resistance capabilities is the Buckfast bee, which tolerates tracheal mites.

Types of Queen Cages

Purchased queens are packaged in queen cages and mailed from breeders to all parts of this country and abroad. Some of the procedures for notifying the postal service when awaiting bee packages should also be followed when queens have been ordered (see Chapter 6). Again, each cage may have a Queen Tab strip for varroa mite control.

In the United States, queen bees are shipped in several kinds of cages. One style is the Benton mailing cage, a small wooden block, approximately 3 × 1 × ¾ inches with wire screening stapled over the length of it to cover the three holes or compartments. Two holes serve as the living quarters for the queen plus several attendants, and the other hole contains candy. At either end of the cage is a small, bee-sized hole. On one end, this smaller hole and the adjacent compartment are filled with candy; the hole at the opposite end, through which the bees were introduced, is plugged with a small cork. The candy serves two purposes: it provides food for the bees inside, and it is also a way to delay the queen's release from the cage into a new colony. The candy plug must be removed by the hive bees before the new queen can be freed.

Other cages used by bee breeders today (see the illustration of queen cages) include the Kelley cage (which has two compartments and measures 2½ × 1 × 1 inches) and the new California Mini Queencage, developed by C. F. Koehnen and Sons. The latter is a narrow wooden block, about 2½ × ¾ × ¾ inches, with one long inner compartment. Inside, it has a large, candy-filled plastic tube inserted into one end. The smaller size of this cage allows more queens to be banked (or stored in queenless colonies), and its insertion between frames does not violate the bee space.

Another available cage is a plastic cage in the shape of the lowercase letter *d*, called JZ's BZ's. This is much smaller than the Benton cage; its overall length, with the stem, is 2½ × 1 × ½ inches. The stem of the *d* holds the candy plug.

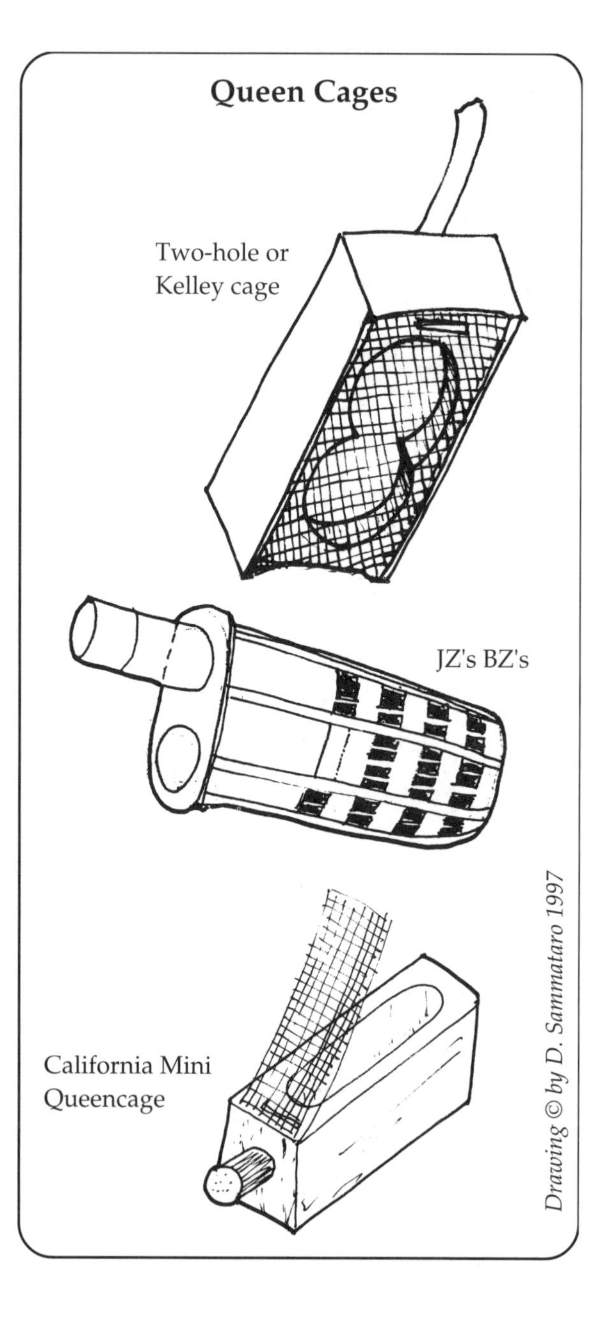

REMOVING ATTENDANTS

When caged queens arrive, they should be properly cared for and placed into a colony as soon as possible. But before the queen cage is introduced, many beekeepers remove the attendants because they are foreign bees and the workers in a colony will try to expel them, biting them (and the queen) through the wire screening of the cage. This action may release alarm odors, alerting other bees to surround the cage, and in the process they could injure or kill the queen.

To remove the attendants from a cage without losing the queen, try removing the staples while holding down the screen. Then place a piece of zinc queen excluder on top of the screen. Carefully remove the screen, taking care not to injure the queen if she is on the wire, and tack the excluder piece to the cage. Now place the cage near some light source (natural or artificial) in a darkened room, and the workers will exit through the excluder. Once the attendants have left, reverse the process, replacing the screen (don't forget to staple it).

Another way of removing the attendant bees is to take out the cork opposite the candy end of cage (or try prying up one the staples and lifting the wire screen) in a darkened room next to a closed window. When the bees exit, they will move toward the light and you can recapture the queen from the window. Pick her up only by the wings or thorax, return her to the cage, and replace the corks or screen. This is a good time to mark the queen (see "Marking and Clipping the Queen" below).

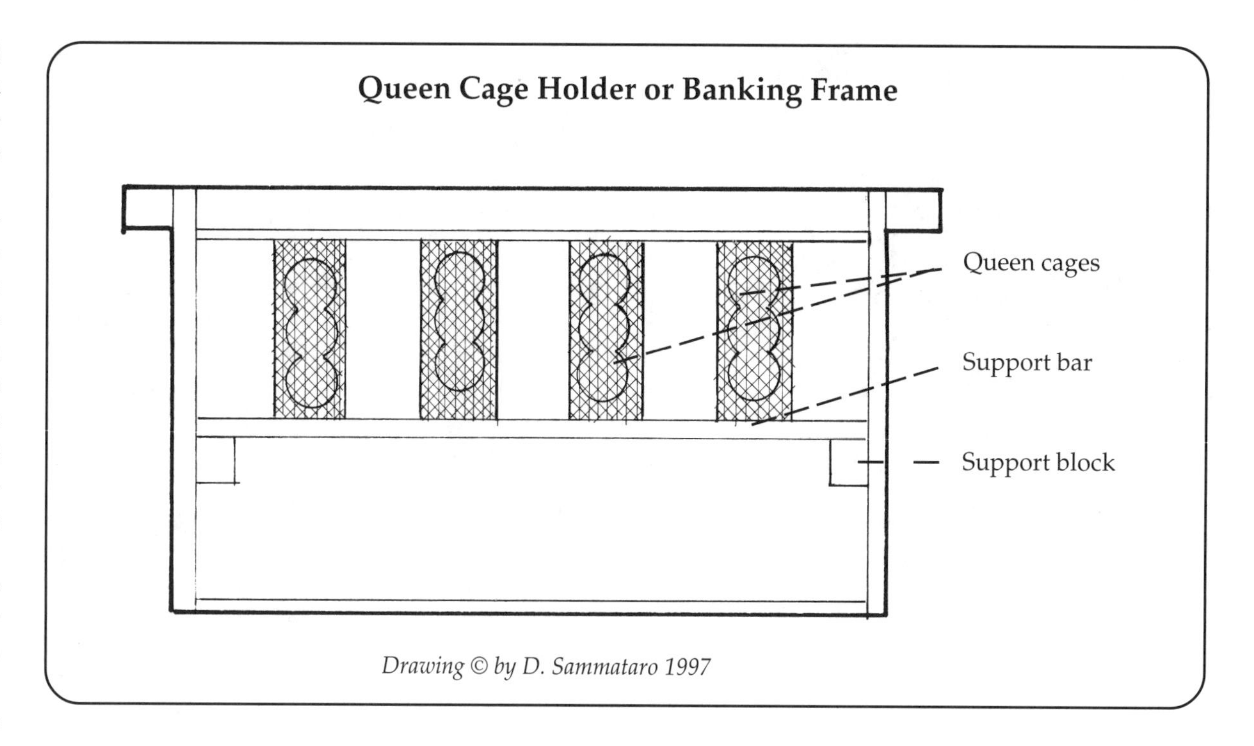

Drawing © by D. Sammataro 1997

CARE OF CAGED QUEENS

If the cages of the newly arrived queens contain candy and attendant bees, they can be kept in a warm (about 60°F [15.5° C]), dark place free of drafts for up to one week. The attendants help keep the queen warm, but cover the cages with a light cloth to keep the warmth in. A very small drop of water should be placed on the screen twice a day. Do NOT get the candy or the bees wet, as they could become chilled.

If the queens are to be introduced by the indirect method, the attendant bees should be removed beforehand, as already described. If you use another method in which the attendants are needed, check the condition of the queen and accompanying bees to see if there are any dead attendants in the cage; if so, replace them with newly emerged workers from one of your colonies. Clean out the old cage, and replenish the candy, if needed, with a ball of candy dough (made of clean honey and powdered sugar) or a marshmallow.

Some beekeepers bank caged queens in colonies, for up to three weeks, if many queens arrive at one time. First remove the attendant bees, then place the cages screen side down between the top bars in a queenless nuc that contains lots of young workers. You can also make a special queen cage–holding frame or banking frame. One end of the cage goes against the underside of the top bar of any empty frame (without comb or foundation), and the other end rests on a bar of wood that has been nailed in to run the length of the frame (see the illustration of the queen cage holder). The bar is nailed to the side bars of the frame; this way, 10–20 cages can be banked in several frames.

Insert the banking frame into a strong, queenless colony or into a queenright hive above a queen excluder. If you surround the cages with frames of emerging bees, pollen, and honey, the caged queens will be well cared for until they are needed. If stored in a queenless colony, add some frames of capped brood once a week and feed the colony. A free queen must not be allowed in the queenless colony or above the excluder; otherwise, the caged queens may be killed.

MARKING OR CLIPPING THE QUEEN

Many beekeepers choose to mark the queen with a spot of color or a color disc (with or without a number) on her thorax, or they clip her fore and hind

Queen with Numbered Tag

wings on one side (see the illustration of the queen with numbered tag). Marking or clipping the queen enables you to keep a record of the age of the queen, and the use of color will also make it easier to find her, especially if she is dark. Clipping the wings of a queen will not control swarming, although this erroneous statement is found in the literature. Today, clipping is considered injuring the queen and may cause the bees to supersede or replace her.

Never hold the queen by the abdomen when picking her up to mark her (see the illustration of the method of holding the queen). If the queen is picked up this way, she can become injured, and her egg-laying ability may be compromised. As a result of this or any injury to the queen, the colony may supersede her. To avoid injuring a queen, first practice on a few drones.

To mark the queen:

Step 1. Grasp the queen by the wings with one hand. Then, with the other hand, hold the sides of her thorax, releasing her wings.

Step 2. If marking with a color spot, use a fast-drying paint, such as nail polish or model paint; mark only the thorax. If marking with a disc, apply adhesive to the thorax, then firmly attach the disc.

Step 3. Allow paint or adhesive to dry before gently returning the queen to the frame from which you took her, or place her on the top bar of a frame and let her walk down the comb. Sometimes the smell of paint will cause the hive bees to attack the queen; cover her with honey so the bees will clean her off instead.

The queen can be removed from the hive without alarming the other bees. In fact, for short periods (5–10 minutes) they will not be aware she is absent.

Some beekeepers requeen every other year to maintain quality stock, reduce swarming, or keep diseases and mites under control. To keep track of the queen's age, record the date and color you marked her in your hive diary. Or use two colors, alternating them every other year.

Proper Method of Holding Queen to Clip or Mark

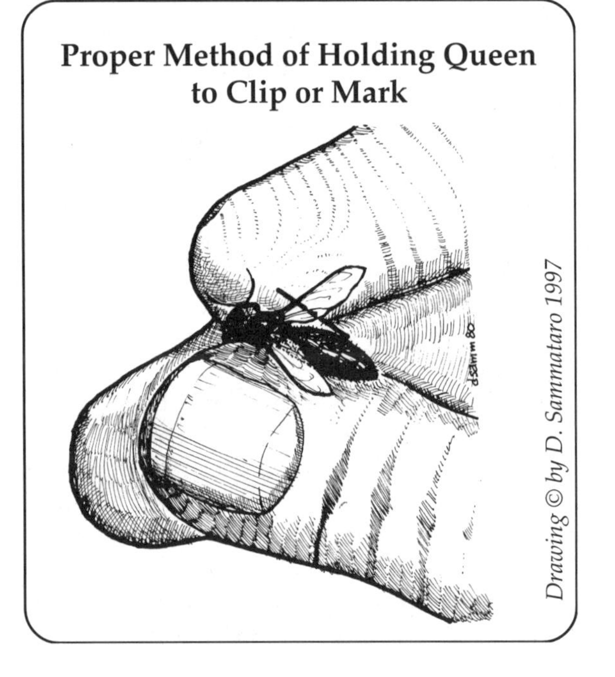

Drawing © by D. Sammataro 1997

You might decide to adopt the European color system, which is on a five-year sequence:

- White or gray—year ending 1 or 6.
- Yellow—year ending 2 or 7.
- Red—year ending 3 or 8.
- Green—year ending 4 or 9.
- Blue—year ending 5 or 0.

Thus in 1998, the queen color is red.

Advantages:

- Queen easily found.
- Queen's age can be determined.
- Queen's absence indicates a change in the colony (e.g., queen swarmed or was superseded).

Disadvantages:

- Bees might supersede "maimed" queen.
- Clipping does not prevent or control swarming.
- Queen could be injured while being handled.
- Queen that was clipped may be a virgin and thus would be unable to fly and mate.
- Virgin queen might sting when handled (but this is not likely).
- Color spot or disc may come off.

FAILURE TO FIND A MARKED QUEEN

If you can't find a marked queen, you can assume that either the colony swarmed with her, she was superseded (replaced), or she was accidentally lost and subsequently replaced. If a queen is not found but eggs and brood are present, you can be assured that there is a queen in the hive (marked or not). If no queen is found after a thorough search or no eggs or brood are present, look for queen cells about to hatch or suspect that a new, young queen or virgin is present but has not yet begun to lay. Check the colony for eggs in another week; if none are found then, requeen.

A virgin queen is more difficult to locate because she tends to move more quickly on the comb than does a laying queen. She might also be a little smaller than a laying queen (more the size of a worker), because she loses some body weight before the mating flight and needs time to regain it before she starts to lay.

The absence of eggs or brood in a colony, therefore, could mean that a virgin queen is present but has not yet mated or has not begun to lay eggs. Before requeening such a hive, be sure that the colony is queenless, since any attempt to introduce a new queen into a hive with a virgin queen or a newly mated queen is likely to fail.

On the other hand, the colony is definitely queenless if laying workers are present. If you find scattered drone brood, scattered capped drone cells, and several eggs in each cell that are attached to the cell walls instead of the bottom, you have laying workers (see "Laying Workers" in Chapter 11).

Spring Requeening

Requeening can be done in the spring, summer, or fall. It is always preferable to requeen during a honeyflow, because a colony is almost certain to accept a new queen when food is incoming. It is easiest to requeen in the spring or early summer.

Advantages:

- Colony less likely to swarm if requeened before the swarming season has started in your area.
- Vigorous egg layer will produce large bee populations, ready for subsequent honeyflows.
- Colony will enter the winter with a large population.
- Old queen easier to find when colony numbers are low.
- Bees are calm and less prone to sting, run, or rob, especially during a honeyflow.
- Less chance of swarming the following year because queen will be only one year old.
- Time to assess queen's performance and to change her if necessary.

Disadvantages:

- Dependent on weather; if you hit a rainy spell, it could be many days before you can get to requeen the hive.
- Queen could be superseded if inclement weather sets in and bees go hungry.

Summer/Fall Requeening

Now that mites infest bees in the United States, many beekeepers here find it better to requeen later in the year. By doing so, you break the bee's brood cycle, especially if you install queen cells. This allows bees to clean out diseased or damaged brood, thus lowering infestation rates (see "Mites" in Chapter 13).

Advantages:

- Less chance of swarming the next year.
- Colony enters winter with a strong population and a young queen.
- Young queens will lay more eggs in the late winter and spring than older queens.
- Colony emerges in spring with a high bee population ready for honeyflow.
- Breaks the brood cycle, thus reducing disease and pest problems.

Disadvantages

- Hive is populous making it difficult to find old queen.
- If no honeyflow is on, bees are prone to sting, rob, and run when hive is opened; robbing could be serious.
- Methods for requeening later in year are more time consuming.
- In fall, fewer opportunities exist to check if queen was accepted.
- Less time to assess queen's performance.
- Could end up with queenless colony and laying workers.

Queen Introduction

Although many methods, including some ingenious ones, have been devised for introducing queens into colonies for the purpose of requeening, none can guarantee absolute success. Often the more time-consuming ones are the most likely to succeed.

It is generally agreed that no matter what method is employed, the most opportune time to requeen is during a honeyflow. If no honeyflow is evident, feed the colony to be requeened at least one week before and after the new queen is installed to simulate a honeyflow. All the methods listed here, except the division screen method, require that the hive be dequeened (the queen taken out to make a hive queenless) for a period of 2 to 24 hours before introduction of the new queen.

During the time you are introducing the new queen, observe how the bees react to her. If the bees in the colony tightly cluster over the queen cage and appear to be trying to bite or otherwise injure her, there may be a queen in the colony that you are trying to requeen. On the other hand, if you observe the bees forming a loose cluster and trying to feed or lick the new queen, they are probably ready to be requeened.

The methods used for requeening can be divided into two categories: *indirect release*, in which there is a delay before the bees have direct access to the queen; and *direct release*, in which the queen is immediately released among the bees. Some of these methods can be combined with swarm control or with making increases in the apiary (see "Prevention and Control of Swarming" in Chapter 11).

After you requeen a colony, check it after a week or so to see if the new queen has been accepted. Look for eggs and young larvae: if present, you have a laying queen. If the colony has a queen other than the one you introduced, you can either assess her qualities for about a month or requeen the colony again later in the year.

Indirect Release: Push-in Cage

The push-in cage is an old style of introduction cage (see the illustration of the push-in cage) first used in the 1900s. It is made by folding all four sides of a square piece of hardware cloth (about a half an inch) at right angles, to form a box with a top, sides, but no bottom. The corners of the box will have to be pinched into a triangle to fold around the sides. After a colony has been dequeened, take out a frame of capped brood, shake or brush off all the bees, and place the queen on the brood. Now place the open end of the cage over the queen and push it into the comb.

The comb under the cage should contain capped brood, a few cells of honey, and no adult bees. After seven days, release the queen by removing the push-in cage. Some beekeepers cut a hole in one of the sides of the cage and plug it with candy made from a small amount of honey mixed with confectioners' (powdered) sugar; the bees will release the queen after eating through the candy. You can also substitute a compressed piece of marshmallow or a mini-marshmallow for the candy plug.

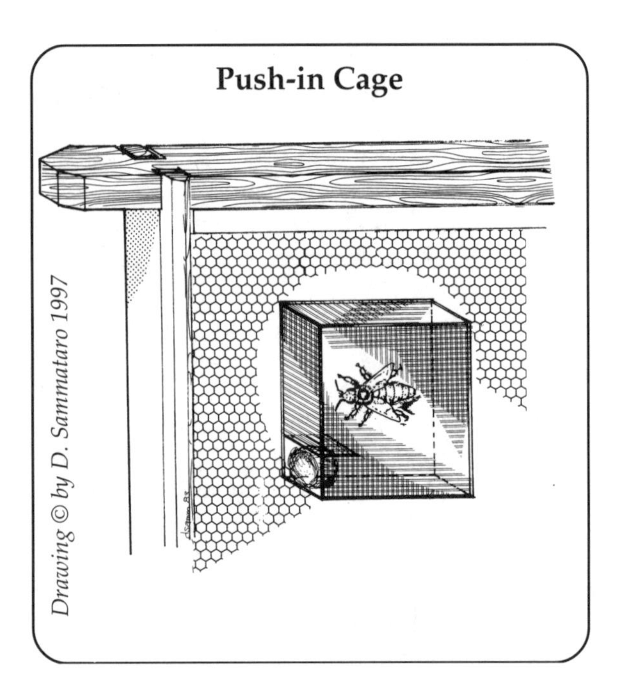

Indirect Release: Shipping Cage

Smoke should be employed in the same fashion as when ordinarily working with the bees. Feed the colony with sugar syrup a few days before and after killing the old queen.

The indirect method of requeening employs a queen cage:

Step 1. Dequeen the colony from 2 to 24 hours before replacing with new queen.

Step 2. Select a caged queen with no attendants.

Step 3. Remove the cork in the candy end and, if the candy is hard, make a small hole through the candy with a nail to make it easier for the bees to free the queen by eating out the candy. The hole should not be too large; one of the purposes of the candy plug is to delay the queen's release and thus enhance her acceptance. Do not make a hole if the candy is soft.

Step 4. Find one or two frames containing uncapped larvae and remove some of the comb to allow the queen cage to fit vertically from the top bar. Push in the cage candy side up, and place the frame back in the hive. Now place a frame of young brood next to the queen, and push frames together so the bee space is maintained. (If you leave a gap, the bees will fill it with comb.) Make sure the screened or open side (if plastic) of the cage is accessible to the bees.

Alternatively, place the cage horizontally between two frames of brood, with the screen side down, so the bees have plenty of access to the queen inside and to the candy. Remove the cage after a few days to delay excess comb from being constructed.

Step 5. Watch how the bees react to the caged queen. If they are defensive, you may have a loose queen in the hive. If they cluster on the cage, trying to feed her and scent fanning, the hive will probably accept her.

Step 6. Examine the colony after one week. If the queen is still in her cage, poke a bigger hole in the candy to let the bees finish releasing her, pull out the cork plugging the other end, or release her directly by pulling up the screen. Close the colony and leave alone for a few more days (see the next section on direct release).

Direct Release

Nucleus Method. This method does not require that the attendant bees be removed from the queen cage. Take two or three frames of capped and emerging brood from a strong colony and shake off all the bees back into the parent colony. Place these frames, now free of bees, in a three- to five-frame nucleus hive (or *nuc*); fill the rest of the nuc with honey and pollen frames. Directly release the queen (and her attendants) onto a frame of emerging brood; any young workers will immediately accept the new queen. Use this method during the summer months so the brood (and the queen) will not be chilled on cool nights. You can later use the nuc with the new queen to requeen an established colony.

To requeen an established colony with the now populous nuc, follow these steps:

Step 1. Dequeen the colony to be requeened at least a day before.

Step 2. Place the nuc containing from three to five deep frames of bees and a laying queen next to the dequeened hive.

Step 3. Apply a small amount of smoke into the nuc

entrance, being careful not to disturb the bees too much.

Step 4. Remove two to three frames from one side of the dequeened hive and replace them with two to three frames of bees and the laying queen from the nuc box; the new queen should be between two of the inserted frames. Any extra frames can be given to a weak colony.

Step 5. Close the hive, checking it after one week.

Another nuc method is similar to the one above, but the attendant bees are taken out. Remove frames of capped brood from a hive and shake off the attached bees. Place frames in a nuc and directly release the queen, without attendants, on these frames; close up the nuc. Now remove two or three frames of young, uncapped brood, making sure the old queen is NOT on these frames. Shake these frames of young bees in FRONT of the nuc, so the bees can enter it.

Honey Method. Dequeen the colony at least one day in advance, then proceed as follows:

Step 1. Open the hive and remove the nearest frame. Check each frame until you find one with young larvae and honey; remove it, shaking off all adult bees.

Step 2. Break the wax seal over some honey, and without injuring the new queen, coat her with honey.

Step 3. Release the queen on the frame with young larvae, and then gently return the frame to the hive. Replace remaining frames and close the hive.

Scent Method. The scent method employs a scented syrup (a few drops of any one per cup of syrup: peppermint, lemon, vanilla, wintergreen, onion, anise oil, or grated nutmeg) that temporarily masks the odor of the introduced queen. As the scented odor gradually diminishes, the queen acquires the odor of the hive, and the bees will accept her. Dequeen the colony at least one day in advance, and feed with the scented syrup. Then proceed as follows:

Step 1. Remove frames containing bees and spray them with the scented syrup, but do not soak the bees. Try not to get the syrup inside uncapped larvae, as it may kill them.

Step 2. Spray the queen, then release her onto the top bar of a sprayed frame. Guide her down between two sprayed frames.

Step 3. After she has crawled down, close the hive and feed with more scented syrup. Check after one week to see if she was accepted.

Smoke Method. Dequeen the colony at least one day in advance, then proceed as follows:

Step 1. Blow four or five strong puffs of smoke into the entrance.

Step 2. Reduce the entrance to 1 inch with loosely packed grass, rags, or newspaper.

Step 3. Close the entrance for one to two minutes.

Step 4. Open the entrance slightly to allow the queen to enter; blow in a few puffs of smoke after she enters.

Step 5. Close the entrance for from three to five minutes.

Step 6. Reopen the entrance to about 1 inch wide (the bees will remove the remaining material within a few days if it is loosely packed).

Step 7. Check for the queen after one week to see if she has been accepted.

Caution should be used with this method if the weather is extremely warm, because a reduced entrance will make it difficult for bees to ventilate the hive.

Shook Swarm Method

Dequeen the colony at least one day in advance, then proceed as follows:

Step 1. Take out frames with bees on them, and spray the bees and frames with syrup (can even be scented).

Step 2. Shake these queenless bees into a screened box, or swarm box (the container should be large enough so bees are not overcrowded and should be screened for ventilation). After the bees have been "shook" off the frames into the container, remove any frames with brood, giving them to other colonies (or use in nucs for new queens, as previously described). Add enough frames of dry comb, honey, or foundation to the hive to make up one deep colony; close up the entrance to prevent bees from robbing the frames of honey.

Step 3. Put the bees in a cool, dark place. Feed them with 1:1 sugar syrup as needed (see "When the Packages Arrive" in Chapter 6).

Step 4. Seven to eight hours later, directly release the queen into the container; she can be sprayed, along with the bees, with a scented syrup.

Step 5. After one hour, reopen the old hive and install the bees as you would a swarm (see "Swarming" in Chapter 11).

Snelgrove, Division Screen, or Screen Board Method

A Snelgrove or screen board, sometimes called a division screen or division board, is a double-screened, rimmed, inner cover–size board that has a small entrance on one side of the rim. It is used to make an increase, to split a colony, or to start a two-queen colony. The screen separates the queen and bees in the lower part of a hive from the queen cells or new queen and bees in a hive body placed above; the smaller colony above can take advantage of the heat generated from the colony below.

To requeen by this method, follow these steps:

Step 1. Reverse the hive.

Step 2. Remove from any strong hive(s) three or more frames of capped brood from which bees are emerging; replace these frames with foundation or drawn comb.

Step 3. Shake the frames so all adult bees fall back into the parent hive.

Step 4. Place these brood frames in the center of an

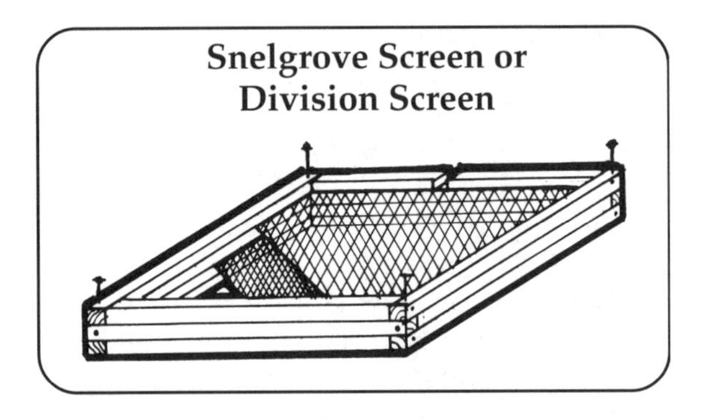

empty deep hive body (3 on the illustration of requeening).

Step 5. Place frames of sealed honey with some pollen on each side of the brood frames.

Step 6. Fill the remaining space with frames of drawn comb.

Step 7. The new hive should include the following parts:

- One deep with original queen (2).
- One shallow, medium, or deep with the new queen (3), as well as an inner cover and an outer cover.
- Super for honey for both colonies as needed.

Step 8. Requeen the top hive body with a queen cell, ready to emerge, or a marked queen (by any method covered previously). As the young bees emerge, they will accept the queen as their own.

Step 9. The entrance of the division screen should be small so only a few bees can pass through at one time (this entrance should be pointed in the opposite direction to the main entrance). Close this new entrance with loosely packed grass for a week until the brood emerges.

Step 10. Check the colony after one week.

Three weeks later, replace the division screen with a queen excluder. The hive can now be run on a two-queen system until after the honeyflow, or the hive (3) can be united (with 1 and 2) by removing the excluder just before a major honeyflow. Find and remove the older queen, or unite the colonies by removing the excluder. If you want to make 3 a separate colony, put it on a new bottom board when the bee population is high.

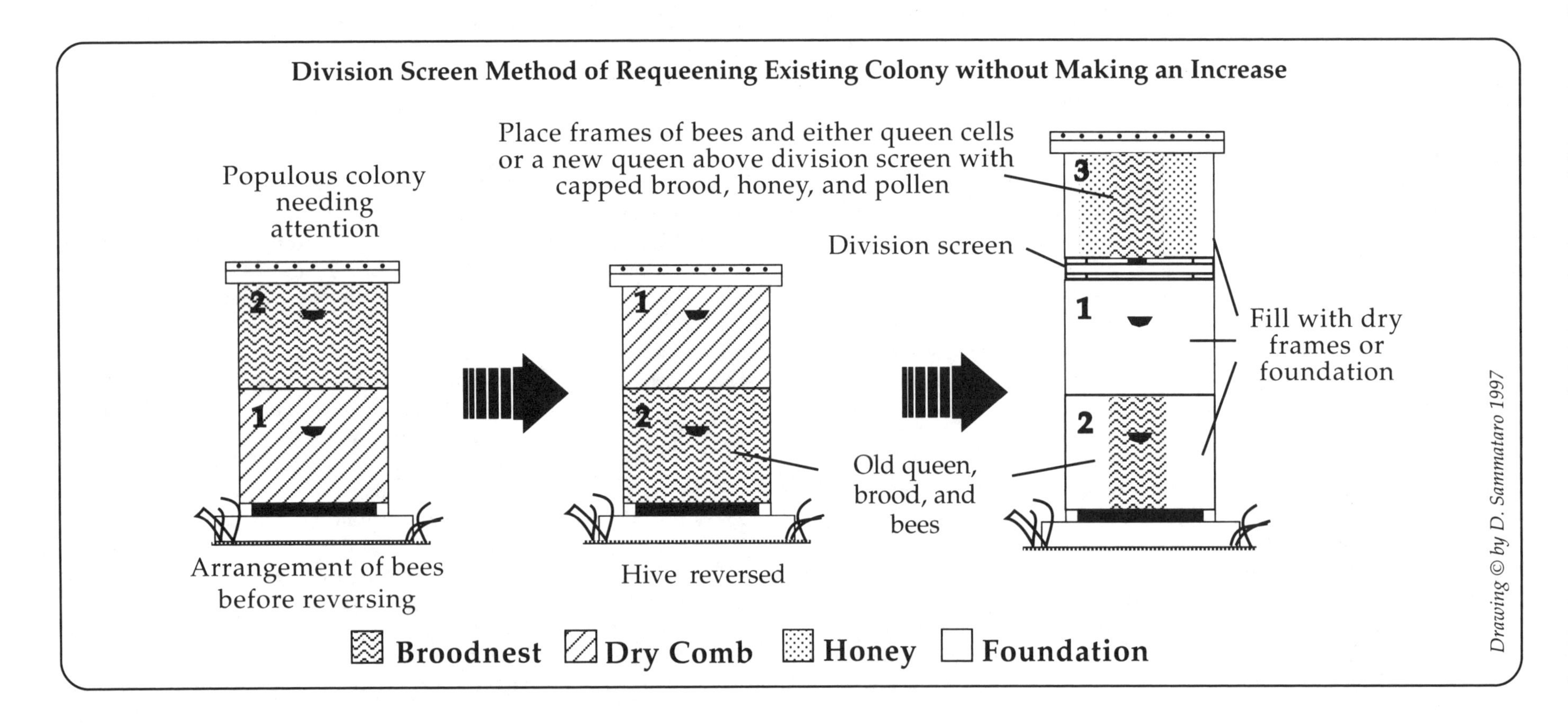

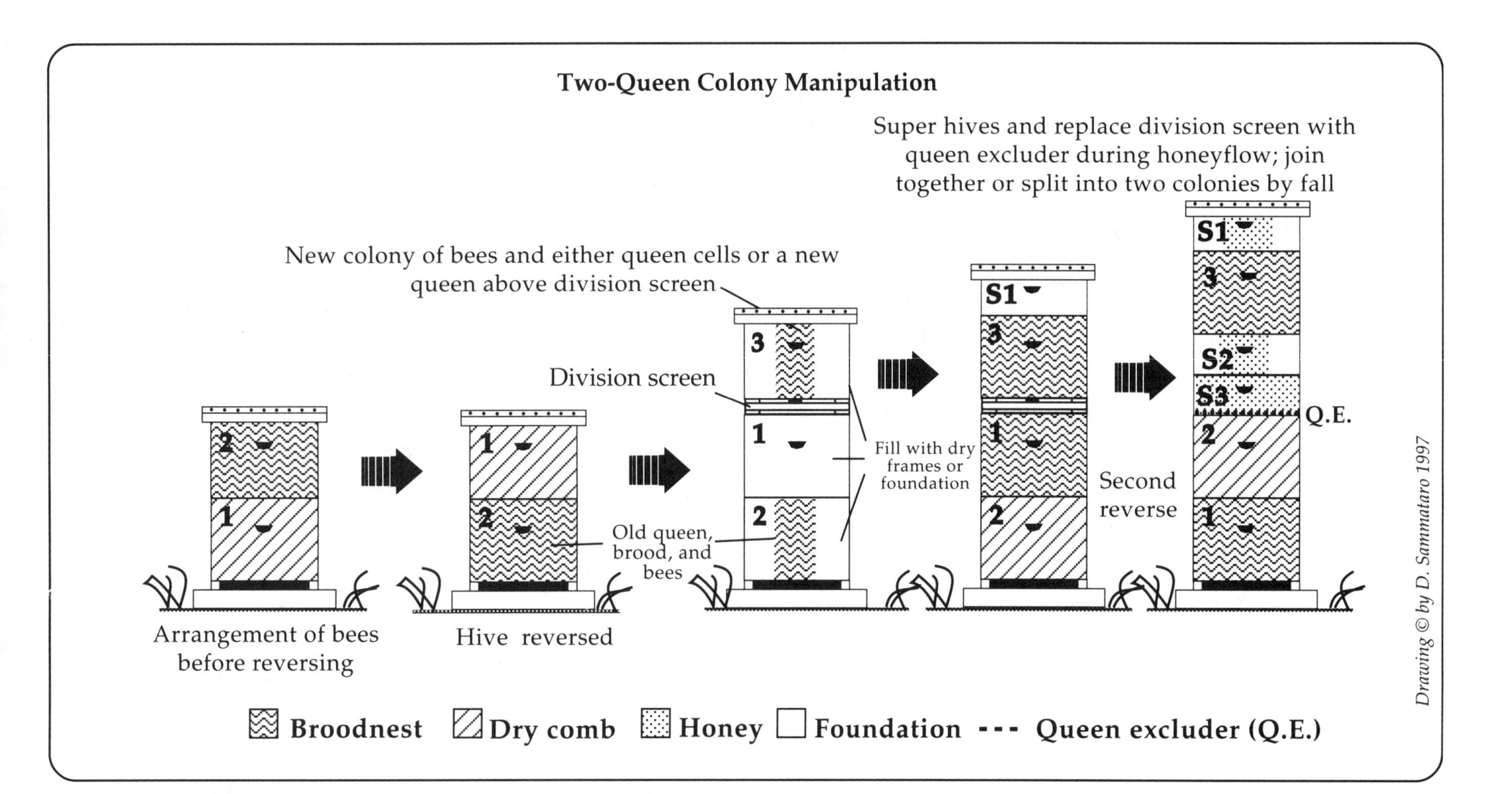

Two-Queen System

Some beekeepers use the two-queen system of colony management to improve honey yields. Two separate colonies, one above the other (each with its own queen) are joined, but a queen excluder is placed between them to protect each queen from the other.

There are various methods of managing hives with two queens; here is one method:

Step 1. Split a very strong colony and requeen the upper queenless portion using the division screen method, placing only capped brood above the screen (see figure on the two-queen colony).

Step 2. Replace division screen with excluder when honeyflow is under way, supering both colonies for honey.

Step 3. Remove the excluder in the fall and either kill the less desirable or older queen or let the two queens fight it out.

Advantages:

- Better yields, since one large colony will produce more than two separate colonies, each having half the strength (and fewer than half the foragers) of the large one.
- More bees, which can be used later for making an increase or for strengthening weak hives.
- Strong hive for wintering.
- If one queen fails, hive has a backup queen. Can be combined with swarm control and requeening techniques.

Disadvantages:

- Even if all manipulations are completed successfully, will not be effective if local honeyflows and colony are not synchronized.

- Time consuming.
- Difficult to manipulate.
- Hard to make work successfully.

Requeening Defensive Colonies

Occasionally a colony exhibits extreme defensive behavior, such as flying at and stinging you with little provocation. These colonies, at peak population, are very difficult to work, and if they release excessive amounts of alarm pheromone when worked, the defensive stance of the other colonies in the yard could be elevated as well. If this happens, you may have to abandon activities and hope for a better day.

First, check for signs of skunk or other pest damage to see if an explanation exists for why the bees are so upset (see "Animal Pests" in Chapter 13). If they can be ruled out, your colony may need to be requeened.

If this colony is close to other homes, schools, or high-use areas (such as parks or sports playing fields), it could become not only a public nuisance but also a liability. Any stinging incident can result in unhappy neighbors and could lead to zoning regulations restricting beehives in residential areas.

If you live in regions where the Africanized bee has become established (parts of Texas, New Mexico, Arizona, and Southern California, as of 1997), defensive behavior may mean that you have serious problems. Only scientists and beekeepers experienced with this bee can absolutely differentiate Africanized bees from European races, but many signs can help you tell the two apart. Africanized bees exhibit extraordinary defensive behavior, with many bees running on the combs and many stinging bees pursuing you (see Appendix G). Requeen such colonies immediately with a purchased queen of a known gentle stock, or destroy the colony outright.

After you successfully requeen, the change in the behavior of the colony will be apparent as soon as the new queen's offspring populate the hive. Thus it will take almost two months for all the existing workers and larvae to be replaced, which may be unacceptable if a more immediate solution is required.

As for any method of requeening, the hive first needs to be dequeened. Finding the queen of a volatile hive is not easy, especially if the colony is populous.

Here is one method of requeening, called the non-shook swarm method:

Step 1. During a favorable day, when most bees are out foraging, move the colony to a new location.

Step 2. If you suspect the bees are Africanized, place an empty deep hive body on a bottom board, with a cover, in the old location. The foragers and drones that return will collect in this hive. In the evening, spray these bees heavily with soapy water, which will kill them quickly.

Step 3. If the bees are European but highly volatile, place a deep hive body with all its complements in the old location, plus one frame of eggs and young larvae from a gentle colony. They should begin to raise a new queen from this frame. You can also replace this frame with a ripe queen cell or a caged queen from gentler stock. Fill the remaining space with dry, drawn comb.

Step 4. Field bees from the aggressive colony will return to the old location and the new hive. Because they will live for only another three weeks, they will be dead by the time the new brood is emerging.

Step 5. Back at the hive you moved, with their population reduced, the bees will be more manageable. Find the old queen and kill her.

Step 6. The next day, introduce the new queen by any method.

Step 7. Wait seven days, then, if a caged queen was introduced. Check to see if she has been accepted.

Step 8. At the end of two months, the undesirable bees should all have been replaced by the new stock. You can return the original hive to its old location and unite it with the small hive, or use the new colony as an increase at its new location.

This method, while effective, is very time consuming, so if possible try to requeen the problem colony at its original location, if the queen can be found quickly. Or you can separate the brood chambers with one or more queen excluders. Return in four days: the queen will be in the box with eggs. If the hive is too populous, split it (separating the hive bodies and supplying each its own bottoms and tops), wait four days, then dequeen the split or splits. Requeen one or both splits, or unite them after one has been successfully requeened.

Queen Rearing

Natural Queen Rearing

The simplest way to rear queens is to let the bees make their own. Kill off the older queen so the bees can make emergency cells. This method has risks, however, for the bees may fail to rear a new queen, or she could be inferior or not as good as her mother. A superb queen can probably be found in any apiary with two or more colonies. Obviously, if queens could be raised from the larvae of such a colony and later be introduced successfully to other colonies, the entire apiary could be upgraded. This leads to inbreeding, however, which you should avoid.

Good queens are reared by bees in a strong colony when there is an abundance of food (honey or sugar syrup and pollen) available to the nurse bees. Queen-rearing operations can coincide with swarm control manipulations (see "Prevention and Control of Swarming" in Chapter 11).

Once you have killed the old queen, provide the queenless colony with a frame of larvae less than three days old taken from a colony with desirable qualities. You can also take extra queen cells from colonies preparing to swarm.

For more control, split a populous colony into two or more smaller hives, each with its own top and bottom and with entrances facing a direction different from those of the parent hive. Let the bees raise their own queens, making sure that either some queen cells (swarm cells) or the right-aged larvae or eggs are available in each nuc you start. The chances are high that one of these splits will take, and you can always choose one daughter to replace the mother.

Some beekeepers prefer to raise their own queens rather than purchase them from a commercial breeder. While educational and exciting, rearing queens can be tricky, time consuming, and often unsuccessful but, with practice, can be accomplished.

Advantages:

- Can be easy.
- Inexpensive.
- Will usually succeed in obtaining queens.
- Few manipulations needed.
- Can coincide with mite control, as it breaks the brood cycle.

Disadvantages

- Queen could be inferior or may mate with inferior drones.
- Will disrupt dequeened colony; no new brood emerges for 43 days.

Conditions Needed to Rear Queens

Before you begin to rear queens on a larger scale, read books on the subject, talk to beekeepers, and take a class (see "Queens" in the References). Then you should choose whether you want to graft larvae into artificial queen cells (Doolittle method) or cut out such larvae and let the bees make their own queen cells (Miller method).

Next, ask yourself three questions: Why, What, and When? We have already covered "why" one rears queens—to replace inferior queens or increase your colony numbers.

The next question is, What is a good queen? Pick your breeder queens carefully, from stock which you can trust and which has the characteristics you want. Decide if you want to instrumentally inseminate the queens or open-mate them and what apiary has the best drones with which to mate these new queens. Then make sure your queen larvae are fed copious amounts of food, because the number of ovaries is directly proportional to the size of the abdomen: bigger queens are better queens.

Finally, when do you want to rear them? Prepare and organize your bees and yourself so you can have abundant workers, food, and drones at the right time.

Conditions needed for successful queen rearing include:

- Abundance of young workers.
- At least 20 pounds of honey.
- Plenty of pollen or substitute.
- Heavy nectar flow or syrup (1:2) three days before you start rearing queens.
- Plenty of mature drones (14 days old) available one week after queen emerges.
- Superior queen mothers.
- Large populations of young "nurse" bees.
- Correct-aged larvae from which queens can be reared.

There are numerous methods of queen rearing (see the References). Although it falls outside the scope of this book to go into great detail on this subject, some simple methods are given below, from which you can start experimenting. You will need breeder queens; starter, cell-builder, and finisher colonies; mating nucs; mating yards; and drone mother colonies.

Breeder Queens

First, you must select your best queens from which to raise new ones. Carefully choose these breeder queens by testing from 3 to 20 colonies (depending on how many colonies you have) that possess the characteristics you wish to perpetuate in your bees. You might begin the screening and selection process one year and rear queens the next.

To choose which queens are best, give some sort of test or criteria ratings to a group of 10 hives at one time, such that weather conditions, nectar/pollen flow, and temperature are the same. Grade the bees, with a letter or number grade for each option, and select at least the top five colonies with the highest grades (see the sample preselection data sheet on p. 94).

Here are some characteristics by which to score your colonies:

1. Brood production (a compact pattern means good brood viability; spotty brood indicates larvae were removed because of poor mating or disease).
2. Disease and pest tolerance (hygienic behavior means bees uncap and remove dead or infested larvae in less than 24 hours; grooming behavior means some bees remove varroa mites from one another; tracheal mite resistance; wax moth resistance; chalkbrood removal).
3. Overall population (how populous the colony is compared with others).
4. Propolis (whether the colony uses excessive or low amounts).
5. Temperament (test without smoke, size of colony not important; note whether bees do nothing or act defensively).
6. Composure on comb (how quiet or runny the bees are when you examine frames).
7. Pollen arrangement and hoarding (whether there is a clearly defined ring around broodnest

Preselection Data Sheet for Breeder Queens

Date ______________ Queen # ______________ Queen origin ____________ Date mated____________

	5 Exceptional	4 Excellent	3 Average	2 Fair	1 Poor	0 Unacceptable
Brood viability						
Temperament						
Spring buildup						
Pollen hoarding						
Cleaning/hygiene						
Honey production						
Disease	-1	-2	-3	-4	-5	Eliminated
Type						
Mites	-1	-2	-3	-4	-5	Eliminated
Type						

Color					
Queen	Black	Dark Striped	Striped	Light Striped	Unacceptable
Daughters					
Total score					

Adapted from Susan Cobey, Apiarist, Rothenbuhler Bee Lab, The Ohio State University, Columbus.

frames or pollen is scattered; gauge amount of pollen stored in frames, especially in the fall).

8. Honey produced or collected (note the number of forager bees during the honeyflow and their plant preferences; note ability of bees to move up into the supers).
9. Beeswax (note speed of drawing out foundation; note color of cappings: white or dark).
10. Swarming tendency (whether the colony swarmed, and if so, how many swarms it cast).
11. Robbing tendency (score the number of times robbers were seen at other hives; conversely, is the colony easily robbed or nondefensive?).
12. Wintering ability and spring buildup (note how many frames of bees are found in the spring and the rate of buildup under early spring conditions).
13. Flight time (how early and at what temperature the bees fly out).
14. Color (whether you want light or dark bees; evenness of the color of workers and drones; if offspring all the same color, indicates purity of mating).

Remember, these records should be kept on each queen you choose to be a breeder. It would be helpful if you had similar records for your drone mothers too, as the worker offspring of your new queens will carry the genetic traits of both the breeder queens and drone mother colonies. For more information on this topic, see H. H. Laidlaw Jr., "Organization and Operation of a Bee Breeding Program," *Proceedings of the 10th International Congress of Entomology, Montreal, August 17–25, 1956* 4:1067–1078 (1958).

Grafting Methods

Equipment for Queen Rearing

You will need to purchase or make grafting needles (to pick up young larvae), queen cups (from which the bees make queen cells), and grafting bar frames (a frame that holds queen cups) or try a plastic queen-rearing system (such as the Jenter, New Zealand, or French systems). *Grafting* means to physically remove a three-day-old larva (age is extremely important) with a grafting tool, and place it into a prepared cup with or without royal jelly. This procedure is tricky to do and takes much practice. You must develop nimble fingers and keen eyes to see what you are grafting.

Doolittle Method

To graft worker larvae into artificial queen cups, you might employ the Doolittle method, named after G. M. Doolittle, a beekeeper who wrote extensively on the subject around the turn of the last century. He found that priming the queen cups with

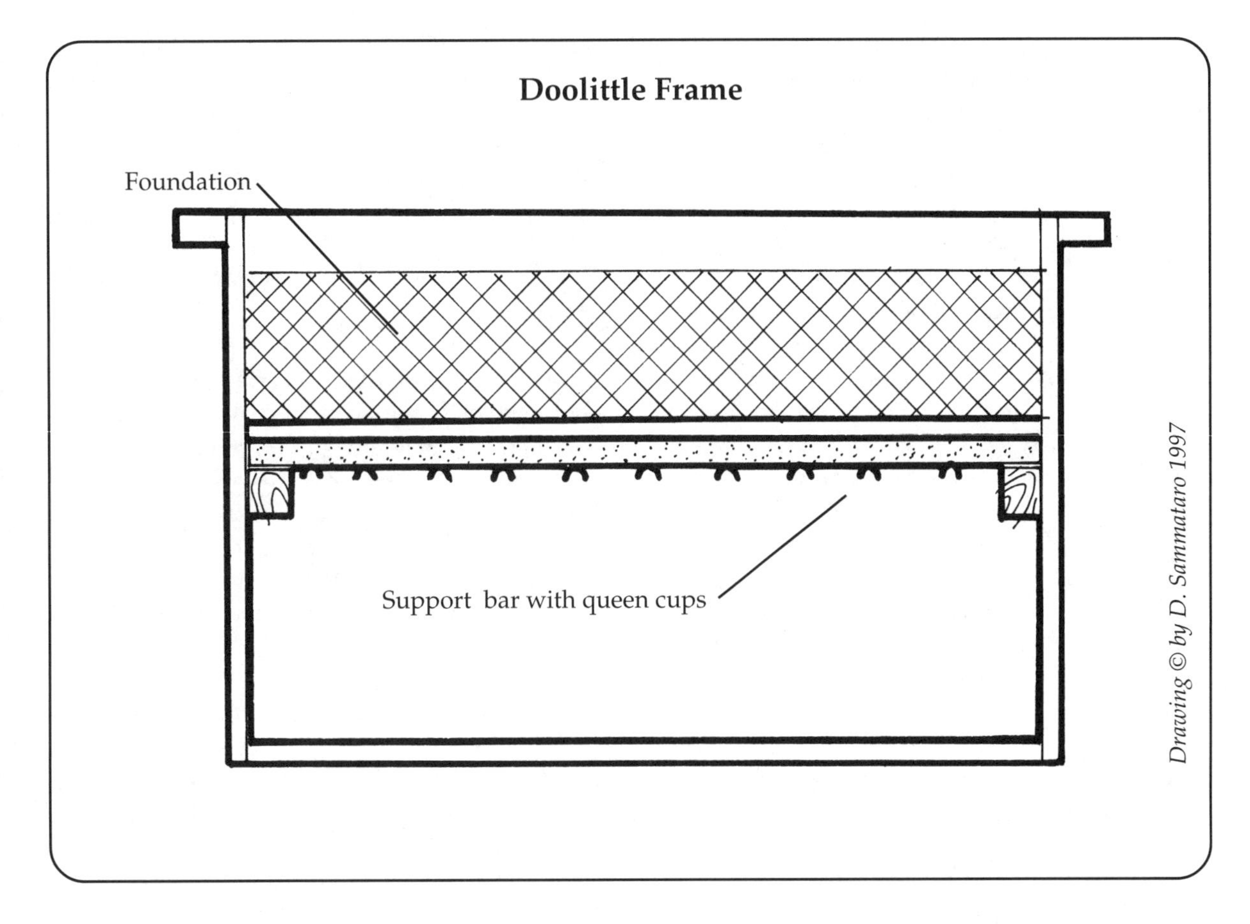

royal jelly and feeding the starter colonies was imperative in rearing good, healthy queens. Here's how to do it:

Step 1. Fit an empty frame with two or three bars to hold artificial queen cups. Some frames have a 3-inch strip of foundation above the bar (see the illustration of the Doolittle frame).

Step. 2. With melted beeswax, attach the wooden cell bases with wax cups to the underside of the wooden bar. If using plastic equipment, drill holes in the bars in which the cups will fit.

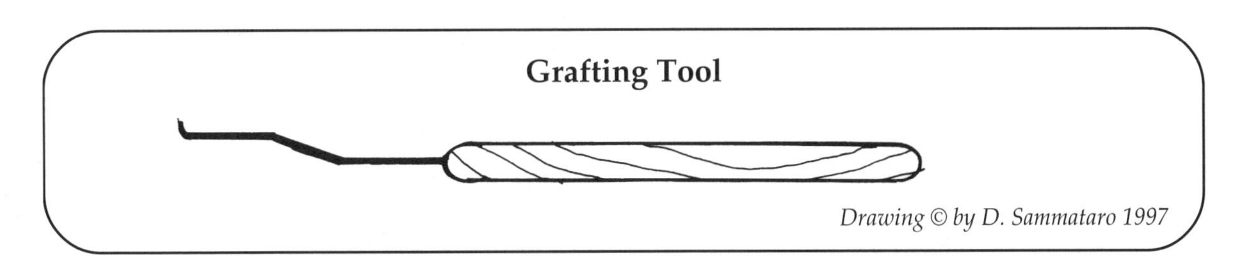

Step 3. Two days before transferring larvae to these cups, dequeen a strong, two-story hive and feed it with syrup.

Step 4. The next day, shake the bees off every brood frame in the dequeened hive and remove all queen cells; these cells will provide you with royal jelly. Collect the jelly with a small spoon, eyedropper, or siphon; you can store the royal jelly in a sterilized jar in the freezer.

Step 5. Once you have enough jelly, remove a frame of larvae less than three days old (larvae 24 hours old or younger are ideal) from one of your breeder hives; these larvae are about the size of an egg.

Step 6. Prime the queen cups with a small drop of the stored royal jelly; place it in the queen cups, covering the bottom. Dilute the jelly, if it is too thick, with about 25 percent sterilized, distilled water.

Step 7. Transfer the larvae with a grafting tool (see the illustration of this tool) or a toothpick (carve one end flat and curve it slightly upward) from worker cells into the primed queen cups. The larvae should be placed on top of the royal jelly, being careful not to drown them in it, and in the same position they were found before the transfer. Because these larvae are about the size of an egg, it may help you to use a lighted magnifying glass to see them. Grafting should be done in a hot, humid room to keep the larvae alive. Cover frames with a damp towel.

Step 8. Insert the finished frame, with queen cups containing transferred larvae, into the dequeened starter colonies.

Starter Colonies

Make up a super full of queenless young bees, capped brood, and lots of food about a day before you add queen cups.

- Rearrange the frames in the dequeened colony so that the lower chamber has mostly sealed brood; the upper chamber should have (in order) a frame of honey, two frames of older larvae, a frame of young larvae, space for a frame with queen cells, a frame of pollen, one frame of older larvae, and one frame of honey.
- Feed pollen patties and syrup to make up any shortfalls. Give 40–50 queen cups to the starter, keeping them there for 24–36 hours.

Because this colony is queenless, it will not be strong enough to cap all these cups: it just fills them with royal jelly. If you are planning on reusing the starter colony, join it to a queenright colony until you need it, or give the colony three or four queen cups to make its own queen.

Once all the queen cups you introduced are full of jelly, take out the frame of cups and divide up the queen cells into cell-builder/finisher colonies.

CELL-BUILDER/FINISHER COLONIES

Each cell builder/finisher colony should be a queenright colony of two or three deeps, full of healthy bees. Prepare this colony a few days ahead of time by confining the queen to the lowest super below a queen excluder. Give each cell builder about 20 cups, right above the broodnest over the excluder. Feed the colony heavily with both pollen patties and syrup, and place the frame holding the queen cups between frames of capped, emerging bees and young nurse bees. Pollen is critical for the nurse bees to produce enough royal jelly. Check in a few days to see if the cells are getting enough jelly. Sacrifice a few capped cells to check jelly levels—there should be extra jelly in the cup and the queen larva should be strong and fat.

After five days, the cells should be capped. Queen pupae are EXTREMELY DELICATE at this time. DO NOT bump, cool, or overheat them or your queens will be deformed. Know the age of the queen cells so that one will not emerge too soon and kill all the others. Do not handle the cells until they are 9 or 10 days old (after grafting). Place ripe cells in a mating nuc (or in an incubator if you are instrumentally inseminating them).

MATING NUCS

Because the first queen can emerge 11 days after grafting, make up nucs a few days before they are needed. Bees added to or making up the nucs should have been treated beforehand to ensure that they are mite-free; finish Apistan treatment about 30 days before adding a new queen. Do NOT treat nucs with Apistan strips for at least 30 days. (Handle queens as you would honey: medicate or treat 30 days before supering for honey.) Nucs can be moved to the mating yard, at least 1–2 miles away, if they have the following:

- No queen.
- Two or more frames mite-free brood/bees.
- Two or more frames disease-free honey/pollen.
- An entrance that can be closed or restricted.
- Sugar syrup, if honey is not incoming.

Transfer one or two queen cells into each mating nuc by cutting the cells from the frame bars. Using a heated knife, cut the cell bases away from the wooden bars. Place cells into an insulated box with a warm water bottle inside to transport them safely to the mating yard. After you select a nuc, wedge the cells between the top bars of two frames, making sure they get covered by bees immediately. Check food stores of each nuc, and feed if necessary. Close the entrance with grass to prevent robbing, and make sure the boxes will not get chilled or overheated.

MATING YARD

Put your nucs into a yard near your drone mother colonies. These colonies must be free of varroa and tracheal mites before the queens are placed in the nucs. Feed your nucs with fumagillin to control nosema if it is a problem or if the weather turns wet and cold and bees are confined. Do NOT apply mite control measures, such as Apistan strips, to your nucs; but you may use oil patties, if tracheal mites are a problem.

Drone Mother Colonies

With the reduction of feral bees, a result of the parasitic bee mites, it is important to have enough drones available to mate with your virgin queens. Because half the genetic makeup of your workers comes from their fathers, select drone mother colonies using the same criteria as you use for your breeder queens. Choose only those colonies that show good characteristics (use only the highest-scoring colonies) but are of genetic stock different from that of your breeder queens, to avoid inbreeding.

Your goal is to produce healthy drones, free of varroa mites. Timing is everything: plan on having mature drones in your mating yard when the virgin queens rise to mate. Drones mature 10–15 days after they emerge.

Encourage drone production by placing from one to three frames of drone comb (if you can find it) into these colonies. Ensure that there is enough incoming forage, and if not, supply extra pollen and syrup. Provide a frame of drone-sized comb in each deep brood super of your drone mothers.

If drone foundation is difficult to find, try inserting frames one-half filled with regular worker foundation and allow the bees to draw out the rest of the frame. Most likely, they will draw out drone cells. Or add one shallow frame in the deep brood supers to achieve the same effect. Some queen breeders also maintain queenless colonies to encourage laying workers to produce drones. If you provide such colonies with frames of capped worker brood once a week, as well as drone foundation and an abun-

Modified Miller Frame

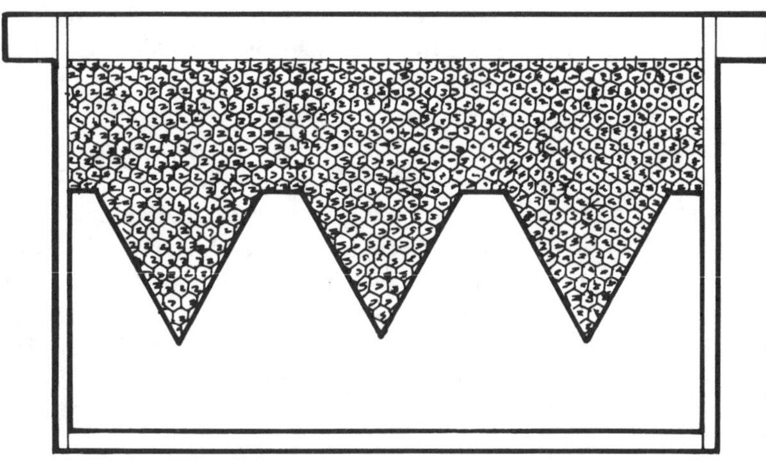

Start with unwired foundation, cut into wedges.

Place frame into your breeder colony so queen can lay eggs in the cut foundation.

After one week in a queenless colony, the frame should look like this:

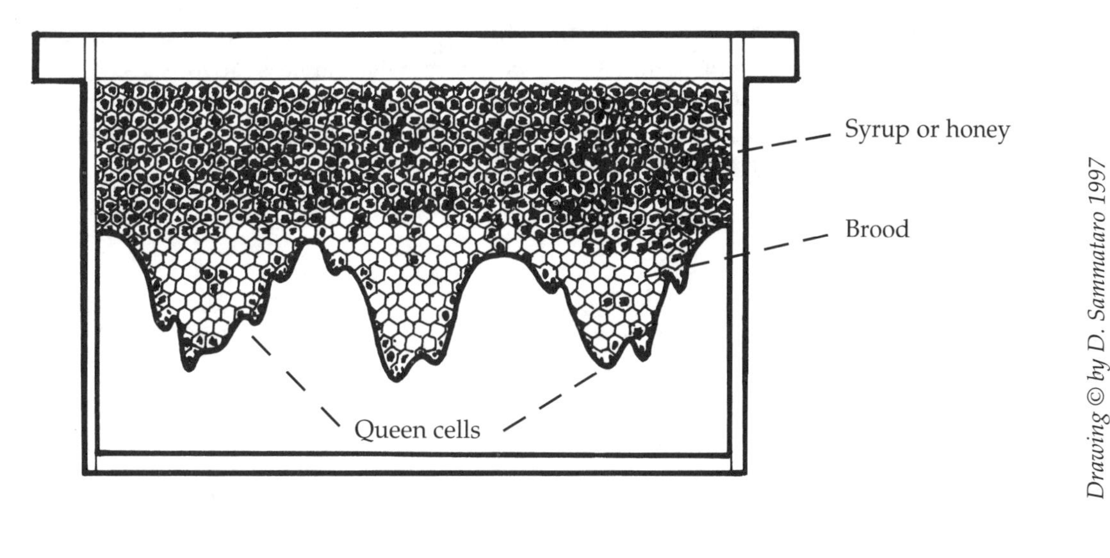

Drawing © by D. Sammataro 1997

dance of food, you can raise large, healthy drones from these colonies too.

Nongrafting Methods

Miller Method

The Miller method of queen rearing, named after C. C. Miller, may be the easiest for the beginner. Prepare an empty brood frame by fitting it with four pieces of foundation, 2 inches wide and 4 inches long (see the illustration of the Miller frame). Cut the unattached lower half of each strip of foundation to form a triangle with its apex pointing downward. Do not wire the foundation to the frame. An alternate method is to cut triangular strips of foundation (as illustrated with the alternate Miller frame).

Now, follow this procedure:

Step 1. Remove all but two frames of sealed brood and the queen from a breeder colony into a nuc or specially prepared breeder hive.

Step 2. Insert the prepared frame between the two frames of the sealed brood.

Step 3. Make sure the queen is on one of the frames.

Step 4. On either side of the brood frames, fill the hive with frames of honey and pollen (there should be no empty cells in these frames, or the queen may lay in them).

Step 5. The queen will be forced to lay in the prepared frame as soon as cells are drawn.

Step 6. About one week later, remove the prepared frame. Trim away edges of the newly drawn pieces of foundation until you encounter cells with small larvae (preferably less than one day old, but never more than two days old).

Step 7. Give this frame to a cell-builder colony. You can insert another Miller frame into the breeder colony, or return the queen to her original hive. Or, use this special hive to start a new colony, requeening the parent hive.

Step 8. Nine days after inserting the prepared frame,

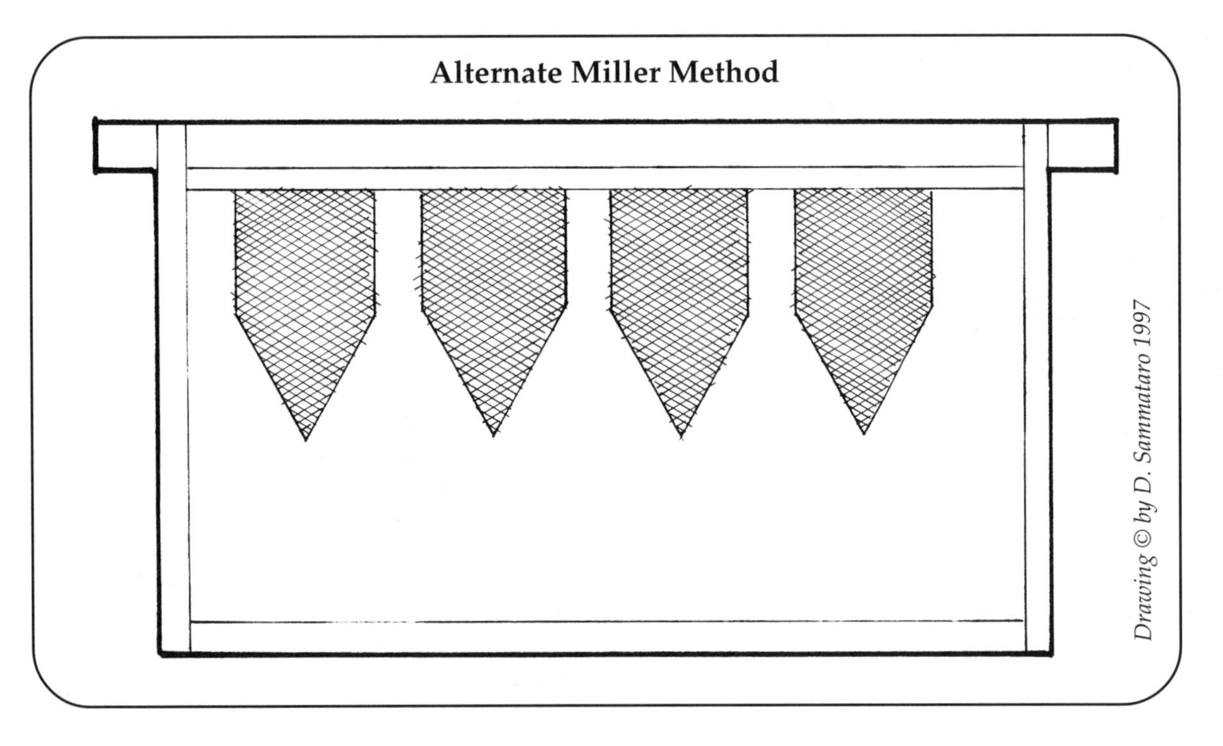

remove the sealed queen cells by cutting them from the Miller frame and attaching them to combs in queenless hives or nuc boxes.

Queens from the nuc hives will emerge and mate. These queens can then be left in the hives from which they mated or used for requeening after they have begun to lay eggs.

Queen Rearing Kits

Several bee supply companies sell queen rearing kits that do not require grafting bee larvae. These come with good instructions, which you should read first. The kits are composed of a plastic box, the size of a comb section square, with a plastic queen excluder lid on one side. The base of the box is plastic comb evenly perforated with holes and covered with a plastic lid. Fill the holes of the comb with the plastic cell cups that come with the kit; the cover keeps the cups from falling out. Once the box is fitted into a regular deep frame (with or without foundation), wired or nailed in place, it is put into a colony so the bees can draw out the comb and help dissipate the plastic odor. After the plastic cells have been worked for a few days, the frame is then ready for your breeder queen.

Place the queen inside the box underneath the queen excluder cover, and return her to the colony's broodnest so she can lay eggs in the plastic cell cups. After this is achieved (usually in a day or so), release the queen back into her hive and take out all the cups containing eggs or young larvae. Attach these cups to a cell bar and place them into a starter colony.

The advantage of this system is that you do not need to hunt for the frames containing larvae of the correct age. All you need to do is check the plastic cups to see if the confined queen has laid in them. Be aware that you have the potential to rear over 100 queens at once, which could tax your facilities if you are not prepared. The down side of this system is its cost and the many little extra pieces of equipment you need, which are easy to lose. Ask more experienced beekeepers about their successes or failures with such a system before you spend the money.

Special Management Problems

Uniting Weak Colonies

Weak colonies (those with small populations), including nucs, splits, and swarms, can be strengthened to make them strong. You can strengthen them by adding adult bees (such as a swarm) or sealed brood. The population in the weaker hive will dictate how many frames of sealed brood to add (no more than the existing bees can cover and care for). If you do these manipulations by early summer, the colony will probably gain enough strength and reserves to overwinter.

If, on the other hand, the weak colonies have not developed into strong ones despite previous manipulations, the reason for their weakness must be investigated. If mites, disease, or other factors are keeping a colony from becoming strong, the colony should be destroyed to prevent the spread of pathogens or mites. If the failure is due to a poor queen, eliminate her and either requeen the weak hive, if it is early enough in the season, or join the weak colony to a stronger one.

Attempts to carry weak colonies through the fall and winter months have never been very successful.

Newspaper Method

Although two weak colonies do not usually make one strong one, a weaker, healthy colony can be united with a stronger one. If you have a preference for one of the queens, eliminate the less desirable one; if both queens are healthy, either introduce one into a nuc or split or simply unite, knowing that only one queen will survive.

The best use one can make of the bees in a weak colony is to unite them with a strong one just before the honeyflow. The united bees will be able to gather more honey.

If you need to unite colonies later in the year—in early or late fall, for example—be sure the joined colonies have ample stores of honey, because they will not be able to collect nectar. Either feed them or provide frames of food from other colonies.

Whenever two colonies are being united, remember that each colony is capable of distinguishing between its members and those from another colony. Hive odors will be different, and unless some precautions are taken the bees will fight. During a honeyflow, this ability to distinguish foreign bees is diminished, but in the absence of a flow, other methods are necessary. If after these manipulations the existing queen fails to improve, the colony should be requeened. Also check for the presence of mites or other pathogens.

The most successful and least time-consuming method of uniting colonies is the newspaper method. This method works because the paper separates the hive odors. As bees chew away the paper, the hive odors gradually blend, and by the time the paper is gone, the bees act as one hive.

To unite colonies with newspaper:

Step 1. Put a single sheet of newspaper over the top bars of the stronger colony (see illustration of uniting colonies).

Step 2. If the weather is warm, make a few small slits in the paper with your hive tool to improve ventilation.

Step 3. Set the weak hive of top and cover; field bees from the weak colony's original hive site will probably drift to other hives.

Step 4. The paper will be eaten through slowly by bees, with most being chewed up within a week. Shredded paper will appear at or near the hive entrance.

Step 5. If the weather is extremely hot (day temperatures over 90°F [32°C]), wait for a cooler day, or unite the hives during the late afternoon.

Step 6. If there is a dearth of nectar and pollen and the bees are unusually aggressive, decrease the possibility of fighting by feeding syrup to the stronger hive for a few days before uniting.

Step 7. If the weaker colony is to be sandwiched between hive bodies, place a single sheet of newspaper below and above the weaker hive, and slit both sheets.

If after these manipulations the colony fails to improve, check for the presence of mites or other pathogens. If such are present, either destroy the colony or treat it accordingly (see Chapter 13).

Moving an Established Colony

Preparing the Colony

Sometimes it is necessary to move a colony of bees. Follow the procedures described below to move an established colony safely. It is general practice to move a hive at least 3 miles (4.8 km) from its old site; if it is moved less than that, many field bees will return to the old location and become lost. The best time to move hives is in the spring, when populations are the lowest and the hives are light in weight.

You must first check and comply with all legal requirements pertaining to moving and selling bees. Have the state inspector certify that the bees are free of diseases and mites, especially if you are moving bees out of state or selling or buying them.

If the colony consists of more than two hive bodies, remove the extra supers (providing they are free of bees) to make it easier to lift. Other methods for moving colonies larger than two deeps are to re-

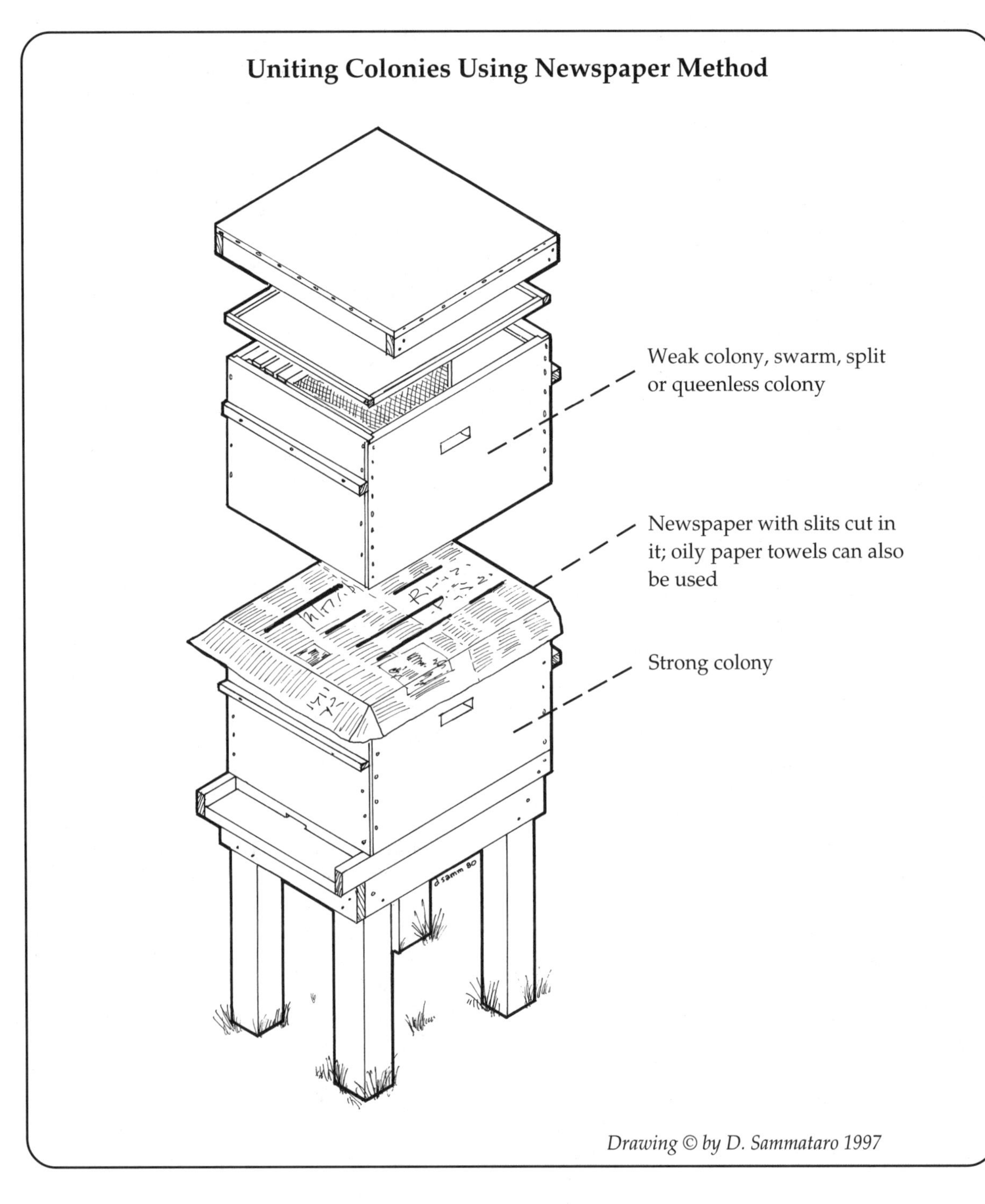

Drawing © by D. Sammataro 1997

move bees by using a blower and to shake or brush the bees from the frames onto the grass in front of the hive. Of course, these manipulations must be done a few days before the colony is moved. You can also rid bees by using bee escapes in the inner cover or an escape board (see "Removing Bees from Honey Supers" in Chapter 9). If a colony is very congested, the bees may suffocate or overheat if confined without proper ventilation. Move a heavy colony with the help of a mechanical lift, extra strong (willing) volunteers, or in manageable pieces, leaving extra hive bodies at the original site to pick up stragglers (or leave a weak hive to collect the bees and move it later).

On the day of the move:

Step 1. Tape or screen all holes and cracks in the two remaining deep hive bodies.

Step 2. If the hive is very populous despite using the methods above or if the weather is very hot, add a shallow super with empty frames above the top hive body to collect the overflow of bees; otherwise, bees might be hanging outside the hive when you return to move them.

Step 3. If the weather is warm, place a screened board (like a division screen) or screened inner cover on top of the hive under the outer cover.

Step 4. Using smoke as needed, tie the bodies together with metal or plastic ratchet straps or bands. This method is far superior to fastening hive bodies together with nails or staples, which do not adequately secure the hive bodies.

Step 5. If moving during hot weather, take off the outer cover and secure the screened top with the bands. Replace the outer cover until ready to move, then remove it when moving the hive.

Step 6. In the evening, near dusk, smoke the entrance to drive the bees inside and use a piece of screen or hardware cloth the length of the entrance and about 5 inches wide to close off the entrance (see the illustration of moving). Slide a V-shaped piece of screen into the entrance so it will spring against the bottom board and hive body, and secure it with nails or staples. Entrances can be closed completely if a screened top replaces the inner and outer covers. You can move the hive that evening or early the next morning.

Loading and Unloading Colonies

Once they are strapped together, load the hives onto a truck to move your bees.

If the weather is hot, remove the outer cover while in transit so the screened top is exposed. The hives should be packed close together on the truck, with the frames parallel to the road; this position will prevent the frames from sliding together if the truck stops suddenly. While loading and unloading, keep the engine running, because the vibration of the vehicle will help keep the bees in their hives.

DO NOT move a hive in the trunk of your car or in the back of a closed van—loose bees inside a vehicle are dangerous to the driver. Once all the hives are in the truck, secure them to the truck with ropes or other tie-downs. The object is to keep the hives from shifting while you drive them to their new homesite. Colonies that move, especially off their bottom boards, make unloading difficult. If moving great distances, purchase a special net that confines all the bees.

Smoke the entrances just before you unload the hives and then just before you remove the entrance screen. Fill the hive entrance loosely with grass to slow the bees' exit and to keep them from drifting. If they exit slowly, a few bees should be able to come out and scent at the entrance; this will help any loose bees to relocate at the hive.

If this site is the final one for these hives, replace the top screened board with the outer cover. Even if this spot is only temporary, replace the outer cover to keep rain out. Inspect the hive after a few days to see if all is well.

Moving an Established Colony

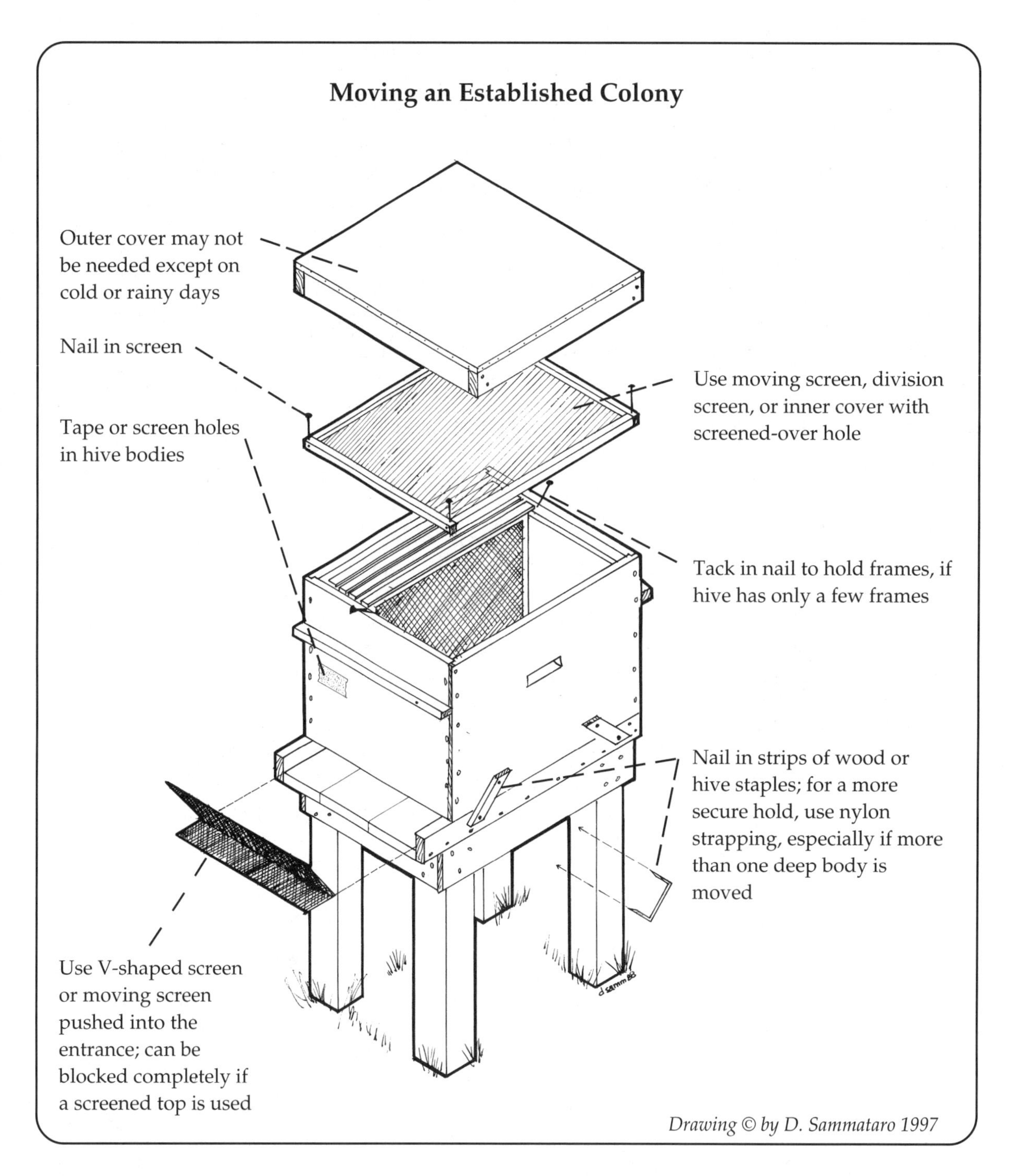

Drawing © by D. Sammataro 1997

Problems associated with moving hives are as follows:

- Hives could shift off their bottom boards or hive bodies could break, permitting bees to escape. If moving old, leaky equipment, cover all the hives with a traveling screen or other netting to contain loose bees.
- Bees can suffocate if the weather is too hot.
- Queen could be killed, injured, or balled.
- Combs could break.
- If moving in winter or very early spring (when temperatures are below 50°F [10°C]), the winter cluster could be broken. Bees might then recluster on empty combs and starve, or existing brood could become chilled before the bees have a chance to cover the brood.

Moving Short Distances I

Follow this procedure to move an established hive less than three miles:

Step 1. During the day, move the original hive off its stand to the new location.

Step 2. In its old location place a nuc box (a small hive with four or five frames) or one deep hive body with bottom board and top cover.

Step 3. Fill the hive or nuc box at the old location with dry comb frames and one frame of brood from the original hive, with or without a caged queen or queen cells. Field bees from the original hive will return to the new box at the old location; this becomes a new, smaller colony that can be moved in about one week.

Step 4. When ready, move the small hive from its original location to a new site at least three miles away.

Step 5. After about two weeks, this hive may be moved to the desired location where it may be united with original colony. It can also become a new colony; in other words, you have split the old colony and created another.

Moving Short Distances II

Move all the hives, leaving no stragglers behind, over three miles away for three weeks. Then move them again to the desired location.

It is often recommended that when moving established hives very short distances, each hive be moved between 1 and 4 feet every few days until they are at the desired location (few if any bees will return to the original location). But this process is slow and not recommended unless the distance to be moved is less than 30 feet.

Robbing

Occasionally, bees will collect nectar and honey from other colonies. This type of bee behavior, referred to as *robbing*, usually occurs when bees are unable to obtain food from flowers. Whenever weather is suitable for flight, foragers will set out in search of food. If plants are dormant or blooming but not yielding nectar, bees continue to search and are attracted to the odor of ripening nectar and honey stored in other colonies. You can recognize robbing bees by their flight activity at the hive entrance (see "Robbing Flight" in Chapter 2) and fighting bees on the landing board.

Nucleus hives, queen-mating boxes, and other colonies low in population or with openings resulting from hive disrepair are the most at risk of being robbed. If you observe increased activity at the entrances of these hives, you may have a robbing problem.

The robbing bees return home, communicate the location of the target hive from which the food was taken, and soon additional foragers, recruited to the unlucky colony, remove the remaining stores. The weakened colony, fighting off the thieves, may be severely affected, even killed. Some robbing likely takes place in apiaries whenever there is a dearth, after which you may notice a weakened colony when you next return to the yard. Reduce the entrance of such a hive, and move it to a location with fewer strong hives.

The best way to prevent robbing activity is to maintain colonies of equal strength, keep the entrances of all weak or small colonies reduced, avoid open feeding of sugar syrups (which stimulates robbing), and minimize or avoid hive inspections during a dearth. If examination is essential, cover any exposed super with a robbing cloth, or place removed frames in a covered, empty hive body. Moving frames also allows honey from burr or brace comb to drip, attracting more bees. Quickly finish your examination to keep robbing bees to a minimum.

If you must feed, do so inside the hive (see Chapter 7), and use sugar syrups rather than honey. If newly extracted combs are to be cleaned (see "After Extracting Honey" in Chapter 12), place them on strong colonies only in the late afternoon. In the fall, supers of freshly extracted combs should be removed after they are cleaned, because the bees may move up into them and remain there until they starve.

If a robbing frenzy has started on a weak or smaller hive, close up the weak hive and dust robbers with flour. By following the dusted bees, you find the robbing colony (or colonies). Reduce the entrance severely, changing the direction of the entrance of the offending colony. Remove the weaker hive or hives to another area if robbing persists. Many beekeepers have their nucs or other weak hives in a separate yard.

Some beekeepers employ a robbing screen, a device that forces bees exiting their colony to fly upward instead of straight out. Robbing bees tend to land on the flight deck of the bottom board and won't figure out how to get into the colony. Another way to help control robbing is to maintain bee strains with low robbing tendencies, such as Carniolans or Caucasians.

Swarming

By reproducing, organisms perpetuate and protect their kind from extinction. Social insects such as honey bees can reproduce new individuals within the colony unit, but this is not sufficient for their continued survival. If bees were to maintain themselves solely by producing young, their colonies—without human intervention—would decrease as a result of mites, disease, fire, predators, or adverse environmental conditions.

Honey bee colonies perpetuate themselves by *swarming*. Swarming is a natural process of reproduction at the colony level. A colony divides, and part of it leaves for a new homesite, usually with the old queen, while the remaining members continue at the original site with a newly emerged—and later mated—queen. In this manner, a single unit becomes two.

An abundance of queen cells, often called swarm cells, indicates that swarming preparations are under way. Shortly after the swarm cells are sealed, the colony will cast a swarm. Bees will exit as a swarm with the old queen, on any warm, windless day, usually between 9:00 A.M. and 3:00 P.M. (earlier or later if the weather is favorable). Occasionally, bees will swarm when the weather is less than favorable.

After the swarm issues, some of the bees will alight on a nearby object and begin fanning with their scent glands exposed to attract the remainder of the swarm and the queen. Soon a "cluster" of bees forms. It is this cluster—readily visible to the casual observer—that is correctly called a swarm (see the illustration of a swarm). Scout bees will then dance on the cluster to communicate a new homesite. When one has been agreed on (within a few hours to a few days), the swarm flies to the new site, guided by scenting bees.

Bees in a swarm are usually quite gentle. Before leaving their old hive, they engorge honey, which seems to contribute to their gentleness. Another reason for their gentleness might be that, because the homeless cluster is only a temporary situation, the division of labor—including guarding—that prevails in a normal hive is either nonexistent or not as prevalent.

Swarm

Drawing by Jan Propst

Swarming versus Productive Hives

Swarming was once considered a sign of "good and productive" beekeeping, for beekeepers could increase their holdings from the numerous swarms available. Straw skeps, logs, and other types of cramped hives have been used for bees since the 1600s, but these containers became overcrowded quickly and thus promoted the swarming of bees.

Today, swarming can be viewed as a sign of the beekeeper's negligence, because it means a loss of both bees (unless the swarm is captured) and the production of honey. Although most beekeepers make efforts to prevent or control swarming, it is not an easy task. The picture is further complicated by the fact that most methods used for controlling or preventing swarming result in manipulations that reduce the colony size (which is what happens when the colony swarms).

Thus, although swarming can be controlled or prevented, in doing so the goal of maintaining populous colonies for the honeyflow is somewhat sacrificed. Nevertheless, this sacrifice is far better than having the colony cast a swarm that may leave the apiary site before you can recapture it.

During the 1990s, swarms have become less numerous as a result of predation by the parasitic bee mites. Colonies weakened by mites are not strong enough to swarm, and feral colonies, which were once common, have been killed by mite infestations.

Reasons for Swarming

Honey bee colonies swarm for any one or more of these reasons:

- Congestion.
- Unbalanced numbers of differently aged workers.
- Overheating (perhaps due to lack of noontime shade).
- Defective combs (those with too many drone cells or cells that are irregular, thick, damaged, or otherwise not suitable for the queen to lay in, reducing broodnest capacity and increasing congestion).
- Egg laying becoming restricted as empty cells are filled with honey.
- Inclement weather, which keeps bees confined to the hive and causes congestion (bees hanging out of the colony).
- Failing queen—instead of superseding the queen, the colony may swarm.
- Decline in queen pheromone production—the level of pheromone being distributed throughout a highly populous colony is insufficient to control swarm preparations.
- Genetics or race of bees.
- Idle nurse bees.

Other Reasons That Bees Leave

Under certain conditions, the entire original colony may depart their home. This is called *absconding* and could be caused by:
- Starvation.
- Disease/mites.
- Wax moth (or other pest) infestation.
- Fumes from newly painted or otherwise treated hives.
- Poor ventilation.
- Excessive disturbance of the colony by the beekeeper or vandals.
- Excessive disturbance by animal pests such as skunks or bears.

Signs of Swarm Preparation

Signs that a colony is in some stage of swarm preparation are clearly visible during routine hive inspections. The list below is a rough chronology of the various signs you may see in a colony that might ultimately swarm.

1. Rapid increase in worker population occurs (especially in spring, after a minor honeyflow, and before a major honeyflow).
2. Drone rearing begins as worker numbers increase.
3. Broodnest (area where eggs, larvae, and pupae are located) cannot be expanded because combs are already occupied with brood or honey or both.
4. Queen cup construction at lower frame edges becomes evident.
5. Queen deposits eggs in these queen cups; larvae are present in cups.
6. Queen's egg laying tapers off, and amount of young brood decreases.
7. Queen is restless.
8. Many queen cells are present, containing larvae that vary somewhat in age.
9. Field bees are less active and beginning to congregate at hive entrance; this can also happen if the weather is hot or the colony is congested.
10. Swarm cells are capped or sealed.
11. Swarm is cast.

Signs of Imminent Swarm Issuance

A colony that has been making swarm preparations can be expected to issue a swarm:
- After queen cells (swarm cells) are sealed over.
- When wax has been removed from the tips of queen cells, exposing the cocoon (referred to as a "bald spot").
- When few bees are foraging (little flight activity of bees at hive entrance) compared with other hives of the same strength.
- When bees are clustered near the entrance, but not because of hive congestion or warm temperatures.
- Usually on the first warm, sunny, calm day following a short period of cold, wet, cloudy days when congestion in the hive is aggravated.

Clipped Queen Swarms

A clipped queen will attempt to leave the hive with a swarm, but being unable to fly, will not accompany the other bees in flight and will be left behind, usually on the ground near the hive from which she attempted to swarm. The swarming bees, without a flying queen, may return to the hive while they are still airborne, cluster on the ground with the queen, or cluster on a branch nearby. After a brief time they will return to the hive. Eventually, the bees will swarm, accompanied by a virgin queen that can fly. Sometimes swarms will issue but the queen remains inside the colony. These bees may remain airborne or cluster temporarily before returning to the colony; bees without a queen will return.

If you witness any of these events, take these steps to discourage their recurrence:

Step 1. Find and cage the queen either before or after the swarm returns.

Step 2. Move the parent hive from its stand and replace it with a new hive of foundation or dry, drawn comb.

Step 3. When the swarm returns, let the queen walk in with the bees. If the swarm has already returned to the parent hive at the old location, shake off half the bees in front of the new hive; the bees will enter this hive. Release the queen so she can walk in the hive entrance with the bees.

Step 4. Check the hive after 10 days.

Step 5. Requeen the colony with new stock; the queen may have swarming instincts and is probably old. Any virgins queens emerging from the original colony will likely also have this swarming instinct.

Another method would be to let the swarm return to the original colony after removing the queen cells. Check the colony after 10 days to remove any queen cell construction, or Demaree the hives (see "Demaree Method," below). Requeen the colony later with nonswarming stock.

Prevention and Control of Swarming

Successful swarm prevention means that you are able to keep bees from initiating the queen cup construction that may lead to swarming. You practice swarm control when you find and remove queen cups and cells and other signs of swarm preparations already evident. Although the times for initiating swarm prevention and control are different, the manipulations are the same and include:
- Relieving congestion by adding more room in which the queen can lay eggs.
- Providing storage space for the growing bee population.
- Separating the queen from most of the brood.
- Interchanging weak colonies with strong ones.

REVERSING

Reversing the brood chambers, or lower hive bodies, at regular intervals or as needed beginning in the spring, is one method used to relieve congestion in the hive. Through the winter, the colony and its queen move upward through the hive bodies (see "The Winter Cluster" in Chapter 8). By spring, the cluster is usually in the topmost super (or supers), and because the queen may not move down, the brood will be confined there. Unless the queen, broodnest, and bees are put on the bottom, with the empty hive bodies placed on top, the colony is likely to become congested and will probably swarm, despite the expansion space below. Even if the broodnest is not congested, still reverse the hive bodies such that empty combs are available for the queen.

Here is a quick outline for reversing hive bodies (two deeps and one shallow; see the figure on reversing):

Step 1. Take an extra bottom board to the bee yard. Move the hive off its stand or from its location and place the extra bottom board in its place. You can also put the bottom board down next to the old location. Take the hive body containing the queen, most of the bees, and brood (S1 and 2) and put them on the extra bottom board, shallow first.

Step 2. Place at least one deep hive body (1) above the broodnest.

Step 3. Clean the original bottom board, and go to the next hive.

Step 4. Repeat the procedure until all the hives are reversed.

Step 5. If the queen is reluctant to move up after one week, exchange a frame of brood from the broodnest with an empty frame, and move it up into the second hive body.

If three deep hive bodies were present, the order after reversing is 1 (top), 2 (middle), and 3 (bottom), as shown on the illustration.

Reversing Hive Bodies

Reversing with Two Deeps and a Shallow

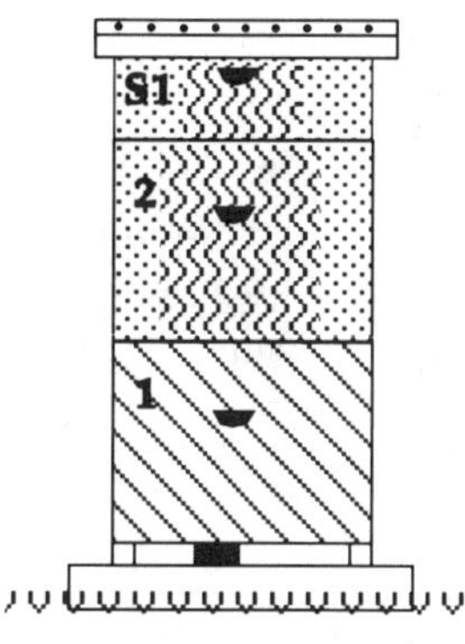

Spring arrangement of bees before reversing

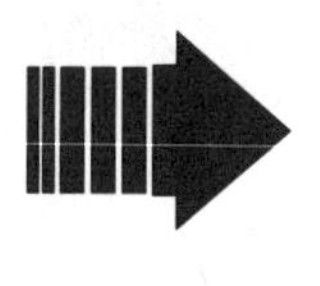

Spring arrangement of bees after reversing

1
2
S1

Reversing with Three Deeps

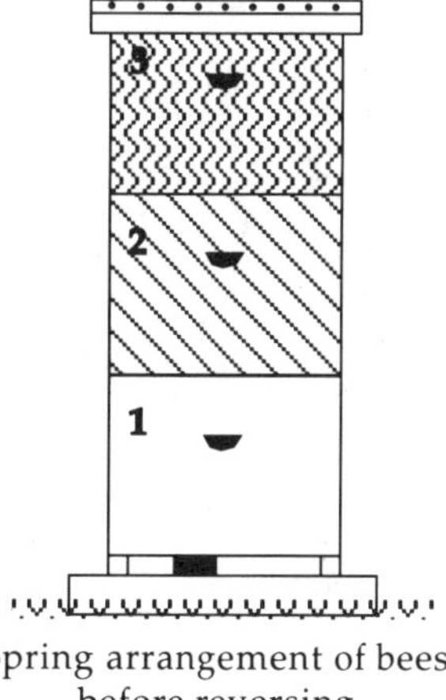

Spring arrangement of bees before reversing

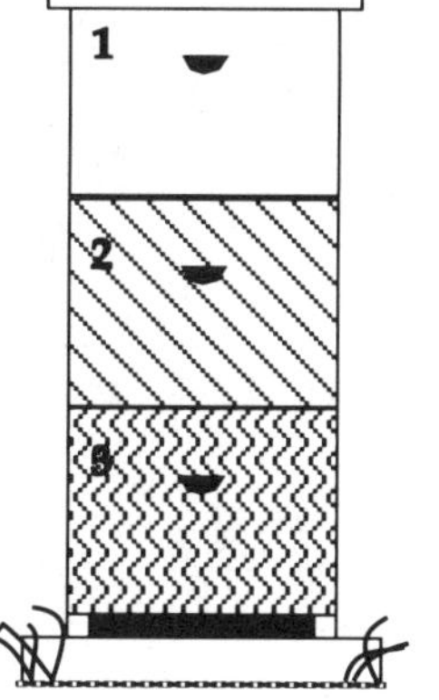

Spring arrangement of bees after reversing

Honey Broodnest Dry Comb Foundation

Drawing © by D. Sammataro 1997

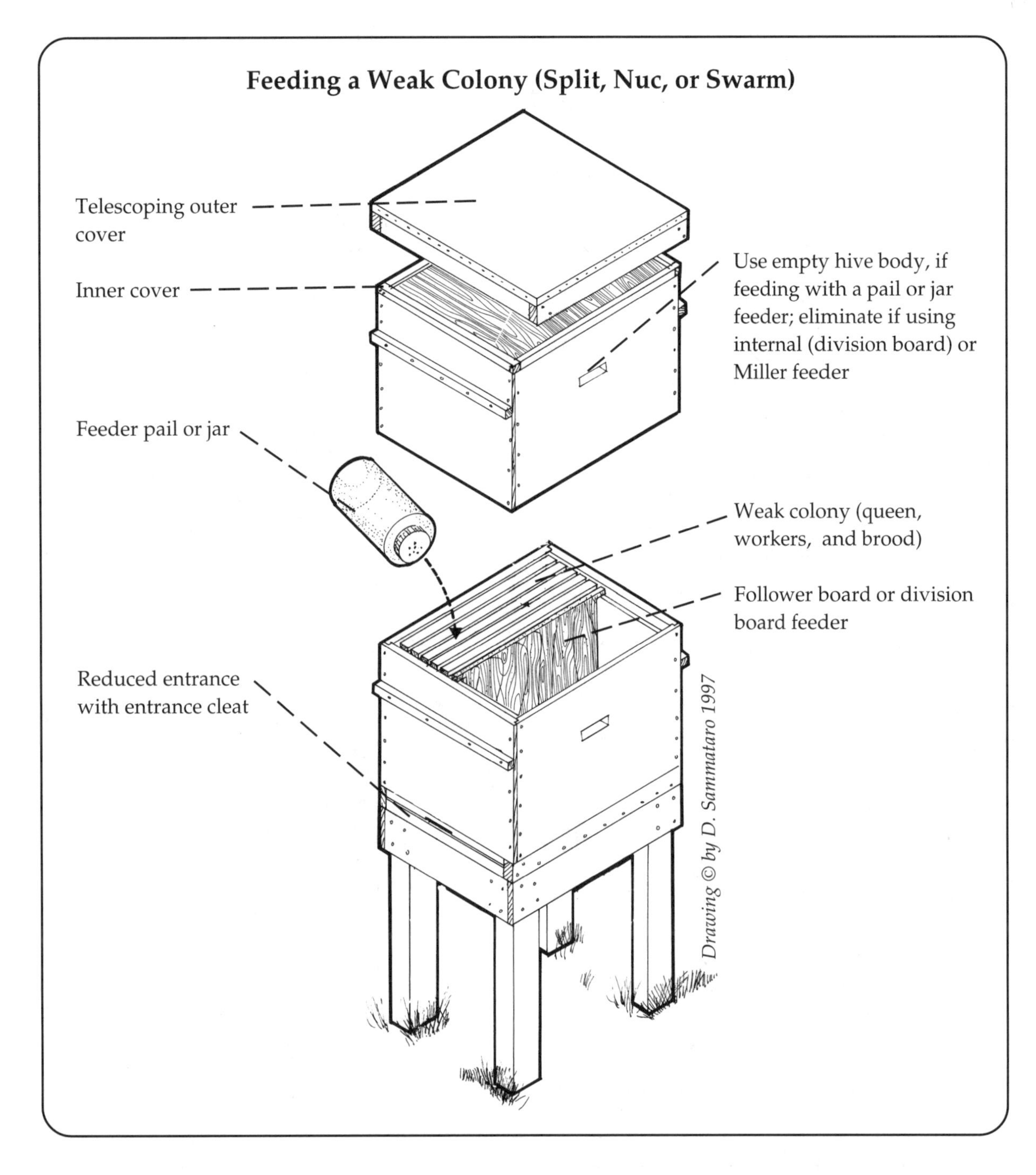

RELIEVING CONGESTION

Hives that are very congested due to poor combs or inadequate space for brood are more likely to swarm. Listed below are some techniques for relieving such conditions:

1. Add extra frames or supers full of foundation.
2. Stagger supers slightly to allow for more ventilation.
3. Separate brood and queen:
   - Place the queen, along with unsealed brood, eggs, and bees, in the lowest super.
   - Above this, place a super with foundation.
   - Above this, place a super filled with capped brood and the rest of the bees.
4. Decrease the number of bees or brood in the hive by splitting the hives to make additional ones, called *increases* or *splits* (see illustration on feeding a weak colony):
   - Place frames of capped brood, honey, and bees from the congested hive into a new box (e.g., one deep body). You can also put these frames into nuc boxes and requeen to augment the number of your colonies or to sell to others (see Chapter 10).
   - If you are combining frames of capped brood and bees from different colonies, spray each frame of bees with syrup to reduce any fighting among the bees.
   - Give the new hive a frame of eggs or newly hatched larvae from your best hive so the bees may make their own queen; requeen the split with a new queen; or provide some queen cells (usually swarm cells). Whatever is provided, place it in the middle of frames of emerging brood.
   - If you made an increase in a deep hive body, it should have frames of capped, emerging brood; frames of foundation; and frames of honey and pollen or empty drawn comb filled with syrup. Use your judgment on how to

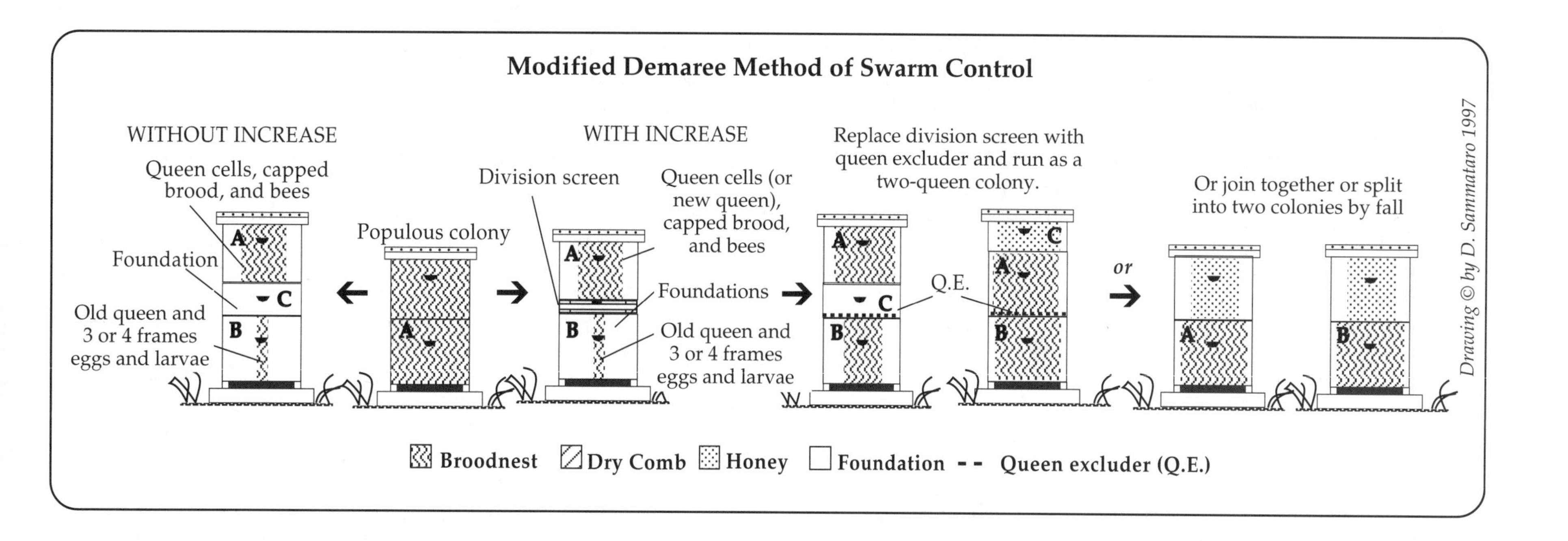

combine the three categories of frames; if no honeyflow is anticipated, put in an extra frame of honey.

- Reduce the entrance to discourage robbing; check after one week.
- If using nuc boxes for your increase, a good rule of thumb is to have each contain two frames of bees and brood and two of food.

INTERCHANGING HIVES

In an apiary in which the hives are in long rows, the bees tend to drift toward the row ends. As a result, the colonies in the middle may be weaker than the colonies on the ends (see "Hive Orientation" in Chapter 4).

If a hive is very populous and seems likely to swarm at some point but has not yet made preparations to do so, interchange it with a weaker hive. More incoming, food-bearing foragers will return to the stronger hive's location but will enter the weaker hive, augmenting its population. Conversely, the strong hive will have a sudden decrease of incoming field bees, and any idle bees that normally might have initiated swarm preparation will begin foraging. Foreign bees entering the switched hives are unlikely to fight if there is a honeyflow in progress. To decrease chances of fighting, wait for a good honeyflow before interchanging the hives. Very weak colonies or nucs should not be strengthened in this manner unless the queen is caged and the candy plug is exposed; otherwise, the incoming strange foragers can overwhelm and kill her.

DEMAREE METHOD

The Demaree method, described first by George Demaree in 1884, makes it possible to retain the complete population of a hive while practicing swarm prevention and control. Basically, it separates the brood from the queen, thus decreasing the hive congestion, without making two different colonies. Here is one way to Demaree (see the illustration of the modified method):

Step 1. Select a populous colony (A). If you are not making an increase, follow the left side of the diagram. If you are, follow the right side of the diagram (see step 9).

Step 2. Have ready a hive body (B) filled with frames of dry drawn comb, foundation or a combination of both drawn comb and foundation. If no honeyflow is on, use less foundation, because the bees will chew it. If you have only foundation, feed bees with syrup so foundation will be drawn out.

Step 3. Find the queen in A and place her on a frame containing very young larvae and eggs. Make sure there are no queen cups or cells on the frame with the queen; if present, remove them or replace the frame.

Step 4. Remove some dry comb frames or foundation from the middle of B and place the frame with the queen and clinging bees there.
Step 5. Remove A from bottom board for the moment, and set B in its place.
Step 6. Add two or three frames of honey and pollen from A to B and fill the remaining spaces with dry comb or foundation.
Step 7. Now add super C (full of foundation or dry comb) above B.
Step 8. If you do not want to make another colony (an increase), add three or four frames of eggs and larvae plus two or three frames of honey in A and place on top of C. Fill in the rest of the space with frames of foundation.
Step 9. If you are making an increase or split, separate old queen in B below (fill up the space as described in step 8) with a division screen. Above the division screen place A, filled with queen cell (or a new queen), capped brood, and bees. If you are requeening A with a purchased queen, remove all queen cells from the brood frames. Any extra frames can be given to other colonies.
Step 10. Once you have two queens, you can separate them with a honey super (C) placed above a queen excluder, and run the colony as a two-queen unit. If, after two weeks, B (with the original queen) is becoming too congested, you may have to Demaree again.
Step 11. You can either use the new queen to replace the older one in A by joining the two colonies together and let the queens fight it out (or kill the older one in A), or split the colonies into two separate hives (A and B) making an increase. Any remaining frames of brood or honey without the clinging bees can be given to other colonies; any empty frames can be stored or placed in a super and added to a populous colony for additional room.
Step 12. After one week, cut out any new queen cups in the upper story.
Step 13. Two weeks later, if the queen's hive body (B, below the excluder) is congested and full of queen cells, remove the queen cells and Demaree again.
Step 14. One week later, remove any queen cells above the excluder.
Step 15. Fifteen days after the last manipulation, because the queen can't get above the excluder to lay, the top supers will be free of brood and will be used for honey storage or remain empty.

Variations of this method are used to rear queens (in warm weather), run a two-queen colony, or make increases (see the illustration); a division screen can be used in place of a queen excluder.

Advantage:

- Population kept at right size to take advantage of any honeyflow.

Disadvantages:

- Must find queen.
- Many manipulations necessary.
- Time consuming.
- Many trips to the apiary needed.

OTHER FACTORS

The following factors may also be of importance in helping to decrease swarming in some hives:

1. Young, vigorous queens.
2. Queens raised from nonswarm stock.
3. Hybrid queens with nonswarming tendencies.
4. Ventilation to increase air flow within a hive:
    - Hive bodies can be staggered.
    - Inner or outer cover can be propped up.
    - Slatted rack can be placed on top of the bottom board; this gives bees lots of room to cluster and can be kept on the hive year-round.

The first two ventilation techniques might encourage robbing when the honeyflow is over; thus, only strong colonies should be manipulated in the way described. To avoid all this extra work, we suggest requeening any colony that has a two-year-old queen.

Catching Swarms

SWARM TRAPS AND BAIT HIVES

It is generally not possible to check one's apiary on an hourly basis, and such attention to outyards throughout the swarming season is impossible. Despite good management procedures for swarm prevention or control, colonies will swarm, and you should expect some swarming from your apiaries. If you are not able to capture the swarm, the opportunity to increase your hive numbers or to return the swarm to its original home is lost.

Some beekeepers who are unable to visit their apiaries frequently attempt to lure swarms to bait hives or provide sites near the apiary for swarms to cluster. Such swarms can be readily seen and caught if they remain clustered until the beekeeper arrives. Types of swarm traps include the following:

- Decoy or bait hives—with drawn comb or foundation—can be placed at various distances and directions from the apiary. Wax, propolis, and other odors may attract the scout bees and, ultimately, the swarm, but they might also attract mice and wax moths, so any remaining bait hives should be removed at the end of the swarming season. Empty hive bodies are also used. Place these bait hives 4 to 6 feet off the ground in a tree, and face the entrances to the south-southwest.
- Lures, using queen or Nasanoff pheromones, can be purchased and placed inside empty boxes or special wood fiber pots. These are used successfully in trapping Africanized bees in southern states.
- Low, dark objects close to the ground—such as a burlap bag wrapped around a low branch to form roughly a sphere—may attract a swarm to cluster there.

SWARM CONTAINERS

To be prepared, you should always have extra hives full of foundation for hiving swarms. If the swarms have to be collected some distance from the apiary, bring along a single deep hive body or nuc box, with the bottom board nailed on. Large swarms may not fit inside a nuc box, although such large swarms are not common today. Shake the swarm in front of the hive, and after most bees have entered, close the entrance with a piece of screen. The hive can then be either carried off or left there unscreened until the evening so that any stray bees can rejoin the swarm; its entrance should be screened when the hive is retrieved in the evening.

Other containers that can be used for collecting swarms are:

- A screened box, like an old bee package, or larger; shake the swarm into the box, and carry the box to the apiary.
- A cloth bag (not a plastic bag); shake the swarm into the bag and transport it to the apiary. If the swarm is on a tree branch, envelop the swarm with the bag, tie it closed, and cut the branch.
- An old basket that can be covered.
- A cardboard box that can be closed.
- A well-ventilated, 5-gallon plastic pail; add holes and a screened top.

Swarms, especially large ones, need plenty of ventilation and must be kept out of direct sunlight. Often bees are "cooked" or smothered when collected in inappropriate containers (one that is too small or too airtight). A swarm in a temporary container should be stored, like a package of bees, in a cool, dark place until it can be placed in a proper hive. But hive it within one or two days because the bees will begin comb construction, making their removal messy.

COLLECTING AND HIVING A SWARM

Beekeepers are often called by homeowners, other individuals, humane societies, and police and fire departments to retrieve swarms. If you wish to collect swarms to enlarge your apiary or strengthen weak hives, notify these agencies by letter or telephone each spring and ask to be put on the "swarm list." Most beekeepers are thankful to get swarms and consequently pick them up without charge. In today's age of lawsuits, however, when you enter another's property you are liable for damages, real or imagined, that you might cause in collecting swarms. Thus it would be wise to restrict your swarm collecting to your own beeyard.

If, however, you choose to collect swarms, forewarned is forearmed. You will need to ask precise questions about the swarm before you go out. Ask the following questions first. They will save you time and grief later on.

1. How far off the ground is the swarm? If more that 10 feet, forget it.
2. Can I drive to the swarm's location? What is the swarm clustered on (e.g., tree, post, drain, or building), and if on a tree, can I cut off the branch on which it is hanging? Bring pruning or lopping shears.
3. How long has the swarm been there? If more than one or two days, it will either be very hungry and therefore volatile or will have departed by the time you arrive.
4. How large is the swarm? The size of a basketball? A softball? Smaller swarms may be virgin queen or queenless swarms and, again, may not be there by the time you arrive.
5. Are you sure these are honey bees? There is no gray, paperlike nest? If there is, then they are hornets, and a professional exterminator should be contacted.
6. Are these bees yours to give away? Or did they come from a neighbor's hives?

Once these questions have been answered to your satisfaction, get instructions on how and where to find the swarm, and off you go.

Swarms are usually well engorged with honey and therefore gentle. But sometimes a swarm is ill tempered, especially if it has been clustered for several days and the bees are hungry. In any case, it is prudent to wear a veil when collecting swarms.

Some beekeepers carry spray bottles of light syrup, which is often medicated. Bees sprayed lightly with the syrup will engorge the food and become gentle and easier to handle.

These are the basic steps for collecting and hiving a swarm:

Step 1. If the swarm is clustered on a tree limb, with the owner's permission cut away excess branches, leaves, or flowers. Avoid shaking or jarring the cluster.

Step 2. If the swarm is jarred and the bees begin to break the cluster, spray the bees and wait for the cluster to reform.

Step 3. While steadying the limb or branch with one hand, saw or slip it free from the tree.

Step 4. Shake the swarm into a hive or collecting container prepared for it or if possible, put the entire cut limb into the collecting container.

Step 5. If the swarm is on a post or flat surface, brush or smoke the bees into a hive or container, directing them gently with puffs of smoke. Try to capture the queen and put her in a queen cage, which would make collecting the swarm much easier. After you put the caged queen in your container, the bees will soon enter it.

Step 6. A piece of cardboard can be used like a dustpan to scrape bees gently into the container or in front of the hive entrance.

Step 7. Come back in the evening to ensure that the bees are all in the box. Close up the box with a screen, and take it home. Sometimes, bees may be left, or a small cluster of bees will have reformed higher up. If possible, spray these bees with soapy water to kill them, as many homeowners are upset if all bees have not been collected.

Step 8. Back at the apiary, shake the bees from the container into a hive filled with foundation, or unite the swarm with a weak colony. If the collecting box has frames, wait a few days with the entrance open and install it as you would a nuc or unite it with a weak colony. Bees in a swarm are gorged with honey and nectar, which stimulates their wax glands. Put in frames of foundation on a hived swarm and you will be amazed at how fast the bees draw out the comb.

Hive the swarm and attempt to evaluate the queen's performance. After about a month, if she lives up to your standards (evaluate the number of frames of brood, how much honey and pollen was collected), keep her (if she is good) or requeen the colony (see "Breeder Queens" in Chapter 10). If you know this is a year-old queen (the colony she came from has swarmed already), requeen the colony with a queen from new, nonswarming stock. Usually, swarm queens are older and should not be used. If you caged the swarm queen, you can unite the bees with a weak colony that has a young queen, destroying the swarm queen. Swarms united with colonies should be placed in a hive with foundation and then placed over the colony with which they are to be united.

If you know which colony swarmed, check it for additional queen cells. Some colonies can cast several swarms. The first, or *prime swarm,* is accompanied by the original, old queen. Subsequent swarms, called *afterswarms* will all contain virgin queens and are generally smaller in size than the prime swarm.

PRECAUTIONS

Always treat swarms as if they are diseased or are carrying mites. Because you would not usually know this beforehand, install all swarms on foundation and feed with medicated syrup. Sample the bees to see if varroa mites are present. If positive, treat with Apistan strips; or you can also trap the varroa by providing the swarm with a frame of drone larvae, about to be capped over. The mites will be attracted to the open brood; once capped, remove the frame and freeze it. If you are going to requeen the swarm right away, delay the Apistan treatment until the queen is laying well, about 30 days. In the meantime, you can apply nonchemical control techniques if the mites are a problem. See Chapter 13 for information on diseases and mites.

If a swarm is put on drawn comb, the bees may regurgitate drops of nectar containing disease spores from their honey sacs into the comb. By being installed on foundation, the bees will consume the contents of their honey sac first, such that the spores would be consumed and excreted.

DESTROYING A SWARM

If the swarm must be destroyed because it is diseased, it is in a location dangerous to people, it is full of mites, or it is Africanized, a thorough spraying of soapy water (1 cup liquid detergent or soap per gallon) will kill the bees quickly. It knocks them to the ground immediately, even if the swarm is airborne.

Queen Supersedure

Supersedure is how the colony replaces an old or inferior queen with a young queen. The workers in the colony build just a few queen cells, and when a new queen emerges, she destroys the other queen cells and may destroy the old queen (sometimes mother and daughter queens coexist for a short time). Swarming does not usually take place when a queen is superseded.

Reasons for supersedure include:

- Queen is deficient in egg laying.
- Queen produces inadequate amounts of queen substance (pheromone) as a result of age, injury, lack of nourishment when a larva, or other physiological problems.
- Queen is injured as a result of clipping, fighting among virgins, or temporarily being balled by workers when released.
- Queen was injured when removed from or placed into a queen cage.
- Queen is defective, not raised under ideal conditions, or poorly mated.
- Queen did not receive enough nourishment as a larva (which may contribute to some physiological defects).
- Colony and queen has nosema disease or mites.
- Weather has been inclement for extended periods (other than winter).
- After installation of a package, when numbers of adult bees decline and no new ones emerge for 21 days, the remaining older workers may undertake supersedure activities.

Supersedure versus Swarm Cells

Young queen larvae can begin their development in queen cups or in worker cells. Queen cells made from worker cells are called *emergency queen cells.* The sudden loss of a queen usually forces the bees to modify worker cells into emergency queen cells, and as the larval queens develop, the cells' edges are slowly enlarged by the added wax. This forms the peanut shape that is characteristic of all queen cells.

Queen cells that begin from a queen cup are either swarm cells or supersedure cells, depending on their location and numbers. Swarm cells normally hang from the lower edges of a comb, are numerous, and contain queen larvae of different ages and sizes. Supersedure cells, on the other hand, are fewer in number, contain larvae of approximately the same age (little variation in cell size), and are found in the central region of the comb.

A queen that has emerged from a swarm cell usually replaces a queen that has departed with a swarm. A queen that has emerged from an emergency queen cell replaces a queen that was accidentally lost. And a queen that came from a supersedure cell replaces a failing queen.

The state of the colony and the time of the year

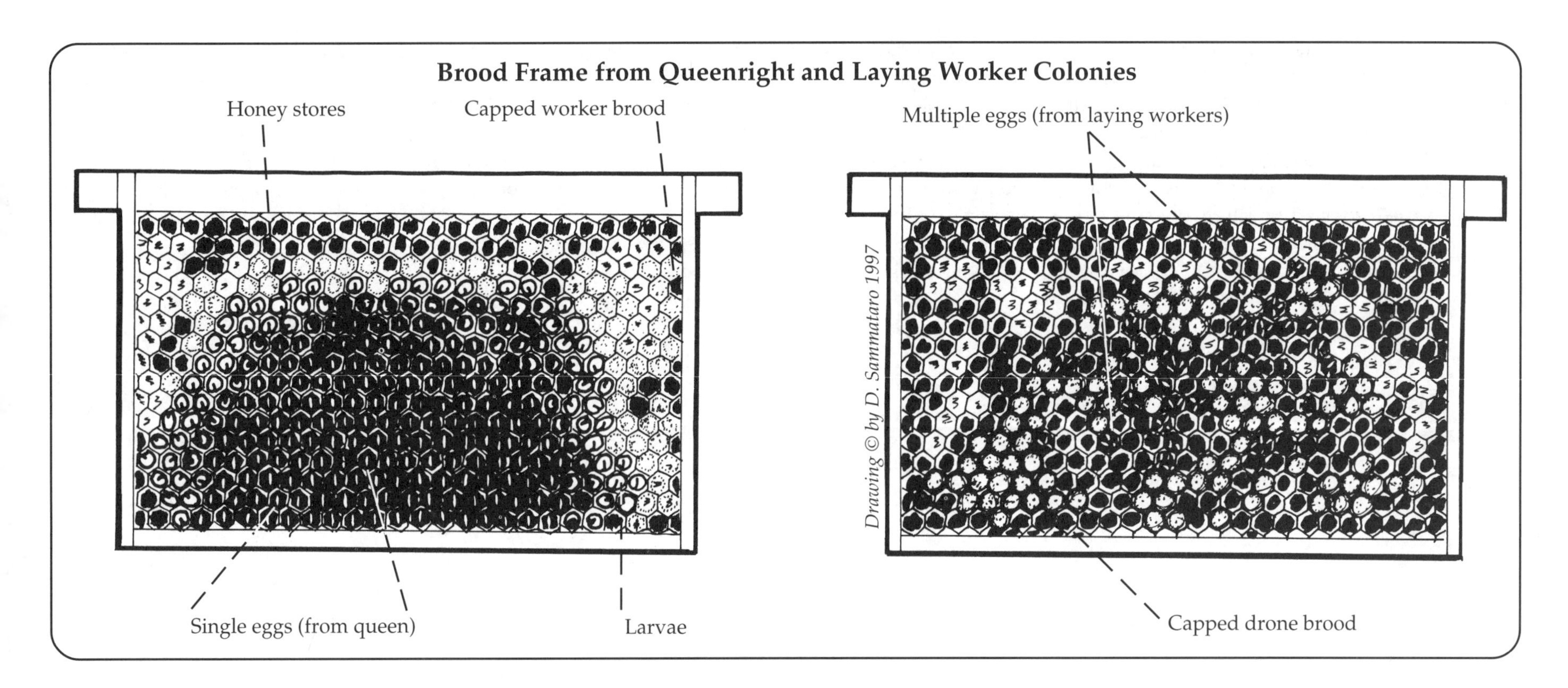

may also indicate whether the colony's aim is to supersede or swarm.

In supersedure:

- Few (between one and five) cells are made.
- The colony is usually not very populous.
- The brood pattern is scattered or almost nonexistent because the queen is injured, diseased, or failing.
- Queen cups or cells are usually in the center of the comb.
- The age of queen larvae is the same, so all queen cells mature at nearly the same time.
- Drone brood often appears in worker cells.
- Queen cells are present after the normal swarming period.

In swarming:

- The colony is populous.
- Numerous frames of capped brood are present, and there is a diminished number of cells with uncapped brood.
- Swarming season is on (early spring and late summer, depending on the latitude).
- Numerous (between 10 and 40) queen cells are present on the lower edge of combs.
- The ages of queen larvae are varied; may result in multiple swarms as different queens emerge over time.

In emergency replacement:

- Queen cells are usually constructed on the face of the comb amid capped worker brood.
- Few (between one and three) cells in a colony are selected from worker larvae that are between two and three days old.
- Attempts to make queens from older worker larvae may result in intermediate forms of queens on maturation.

Laying Workers

When a colony loses its queen and is unable to rear a replacement as a result of a lack of eggs or young larvae (less than three days old), some workers may start to lay eggs. The ovaries of these females will mature, and after these bees are fed the high-protein royal jelly, eggs will mature. Since workers are incapable of mating, the eggs they lay will be only unfertilized or drone eggs. The worker population within a colony with laying workers, therefore, will slowly decline because the rearing of new workers stops with the loss of the queen.

Laying worker colonies can be recognized by the number of eggs seen in each cell. If more than one egg is present and they are on the sides of the cell walls, instead of on the bottom, laying workers should be suspected (see the illustration of brood frame). To correct this situation, several methods have been tried, some of which involve the introduction of a queen, queen cells, or a frame containing larvae less than three days old. Often bees within a laying worker colony are first shaken from their frames at a distance of 100 yards or more from the parent colony just before the introduction of a queen, queen cells, or young larvae. This action supposedly leaves behind the laying workers, too heavy to fly, causing the rest of the bees to accept a new queen readily.

Unfortunately, these attempts at rescuing a colony of laying workers from inevitable doom have never worked to anyone's satisfaction. Frequently the colony will reject the introduced queen or queen cells, or it will rear workers rather than queens from the introduced larvae. You also lose valuable honey-making time. So the best solution is to unite a colony of laying workers with a queenright hive. Experimenting with the other methods, however, may be worth the experience.

12 Products of the Hive

Honey

Nectar: Physical Changes

Nectar contains primarily sugars and water. The water content of nectar is high and has to be reduced as part of the process of converting nectar into honey. The reduction in the water content is one of the steps necessary in making the final product—honey.

Bees returning to the hive with this sugar-water solution (nectar) transfer their load to house bees, who release small portions of the liquid onto the base of their proboscis, or tongue. By stretching out their proboscis, the liquid comes in contact with air movements in the colony. This helps enhance the evaporation of water from the nectar. Then, small amounts of the nectar are placed in the cells, where further evaporation and ripening takes place. As warm air circulates in the hive, fanned by other hive bees, the evaporation rate of the nectar is increased. The elimination of water from nectar represents the physical changes taking place in its conversion to honey.

Staggering honey supers by about ½ inch or so during the honeyflow will increase ventilation and hasten the ripening process in your colonies, especially if they are in a humid location.

Nectar: Chemical Changes

Nectar from flowers generally consists of 60 percent water and from 30 to 35 percent sucrose, or table sugar; nectar also contains other components, in minor amounts. The sucrose is altered by the action of an enzyme called *invertase,* which breaks sucrose into the two simple sugars (carbohydrates) glucose (dextrose) and fructose (levulose).

These two sugars are the principal components of honey, with more fructose than glucose. Other sugars that remain after invertase activity include small amounts of sucrose and other complex sugars, depending on the type of nectar.

In addition to invertase, another enzyme in honey is *glucose oxidase,* which converts glucose to produce *gluconic acid* and *hydrogen peroxide.* The factors that endow honey with antibacterial properties include the hydrogen peroxide and honey's high sugar content (about 80 percent) and high acidity. The enzymatic activity in nectar represents its chemical alteration into honey.

When honey is reduced to ash, trace amounts of minerals are found: these are calcium, chlorine, copper, iron, magnesium, manganese, phosphorous, potassium, silica, sodium, and sulfur. Other components of honey are acids, proteins, amino acids, and vitamins—all in trace amounts (see the table on properties of honey).

Forms of Honey

Honey is packaged and sold in several forms—as all liquid, as a combination of liquid and comb, as all comb, or in a granulated or crystallized form. *Extracted honey,* the liquid form, is removed from the comb and packaged in bottles or jars. Honey is classified into colors ranging from water white (clear) through amber (gold) to dark (black). Light-colored honey tends to be mild in flavor, whereas darker honey has a more pronounced, stronger flavor. Most extracted honey will eventually granulate, or crystallize (becoming semisolid), if not treated (see "Granulation" below).

Comb honey remains in the wax honeycomb, where it too can granulate. Granulated comb honey cannot be extracted except with heat, which will melt the wax and the honey until it is warm enough to separate. Some honey (e.g., canola or rapeseed) granulates very quickly. To keep your comb honey from becoming solid, record the blooming dates of your major honey plants so you can identify which flower nectars granulate quickly.

The basic types of comb honey are:

Section comb, consisting of individual wooden, boxed sections; plastic Half-Comb Cassettes (designed by J. Hogg); or circular plastic sections, now called Ross Rounds (but designed by the late W. S. Zbikowski).

Bulk comb, in which the entire comb on a frame is wrapped or sections are cut out of a frame (cut comb) and packaged.

Chunk comb, in which sections of cut comb are placed in a bottle that is then filled with liquid honey.

Comb honey is the best type of honey; the delicate fragrances and flavors of the floral source are sealed in the comb and not dissipated by heating the honey. Unfortunately, many customers today do not know how to eat honey in the comb. A technique used by many beekeepers to improve sales is to include recipes or have taste tests (see "Selling Your Hive Products," below, and "Honey and Honey Products" in the References).

Honeydew

Honeydew refers to the sugary excretions produced by plant-sucking insects, such as aphids. These insects are found feeding on the tender shoots of conifers, such as pines (*Pinus* spp.), spruce (*Picea* spp.), and larch (*Larix* spp.), and deciduous plants, such as oaks (*Quercus* spp.), maples (*Acer* spp.), willows (*Salix* spp.), and some others. If the secretions are profuse, they are eagerly collected by honey bees because they are rich in sugars normally found in honey. "Pine honey" or "Forest honey" is popular in

Chemical Properties of Honey

Principal Components	Percentage
Water	17.1
Fructose or fruit sugar	38.5
Glucose or grape sugar	31.0
Sucrose or cane (beet) sugar	1.5
Maltose and other reducing disaccharides	7.2
Trisaccharides and other carbohydrates	4.2
Total sugars	99.5
Minor constituents, including vitamins, minerals, and amino acids	0.5
Total	**100**

Enzymes

Invertase (inverts sucrose into glucose and fructose: comes from bees)
Glucose oxidase (oxidizes glucose to gluconic acid and hydrogen peroxide plus water; from bees)
Amylase (or diastase, breaks down starch, and may aid in bee's digestion of pollen)

Aroma constituents

Alcohol, ketones, aldehydes, and esters.

Source: National Honey Board.

some parts of the world, especially Europe. This honey comes from honeydew found on some evergreens in European forests.

Honeydew honey comes in varying flavors and colors, a result of the plant source. One characteristic of some honeydew honey is its rapid crystallization. It is possible that many honeys contain some honeydew, especially if bees are collecting nectar in areas next to plants with heavy aphid populations.

Microscopic components in honey can be used to identify its source: these include pollen for nectar honey, and algae and sooty mold particles from honeydew honey. Dark honeydew honey can be bitter in flavor in part because of its higher mineral and ash content. Proponents of this honey claim it has higher antibacterial activity, and the honey collected from aphids working spruce is especially prized in some countries.

The Honey House

A sanitary honey house for extracting, bottling, or otherwise handling honey is very important and in some states must meet certain health and food safety requirements. It is imperative that it be kept clean—running water and a washable floor are necessities—and be as insect-proof as possible. Bees and other pests attracted to the smell of honey or wax can get into everything, including newly strained or bottled honey. Fecal material from insects and animals is also a problem in any shed that is not tight.

If you have only one or two hives, you can extract honey in the kitchen or a clean basement. Put down some newspaper to keep the floor from getting sticky, and after you cut the cappings off one or two frames of honey, let the honey drip into a pan in a warm room. But if you have more than a few hives, you may have to commit one place to doing all the extracting and associated bee tasks.

If you are moving in this direction, use these guidelines in planning a good honey house to include the following necessities:

- Electricity
- Hot and cold running water
- Restroom facilities
- Washable floor (concrete or ceramic tile) with center drain
- An uncapping and extracting area where sticky, wet frames are handled and honey is strained
- A hot room, to warm unextracted honey supers
- Dehumidifier, if area is excessively damp
- Storage space for empty supers in an unheated portion of building
- Space for bottling, labeling, and storing honey

Though not essential, the following equipment may also be stored in a honey house, if the structure is large enough:

- Uncapping knives
- Extractor
- Capping tank or tray

- Honey pumps
- Straining cloths
- Capping baskets
- Screen to drain cut comb honey
- Bottles and labels
- Holding tanks

If your area is large enough, use part of the room as work areas for constructing and repairing frames, supers, and the like when you are not extracting. Also, store your smokers, hive tools, gloves, suits, and veils here as well.

Some excellent floor plans have been published in various books. Check out the "Books on Bees and Beekeeping" and "Honey and Honey Products" sections in the References; any book by Eva Crane is good, as she lists other references. Also talk with bee supply dealers, most of whom have floor plans for extracting rooms.

Beekeepers with fewer than 20 hives can easily manage by using any sanitary space (such as a garage or basement), an extractor, uncapping knife, and some of the equipment listed above. Or consider getting together with other beekeepers in your area or members of another bee club to buy or share equipment and space. An arrangement whereby beekeepers help one another pull and extract honey supers, dividing the finished product proportionally, should be made ahead of time, to avoid misunderstandings.

If you have 100 or more hives, congratulations, you are no longer a hobby beekeeper. Consider going into this as a business and purchasing more professional, commercial equipment. If you are heading in this direction, work with some commercial beekeepers to see how they run their operations before investing much of your hard-earned money.

Extracting Honey

Once a frame of honey has been at least three-quarters sealed with wax cappings, it can be removed from the hive and processed. Supers with frames of honey ready to be extracted should be placed in a bee-tight room or honey house. If the room temperature is between 80 and 90°F (26.6–32.2°C), the honey can be extracted with ease. Let the supers stand in a warm room until the honey is room temperature. So as not to promote the granulation of honey, avoid storing supers in temperatures below 57°F (13.9°C).

The wax cappings that seal the honey in the cells are commonly cut away with a heated or an electric uncapping knife or rotating chains. Hobbyist beekeepers generally use an uncapping knife, whereas the commercial outfits have more mechanized equipment. Cut the cappings off both sides of the frame, letting them drop into a screened basket or some other container that will permit the honey to drain off. These cappings are later melted down to separate them from any remaining honey (see "Wax Cappings," below).

Place the frames into an *extractor* (radial or basket-type) such that frames of equal weight are opposite each other (see the illustration of honey extractors). If the weight is distributed unequally, the extractor will wobble and vibrate wildly.

If using a *basket-type* extractor (frames are parallel to the sides), start with a slow spin and gradually increase the speed. Spin the frames on one side for

Honey Extractors

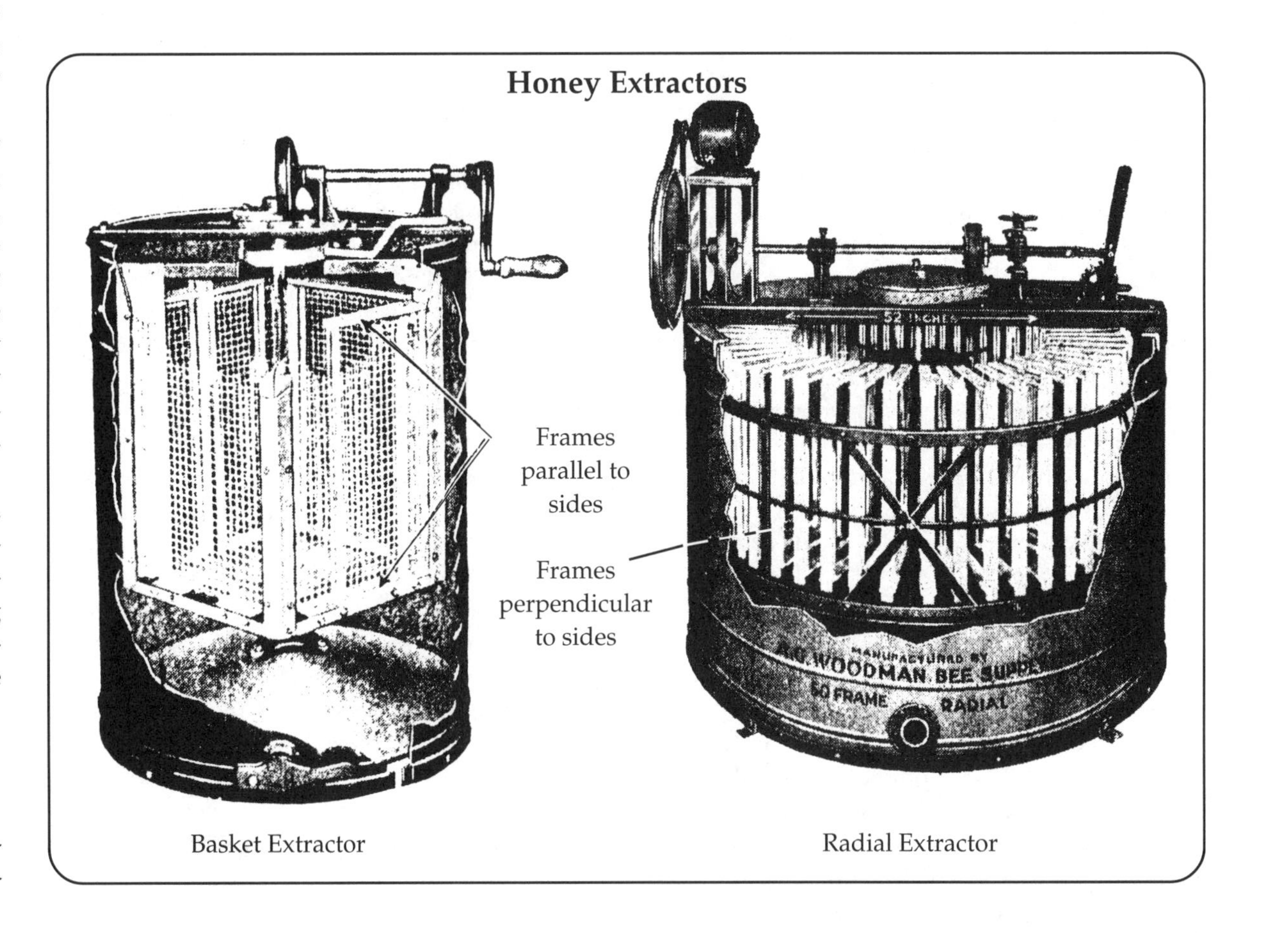

Basket Extractor

Radial Extractor

three minutes; then reverse them and spin on the other side for three minutes.

If using a *radial* extractor (frames are perpendicular to the sides), there is no need to reverse frames—both sides of the frame are extracted simultaneously because of their perpendicular placement. Start with a slow spin of about 150 revolutions per minute (rpm) and gradually increase speed to 300 rpm. Spin at the maximum rate for about 15 minutes. Honey can be draining from the extractor into a strainer on a bucket while it is spinning, but keep your eye on the bucket, as it will quickly fill up and overflow.

After Extracting Honey

Extracted honey should be strained immediately to remove wax, bees, and other debris and may require a second or a third straining. The strainer can be made of nylon or metal screen—any material which is easy to wash and which will not become easily clogged. Remember, warm honey strains much more quickly than cold honey.

After extracting:

Step 1. Place the strained honey into a storage container or a holding tank until it can be put into the final containers. Honey in a storage tank should sit overnight, covered, in a warm room before being bottled. During this time, fine air bubbles and wax particles rise to the top. Skimming off this layer before bottling results in a better, cleaner finished product.

Step 2. Remove the empty, wet frames from the extractor and place them in empty supers; return these to hives at dusk to allow bees to clean the wet frames. To avoid robbing problems from yellow jackets and other bees, tape over any holes or cracks to block such entrances. If no other honeyflows are anticipated, remove the cleaned supers after one or two days and store them. Wax moths will destroy combs that are not properly stored—that is, if not stored at freezing temperatures or otherwise protected (see "Wax Moth" in Chapter 13). Extracted combs should not be stored wet, because any remaining honey will crystallize in the cells, providing the seeds that will hasten the granulation of next year's crop. Wet combs can also ferment and mold, in which case you should melt them down and install fresh foundation.

Step 3. Remove as much honey as possible from the cut cappings and then melt the wax cappings down (see "Wax Cappings," below).

Physical Properties of Honey

Honey is a supersaturated solution of sugar and water and as such has various properties, such as hygroscopicity, fermentation, granulation, and thixotropy.

Hygroscopicity. Hygroscopicity is the ability of a substance to absorb moisture from the air. Honey absorbs moisture when the relative humidity in the storage area is above 60 percent. In low humidity honey will give up moisture to the air. The hydroscopic nature of honey is due to its sugars. This property of honey is beneficial to the baking industry, because it helps keep baked goods that contain honey moist and soft. On the other hand, if honey incorporates too much water (more than 17 percent), sugar-tolerant yeast will spoil the honey by causing it to ferment. Honey that is 19 percent water or over will certainly ferment.

Fermentation. Sugar-tolerant, or *osmophilic,* yeast spores, under high-moisture conditions, are able to germinate in honey and metabolize its sugars. As the sugars of honey are metabolized, the yeasts produce alcohol and carbon dioxide as by-products, which will spoil honey. Probably all honeys contain osmophilic yeasts.

Conditions that determine whether these spores will germinate and multiply include:

- Water content of the honey.
- Temperature at which the honey is stored.
- Number of yeast spores in the honey.
- Granulation of the honey, which results in an increase in the water content of the remaining liquid portion.

Fermentation of honey can be prevented if its moisture content is less than 17 percent, if it is stored at temperatures below 50°F (10°C), or if it is heated to 145°F (63°C) for 30 minutes, which kills the yeast.

Granulation. Most honey will granulate after it is removed from the comb; some kinds of honey granulate just a few days after being extracted (pure canola, aster, or goldenrod honey), whereas other types remain liquid for weeks, months, and even years. Honey consisting of a greater proportion of glucose to fructose will granulate faster. Different flower nectars have different proportions of these sugars. To keep extracted honey in a liquid state, heat it to 145°F (63°C) for about 30 minutes. To keep out any "seed" or particulate matter, air bubbles, wax particles, or pollen, commercial packers add diatomaceous earth to warmed honey, then pump the "flash-heated" honey through several filters. The cleaned honey is then "flash-cooled" and stored or bottled.

Honey that granulates naturally can lead to fermentation or to the production of undesirably large crystals. Partially granulated but unfermented honey, often considered "spoiled" by the uneducated consumer, is in fact perfectly good to eat. Many honey producers use the Dyce process to granulate honey, which they sell as a spread (see the description of this method on p. 117). This honey has very fine crystals and does not ferment. The ideal temperature for honey to granulate is 57°F (14°C). Unless the object is to produce this kind of honey, store honey above this temperature. To prevent granulation during long-term storage, honey may be kept in a freezer.

Thixotropy. A few rare honeys have thixotropic characteristics. A single, particular plant protein im-

Dyce Process for Cremed Honey

Heat honey that has between 17.5 and 18 percent water until it reaches 150°F (66°C), stirring constantly. Strain through very fine nylon mesh to collect all pollen grains, sugar crystals, and other impurities. This straining is very important, otherwise the honey may "seed"on these particles instead of on the seed honey you add. Cool honey rapidly to 75°F (24°C), stirring slowly, but DO NOT add any air bubbles (stir slowly without breaking the surface of the honey).

When the temperature is between 70 and 80°F (21–27°C), add 10 percent by volume of the starter seed (or you can do this by weight: if you have 10 pounds of honey, add 1 pound of starter). The starter should be from store-bought creamed honey that has a very fine-grained consistency. You can also reuse your own granulated honey, as long as the crystals are ground very fine. Incorporate this into the honey without the addition of air bubbles, as bubbles will create a frothy top on the finished product.

Let this settle for a few hours, then pour the mixture into the final, marketable containers. Now store the honey at temperatures between 45 and 57°F (7–14°C) for a week until compleletly crystallized. Label and sell.

According to the National Honey Board, this product is "cremed"honey.

Source: R. A. Morse and P. K. Flottum, eds., *The ABC and XYZ of Bee Culture*, 40th ed. (Medina, Ohio: A. I. Root, 1990).

parts this unique property to the honey. In the comb, the honey appears to be solid but cannot be extracted because of its thick, viscous nature. The honey will liquefy, however, if it is subjected to vibration with a special type of extractor or as it is being spread on bread. As soon as the vibration stops, the honey reverts back to a thick, gel-like solid. The most famous thixotropic honey is from ling or heather (*Calluna vulgaris*), a plant commonly found growing on the moors of Europe. Another is pure grapefruit (*Citrus paradisi*) honey.

Cooking with Honey

Honey is a natural sweetener, often used to replace sugar in cooking. Because honey is a combination of sugars that are broken down by bees into the simple sugars of fructose and glucose, it is very digestible. Some people maintain that honey is helpful in the following ways, although these claims may be exaggerated.

- Retaining calcium in the body.
- Counteracting the effects of alcohol in the blood.
- Deterring bacterial growth, especially on burns.
- Providing quick energy.
- Keeping baked goods moist.

When substituting honey for sugar in any basic recipe, observe the following guidelines:

- Measure honey with a greased utensil.
- Honey has 1½ times the sweetening power of sugar.
- Reduce the liquid in the recipe by ¼ cup for each cup of honey used to replace sugar.
- Use a mild-flavored honey, unless the flavor of the honey is a necessary part of the product.
- Some people add ¼ teaspoon of baking soda per cup of honey to counteract the honey's acid.
- Reduce the cooking temperature of the final product by 25°F (4°C).

If honey is stored in a dry place (do not put in the refrigerator) or is frozen, it will granulate much more slowly and will not ferment. To liquefy granulated honey, place the container in a pan of hot water until the crystals have melted. Do not let the honey overheat, as many of the flavors and aromas of honey are volatile and are destroyed by heat. Another method is to put an uncapped container of honey into the microwave and heat it first for 1 minute and then in 30-second increments until it is liquefied. Never leave honey containers unattended

Other Measurements of Honey

- Acidity: average pH for honey = 3.9 (pH for vinegar, Coca-Cola, beer = 3.0; pH for tomatoes = 4.0).
- Osmotic pressure more than 2000 milliosmols/kg (makes honey a hyperosmotic solution); also contributes to its antibacterial activity.
- Refractive index: 1.55 if water content is 13%; 1.49 if water content is 18%.
- 1 gallon (3.79 liters) weighs 11 lbs. 13.2 oz. (5.36 kg).
- 1 lb. (0.453 kg) has a volume of 10.78 fl. oz. (1 1/3 cup) or 318.8 ml.
- Caloric value = 1380 calories/lb.
- 1 tablespoon = 60 calories.
- 100 g = 303 calories.

Freezing Point Depression

- Honey (15%) freezes at 29.4°F (-1.4°C).

Sweetening Power

- 1 volume honey = 1.67 volume of granulated sugar.
- 1 gallon (3.79 liters) = 9 3/8 lb. (4.25 kg) total sugars.

Source: National Honey Board

when heating. For more information, see "Honey and Honey Products" and "Honey Cookbooks" in the References, as well as the list of other measurements on page 117.

CAUTIONARY NOTE: IT IS NOT ADVISABLE TO FEED HONEY TO INFANTS LESS THAN ONE YEAR OF AGE. Spores of the bacterium *Clostridium botulinum* exist absolutely everywhere and are able to survive even in honey. In the digestive tracts of infants under one year of age, the spores may progress to the vegetative stage of their life cycle and produce toxins that could prove fatal or injurious to infants. This disease is known as infant botulism; see "Honey and Honey Products" in the References.

Preparing Comb Honey for Market

If you are cutting comb honey from a frame (cut comb honey), use a warm, sharp, thin-bladed knife and cut the comb on some kind of screen to allow the honey to drip off. Use a template when you cut so as to keep all the sections the same size, and leave the cut comb to drain overnight in a warm room. Carefully package this honey in clear plastic cut comb boxes, but wrap these with plastic if necessary to ensure they will not leak when displayed. Other leakproof containers are available from a variety of sources; check catalogs and journals.

Section comb honey, whether in wooden boxes, plastic rounds, or plastic cassettes, is relatively simple to prepare for the consumer. It takes no special equipment and is easy to make marketable. One problem is what to do with stained, damaged, or half-filled sections. Half-filled ones can be used the next season to bait new section supers. Stained or damaged sections can be sold as seconds, put in a jar and filled with liquid honey (to make chunk honey), or crushed to extract the honey.

All the sections need to be as clean as possible, free from debris, propolis, and other foreign matter. A sharp knife is used to scrape off the propolis that accumulates on the rim of the sections. This job is both fast and easy, but make sure you do not nick the comb or allow debris to fall inside the finished product. Nicking the comb or otherwise damaging it, no matter how small, will cause the section to leak. Leaky section boxes are not only unattractive but also do not sell well and discourage future customers.

As an extra precaution, place the square wooden sections in a plastic sandwich bag before slipping them inside the cardboard display box (that you have purchased). Print the price and weight on each box, and stamp it with your name, address, and zip code. You can do the same for the round sections or the plastic cassettes, which come with a long, narrow label that fits around the perimeter of the container. Be sure to use indelible ink to keep your stamp from washing off.

Because you have frozen your comb honey sections, make sure you apply the round or cassette labels after the boxes have come to room temperature or the labels will wrinkle on the wet containers. When freezing square wooden sections, place them in the plastic sandwich bags first and freeze in a box to prevent damage to the sections. Then label the cardboard display boxes after they have thawed.

Beeswax

The domestic wax industry can obtain only two-thirds of the beeswax it needs from U.S. beekeepers; the rest is imported. The largest users of beeswax in the United States are the cosmetics and related industries, followed by the candle companies, industrial manufacturing interests, and beekeeping companies (for foundation). Minor users include pharmaceutical and dentistry concerns, followed by companies that make floor polish, automobile wax, and furniture polish. Beeswax is used as a minor ingredient in some adhesives, crayons, chewing gum, inks, specialized waxes (grafting, ski, ironing, and archer's waxes), and woodworking supplies and for arts and crafts projects.

Because wax foundation is expensive and beeswax is a valuable hive product, you should make every effort to save all cappings, old combs, and bits and pieces of extra wax scraped from frames and

More Information on Beeswax

- Between 7 and 9 pounds of honey will produce 1 pound of beeswax.
- 1 pound of wax makes for 35,000 wax cells.
- 1 lb. wax stores 22 lbs. of honey.
- There are 500,000 wax scales/pound (one scale measures $^{1}/_{8}$ inch in diameter).
- Optimally, 10,000 bees produce 1 pound of beeswax in three days.
- Cappings yield from 90 to 97% beeswax, depending on extraction method; these numbers reflect conventional, not commercial, methods.

Beeswax Components

Component	Percentage of Total
Monoesters	30–35
Diesters	10–14
Hydrocarbons	10.5–14
Free acids	8–12
Hydroxy polyesters	8
Hydroxy monoesters	4–5
Triesters	3
Acid polyesters	2
Acid esters	<1
Free alcohols	<1
Unidentified	6

Characteristics of Beeswax

- Melting point: between 141.9 and 150.8°F (61–66°C); wax from cappings: 146.7°F (63.7°C). Paraffin melts at between 90 and 150°F (32–66°C).
- Solidification point (liquid wax becomes solid): between 140 and 146°F (60–63.3°C).
- Flash point (wax vapor ignites): between 490 and 525°F (254–274°C), depending on the purity of the wax.
- Temperature of plasticity: 89.6°F (32°C).
- Relative density at 68°F (20°C) = 0.963 (water is 1.0, thus wax floats).
- Saponification value: 95.35.
- Acid value: ranges between 16.8 and 24.0.
- Ester value: ranges between 66 and 80; ratio of ester to acid is 3 : 4.3.
- Refractive index (light-bending property) at 176°F (80°C) = 1.4402.
- Electrical resistivity: between 5 and 20 × 10^{12} ohm m.
- Shrinkage: if melted to 200°F (93.3°C), wax shrinks 10 percent (by volume) when cooled to room temperature.

Solubility: insoluble in water; slightly soluble in cold alcohol; soluble in benzene, ether, chloroform, and fixed or volatile oils. Beeswax is stable for thousands of years.

other hive parts. Melt these pieces in a solar wax melter, and trade or sell wax blocks to dealers.

Cappings, old combs, and wax scrapings should be kept in airtight containers or frozen until you are ready to process them. Wax moths can quickly infest such wax if it is not covered, or their eggs may already be present, waiting to hatch. Also, melt cappings separately from the old combs, because the latter contain nonwax substances that would impregnate and reduce the value of the almost pure wax cappings. Use extreme caution when melting wax—it ignites easily, and wax fires are difficult to put out (see the information on beeswax on p. 118).

Melting Beeswax

Use extreme care when processing beeswax, as it is highly flammable and can cause serious burns.

Follow these simple precautions when melting beeswax:

- Extended exposure to high heat over 185°F (85°C) will discolor wax.
- NEVER use an open flame; use a double boiler over electric heat.
- Use only aluminum, nickel, tin, or stainless steel containers to melt wax, because other metals will discolor wax.
- Never use direct steam to melt wax, because it permanently changes the composition of wax.
- Separate dark wax from light, and never add propolis to wax.
- Don't melt plastic frames in the wax melter: they can warp.
- Many pesticides are lipophilic and will bind with wax. Never store such chemicals near beeswax or combs. This includes Apistan strips.

Wax cappings, old combs, and scrapings can be melted with one of these devices:

- Electric wax melter.
- Solar wax melter.
- Double boiler made of aluminum or stainless steel; iron or copper containers will darken the wax.
- Steam chests.
- Wax press that is steam-heated.

An old-fashioned and inefficient method of processing beeswax is to place the old comb and scraps in a burlap bag; submerge the bag in a tub or barrel of rain water or acidic water (NEVER in alkaline water, because the wax will become spongy and will not be fit for other uses). Place stones or bricks in or on the bag to help keep it submerged. Heat the water to between 150 and 180°F (66–82°C) for several hours, occasionally poking the bag with a stick to allow the wax to move through the fabric to the surface of the water. After the wax has melted, remove from the heat and allow the water to cool. The wax will solidify on the surface of the water, and most of the debris will fall to the bottom of the tub or barrel. Render and filter the wax cake a second time.

None of these methods will be sufficient to render all the wax found in old combs. The remaining mixture of wax and debris should not be discarded but saved and taken to a dealer who has the special equipment needed to render it.

Solar Wax Melter

The solar wax melter is essentially a box painted black outside and white inside and covered with a piece of glass, Plexiglas, or plastic and made airtight. It is put in a sunny location and tilted at a right angle to the sun's rays. The sun heats the interior of the box, something like it would a greenhouse, melting the wax inside, which collects into a pan. For greater heating efficiency, use two pieces of glass or Plexiglas, separated by a ¼-inch (6.4 mm) gap. The inside of the box

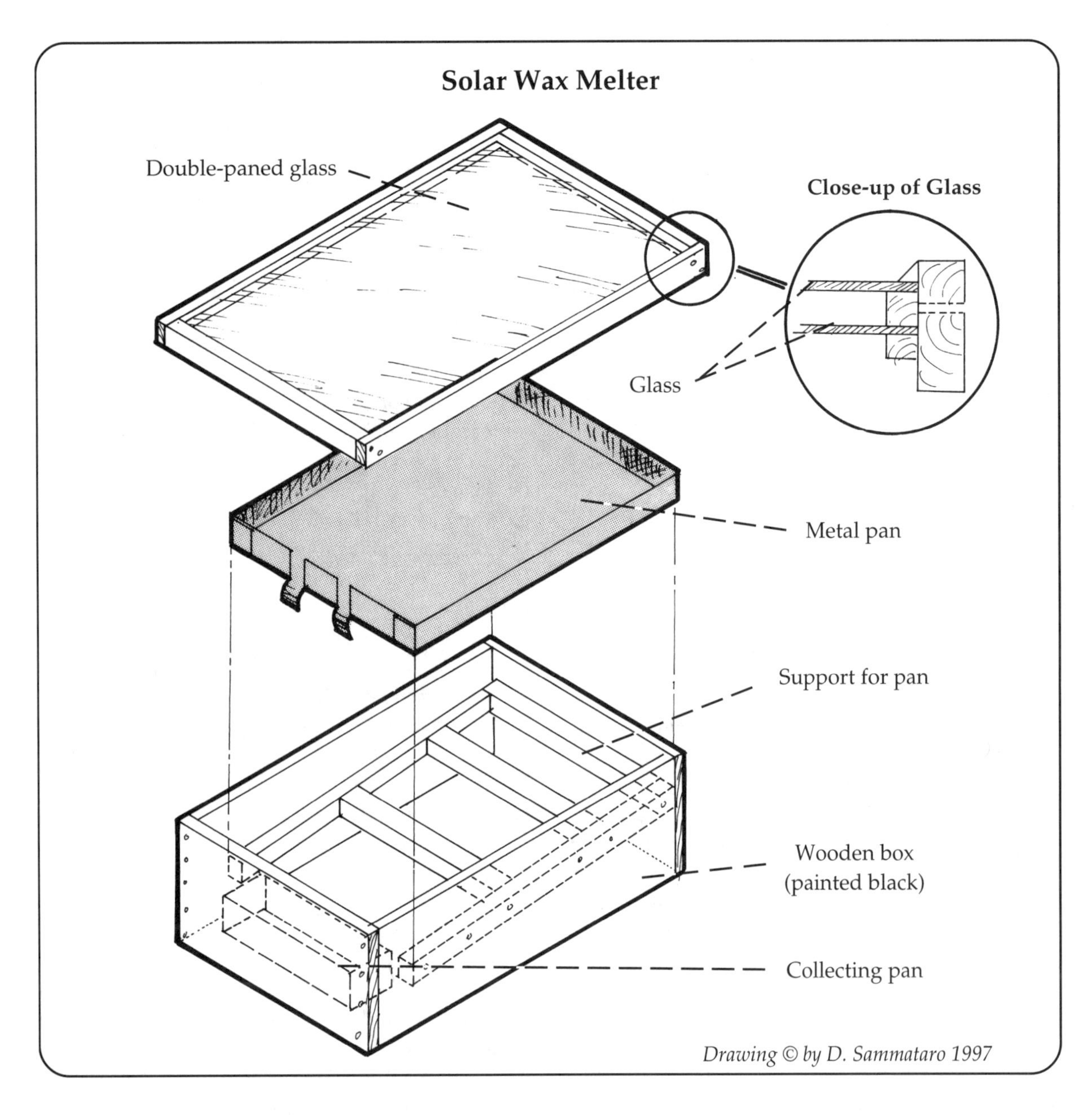

Drawing © by D. Sammataro 1997

contains a metal tray, fashioned from sheet metal, onto which the wax comb and scraps are placed (see the illustration of the solar wax melter).

The melter will render cappings, new burr comb, and old comb, but it will not melt the old comb completely. After old, dark comb has been in the melter for a few days, collect the black, gummy remains (sometimes called *slumgum*) and either compost and use as a soil amenity or discard. If feasible, take the remains to someone with the equipment to extract the remaining wax chemically. The amount of wax one can derive from broken pieces of foundation, cappings, and old or burr comb is significant and well worth the effort.

Wax Cappings

When extracting honey, you are always left with buckets of wet wax cappings and honey. If you have only a little of this, you can place the bucket in a hot room (where it heats up to about 143–151°F (62–66°C). Some beekeepers make an insulated box heated with a lightbulb; this will melt all the wax in 24 hours. The remaining honey, if not overheated, can be eaten (bottled) or fed back to bees (if it contains no foulbrood spores). If overheated, discard the honey, which will be unacceptable to both people and bees, or use it to feed wax moths, if you are rearing them (see Appendix H).

For larger quantities of cappings, you may want to invest in a cappings spinner (an extractor to separate out most of the honey) and some sort of water-jacketed, stainless steel melter. The best way to determine what you need is to visit several beekeeping operations and see how they handle their wax.

Cappings wax is worth a good deal of money and thus well worth the effort to recover. After you separate out the honey, melt the wax and pour into plastic tubs or loaf pans (first coated lightly with talcum powder). Weigh the finished block so you can estimate how many pounds of wax you will end up with. Either sell the wax in blocks or render it for use in salable products, such as candles or ornaments.

Rendering Wax: Final Filtering

Once you have collected the chunks or blocks of wax from your wax melter, you must further refine it to clean out the fine particles of dirt, honey, and other impurities. Use a large double boiler or other wax melter (check bee supply catalogs), and filter through several layers of cotton sacking, heavy cotton, or paper towels.

When melting small amounts of wax to pour into molds or the like, use an electric frying pan, an old coffee pot, or a double boiler arrangement (in which you place the wax in a small coffee can in a pan of hot water). When final-filtering, pour hot wax into a filter-covered can, and to keep the wax liquid, put the can into a warm oven until all the wax has drained. This final wax can then be used for candles, in batik, in cosmetics, or for whatever you decide (see "Beeswax" in the References).

Bee Brood

Generally an unexploited product of the hive, bee brood is rich in proteins, fats, and other substances required in our daily diet. Bee brood consists of over 15 percent protein, 4 percent fat, and mostly water, about 77 percent, compared with beef, which is 23 percent protein, 3 percent fat, and 74 percent water. It has over 100 times the vitamin A present in milk and over 6000 times as much vitamin D.

The value of this hive product does not yet compensate the cost of removing brood from comb and the reduction of the adult colony population that ensues if too much brood is removed. Honey bee brood is currently used, on a small scale, as food for birds, reptiles, and fish. Drone larvae and pupae are often used for fish bait, and if you cut out drone brood for varroa control, you may be able to sell it. Bee larvae also make a good quiche or soufflé.

Honey Bee Soufflé

1/2 cup butter
1/2 cup sifted flour
1 1/2 tsp. seasoned salt
1/2 tsp. paprika
Dash of hot pepper sauce
2 cups milk
1/2 pound sharp cheese, grated
6-8 eggs, separated
1/2 cup marinated bees

Marinate overnight bee larvae or pupae in 1 cup soy sauce, 1/4 cup sake or sherry, 1 clove garlic, 1 dried hot pepper, and 2 tbsp. ginger root, grated.

Melt butter in double boiler over boiling water and add next four ingredients; mix well and gradually add milk, stirring until sauce thickens. Add cheese until it melts; remove from heat. Beat egg yolks until lemony in color and stir slowly into cheese sauce. Beat egg whites until stiff and fold sauce into whites. Layer bottom of 1-quart greased soufflé dish with bees and cover with sauce. Bake at 475°F for 10 minutes; reduce to 400°F and bake 25 minutes longer.

Source: Adapted from R. L. Taylor and B. J. Carater, *Entertaining with Insects* (Yorba Linda, Calif.: Salutek Publishing, 1992).

Bee Venom

Bees require pollen in their diets to synthesize some of the components of venom. The synthesized venom is stored in the poison sac of worker and queen bees (see Appendix B). Venom contains a complex array of chemical substances, such as water, amines, amino acids, enzymes, proteins, and peptides, as well as some sugars and phospholipids. Venom also contains histamine, which reacts adversely with the body chemistry of some individuals. See the table on effects of venom compounds, page 47.

To collect substantial amounts of venom, either for medical or other uses, place an electrical grid near the entrance of a hive. This special grid produces a mild shock, and bees that land on it react by stinging a sheet of nylon taffeta below this grid. The venom is deposited on and collected from a glass plate located below the nylon portion of the device.

Research is still in progress concerning the benefits obtained from honey bee venom for people with rheumatoid arthritis and other diseases. Bee venom therapy, including *apitherapy,* is becoming more popular today. In addition, more recent research indicates that some of the components of venom are much more effective than other serums in desensitizing those who are allergic to bee venom (see Appendix C and "Miscellaneous Hive Products" in the References).

Royal Jelly

Royal jelly consists, roughly, of 66 percent water, between 4½ and 17 percent ash, between 11 and 13 percent carbohydrates, 12 percent protein, 5 percent fat, and 3 percent vitamins, ether extracts, enzymes, and coenzymes. The sole food of queen larvae, royal jelly is manufactured by young nurse bees from two glandular secretions: a white jelly from the hypopharyngeal and mandibular glands, and a clear jelly, which is a hypopharyngeal secretion mixed with honey.

Royal jelly has long been collected and used in Asian countries for medical purposes. Modern uses include cosmetics, lotions, and dietary supplements. But little proof exists that this substance has any miraculous curative or cosmetic powers. Queen breeders often collect and freeze royal jelly to use during queen-rearing operations.

None of its curative properties have been extensively studied in the United States, nor should any be claimed on product labels. In addition, some people can be allergic to this product, and an appropriate warning should be applied to any label.

Propolis

Propolis is a resinous mixture used by bees to seal cracks in the hive. The word derives from the Greek words *pro* (before) and *polis* (city) and refers to its use by bees to reduce the hive entrance ("in front of," or "before," the "city") for winter protection and defense against ants.

The original semiliquid material comes from the sticky exudations of trees and buds—such as the alders, poplars, and some conifers—and is collected by foragers, either in its natural state or from abandoned equipment (such as mite- or disease-killed colonies). Foragers transport the material back to the hive on their pollen-collecting structures. Inside the colony, house bees pull it off, because it is too sticky for the foragers themselves to dislodge, and then mix it with beeswax and a third, unknown substance. It is this mixture that is called propolis.

Propolis is used by bees to:

- Coat the inside of the hive walls.
- Seal the inner cover, bottom board, and outer cover to the hive body.
- Strengthen comb.
- Glue frames together.
- Seal hive bodies together, and caulk any small cracks or crevices.
- Embalm foreign objects too difficult to remove (e.g., dead mice, snakes, or large insects).

To beekeepers, propolis is a gummy mess in the hive and makes cracking the hive bodies apart difficult, even when using a hive tool. Bees fill small, tight spaces and cracks that are smaller than the bee space with propolis.

Studies have documented the antimicrobial activity of propolis, which may help bees keep the hive clean and kill foreign bacteria that would otherwise live in small cracks. Dead mice coated with propolis are virtually mummified and will not decompose.

In Russia, its antimicrobial action has been found effective against certain infections in farm animals. Some of the components identified in propolis include flavonoids, benzoic acid, cinnamyl alcohols and acids, alcohols, ketone, phenolics and aromatics, quercetin, terpenes, sesquiterpene alcohols, minerals, sterols, volatile oils, beeswax, sugars, and amino acids. The compounds in propolis that have some pharmacological (antiviral, antibacterial, antifungal, anti-inflammatory, and antihistamine) activities are quercetin, pinocembrin, caffeic acid, caffeic acid phenethyl ester, acacetin, and pinostrobin (see "Miscellaneous Hive Products" in the References).

Some beekeepers collect propolis by laying a wire screen inside the colony, as one would a queen excluder, and allowing bees to propolize it. Once filled, the screen can be frozen and the propolis knocked off, cleaned, and bottled. Propolis is even appearing in capsule form in health food stores, as a health supplement. None of its curative properties have been extensively studied in the United States, nor should any be claimed on the label. As with other bee products, some people can be allergic to propolis, thus an appropriate warning should be applied to this product.

Because propolis is not water soluble, use acetone or ethyl alcohol to remove it from hands and clothing. Other beekeepers use mechanic's hand cleanser to clean off propolis.

Pollen

Pollen, the protein-rich powder produced by the male parts of flowers, is collected and sold by beekeepers to health food stores, to pollination businesses, to bee dealers (for bee food), and to allergy patients (as a desensitizing agent). Pollen traps are put on hives to collect pollen pellets from foraging bees. Collected pellets should be stored properly (see "Pollen" in Chapter 7 and "Miscellaneous Hive Products" in the References).

Pollen pellets, as a vitamin supplement, are sold in many health food stores. Pollen contains a number of nutrients, such as carbohydrates, proteins,

Pollen Components

Component	Percent of Total
Water	7.0–16.2
Crude protein	7.0–29.9
Ether extract	0.9–14.4
Carbohydrates	
Reducing sugars	18.8–41.2
Nonreducing sugars	0–9.0
Starch	0–10.6
Lipids	4.8
Ash	0.9–5.5
Unknown	21.7–35.9

Other components include organic acids, free amino acids, nucleic acids (plant), enzymes, sulfur, nitrogen, flavonoids, and growth regulators (plant).

Other Facts about Pollen

- Bees collect 33–88 pounds (15–40 kg) of pollen in one season; this is about 26–71 ounces (740–2000 g) per day.
- One colony can eat 44–65 pounds (20–30 kg) per year.
- One pound (0.4536 kg) makes 4500 to 6000 bees.
- One frame of honey plus one frame of pollen makes one frame of bees.

and minerals. The composition varies according to the plant species, and bees are not able to distinguish the more nutritious ones. Pollen constitutes 10–35 percent of the protein in the diets of bees. Remember, it is from pollen that bees also acquire the essential amino acids needed to manufacture their own proteins and enzymes.

Pollen contains such vitamins as A, C, D, E, B_1, B_2, B_6, and B_{12}, as well as minerals such as sulfur, nitrogen, phosphorus, and many minor elements (see the description of pollen components on p. 122). Some substances found in pollen have yet to be identified.

Pollen is also sold for human consumption. Some claim that pollen is a superior food which has curative powers and which provides extra energy. Authoritative reports on the value of pollen as human food are contradictory. To survive the rigors of sun and rain and preserve their precious cargo, pollen grains are protected by a wall of intine and exine. Some pollen grains have been excavated intact from acid bogs over 10,000 years old.

Bees have special stomachs to grind off some of the exine layer and digest the grain's contents. Recent studies indicate that the protective layers of some pollens can be ruptured and proteins released into human digestive systems.

Whatever the claims, there is a market for pollen, and it may be lucrative for beekeepers to collect and sell it. The importation of freshly collected pollen pellets from overseas is not permitted, because they could contain chalkbrood spores.

Before selling or consuming pollen, be certain that it was not contaminated by pesticides. Beekeepers trap pollen while some crops are being sprayed with insecticides so as to reduce the amount of contaminated pollen that enters the colony. Such pollen should never be sold for human (or bee) consumption. Know the source of pollen before buying or selling it.

Some companies will buy bee-collected pollen, so check the bee magazines for advertisements. But remember, none of its curative powers have been extensively studied in the United States, and some people are hypersensitive to this product and could suffer a severe allergic reaction. Add a warning label when selling pollen.

Any bee product can cause an allergic reaction in some people, and you can face potential lawsuits if you are sued. Look into getting product liability insurance if you will be selling bee products. DO NOT RELY ON YOUR HOMEOWNERS INSURANCE.

Selling Your Hive Products

There are many avenues through which you can sell your hive products: the office, local retail outlets, roadside stands or orchards, flea or farmer's markets, fairs, and your local church or other organizations to which you or your friends belong. Keep it simple at first; large grocery chains are expensive and difficult to buy into and require a year-round supply of honey.

Here are some rules, researched by the National Honey Board, that you should follow to be successful in keeping your customers happy:

- Have a reliable supply of quality product.
- Local honey is best. Use slogans such as "produced by bees in ______ County" for regional appeal.
- Keep your jars clean; people do not like sticky honey jars.
- Do not heat your honey higher than 120°F (49°C),

Various Types of Honey Containers

1. Plastic squeeze skep
2. Glass hexagonal jar
3. Plastic squeeze honey bear
4. Plastic queenline jar
5. Glass queenline jar

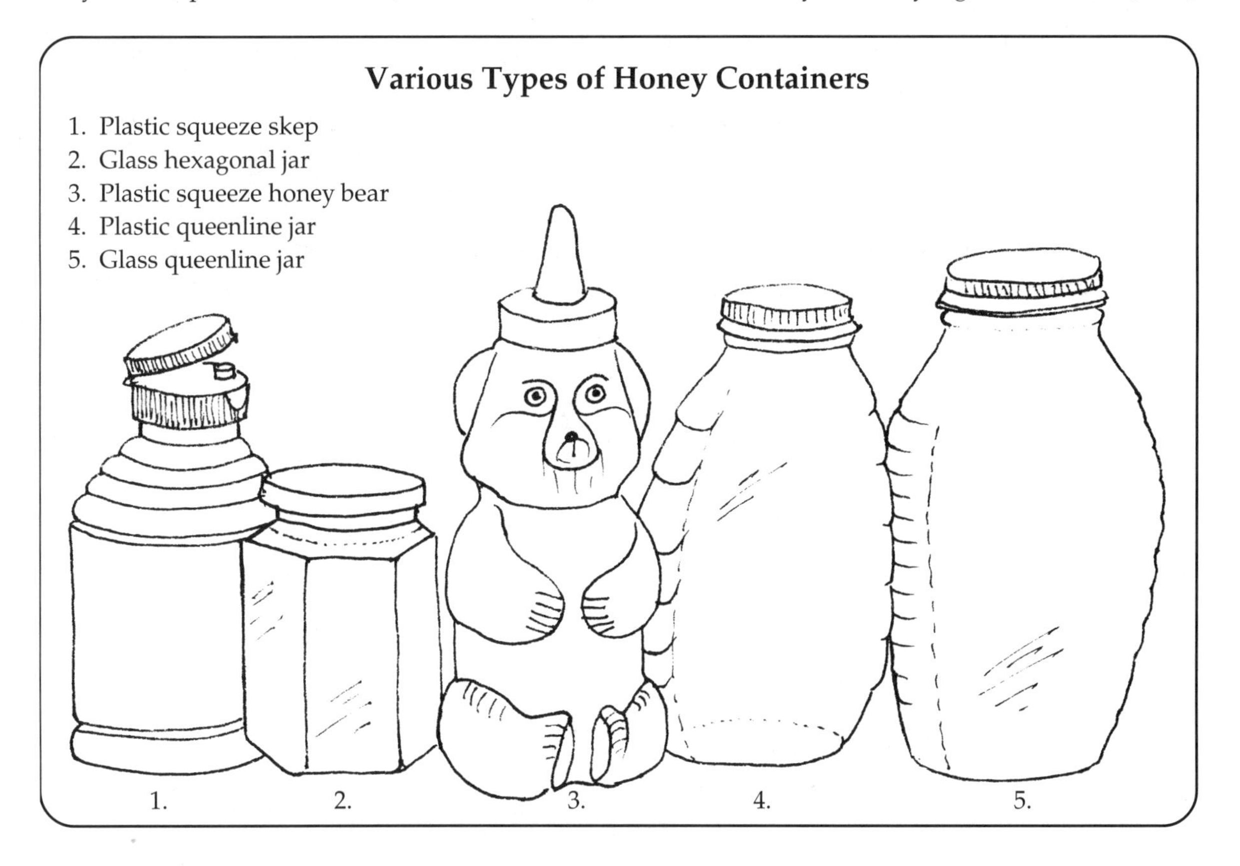

or you will ruin the delicate, floral flavors.

- Make sure your bee product is clean and free from debris.
- Use eye-catching, attractive labels; in general, the public does not like to see bees or bee hives on the label.
- Use consumer-friendly words, such as "pure," "natural," or "organic" on natural hive products. Honey can be labeled "Raw" if it has been minimally filtered or heated.
- Conduct taste tests, give out free recipes, dress honey bears in holiday outfits, or use other "value-added" techniques to help sell your products.
- Try making creamed honey (see the information on the Dyce process on p. 117).
- Use different kinds of honey jars (see the illustration of various honey containers).
- Mix honey with dried fruits, nuts, or flavorings; label these products carefully.

Advertising is also an important part of selling your product. Take advantage of promotions available from the National Honey Board, many of which are free (see "Honey and Honey Products" in the References). Many beekeepers simply sell honey from home, but before expanding your operation to a point at which you need a sign outside, check your homeowners insurance policy and any local zoning codes: forewarned is forearmed. A small classified ad in local papers may be all you need to have a steady supply of customers.

Honey Labels

Jars of honey have to conform to state and federal pure food laws. The guidelines below will meet most requirements, but check with your state agriculture department for more information.

The following information must appear on each jar or section comb of honey you sell:

- Name, address, city, state, and zip code of the manufacturer, packer, or distributor.
- Net amount in lower 30 percent of the display label (e.g., Net Wt. 8 oz. [227 g]).
- Name of product (e.g., Orange Blossom Honey).
- Ingredients, if other than honey, that are in the jar (e.g., fruit, nuts).

Labels can be purchased from some bee supply dealers or label specialists that advertise in the journals, with or without your name imprinted on them. Labels for round or square sections are sold by bee supply companies or label printers that advertise in the bee journals. You can purchase a separate stamp with indelible ink to imprint your name on those labels.

You can also have a printer make up your own label, but the initial cost is rather high. Floral sources can be included on the front label—for example, "Wisconsin Wild Flower."

Nutritional labels are required, especially if you claim a nutrient content or make a health-related statement. Check with the National Honey Board or your state food and labeling regulatory agency for particulars. Some beekeepers put on additional labels giving instructions on how to reliquefy honey that is crystallized. Unfortunately, any such instructions you supply could make you liable for any injury to a person following those instructions. Carefully label your products, but you may want to check with a lawyer about the wording.

For more information on marketing honey, and honey products, contact the National Honey Board (see "Bee Organizations, North America," in the References).

13 Bee Pests and Diseases

Diseases of Adults

Nosema Disease

Nosema, the most common disease of adult bees, is caused by the microscopic protozoan *Nosema apis* Zander. This is a spore-forming organism that invades the digestive cell layer of the midgut. Up to 30 million spores can be found in a single bee one and a half weeks after initial infection. Now found worldwide, wherever bees have been introduced, it is a serious problem in temperate climates. The disease is most prevalent in the spring, especially after winter weather has confined bees to their hive. Nosema greatly reduces the life span of all castes of adult bees, reduces honey yields, and is a factor in the supersedure of package bee (and other purchased) queens, further delaying the growth of a colony.

The spores are viable for up to one year in the bee's fecal material and are found outside the colony environment, especially at drinking areas commonly used by bees. The disease is spread at these water holes, by drifting and robbing bees, and from feces on frames and combs. Package queens or caged queens may have nosema and should be treated when installed. Some effects of severe nosema are:

- Reduced longevity of workers (by 50 percent).
- Reduced honey yield (by 40 percent).
- Queen supersedure, as egg-laying ability in queen is adversely affected.
- Hypopharyngeal (food) glands of workers are affected, and poor brood-rearing ability will result because the nurse bees are unable to produce enough brood food.
- Hormonal development is disrupted, causing bees to age faster and forage earlier in life than normal.
- Secretion of digestive enzymes is disrupted, causing bees to starve to death.

Although bees with the disease display no specific symptoms, listed below are some signs that could *also* be associated with pesticide poisoning, mite infestation, or mite-associated pathogens. If most of these symptoms are observed in the spring of the year, after winter confinement, however, nosema should be suspected:

- Bees are unable to fly or can fly only short distances.
- Bees are seen trembling and quivering; colony is restless.
- Feces found on combs, top bars, bottom boards, and outside walls of hive; also correlated with dysentery or diarrhea (see "Dysentery," below).
- Bees are seen crawling aimlessly on the bottom board, near the entrance, or on the ground; some drag along as if their legs are paralyzed.
- Wings are positioned at various angles from body, or bee is said to be *K-winged*—that is, wings are not folded in the normal position over abdomen but with the hindwing held in front of the forewing.
- Abdomen distended (swollen).

The only way to diagnose nosema disease is to dissect the bee and look for spores. A field test, which is not very reliable but good in a pinch, is to pull apart a bee until the viscera are visible. If the midgut (ventriculus) is swollen and a dull grayish white, and the circular constrictions of the gut (similar to constrictions on an earthworm's body) are no longer evident (normal gut color is brownish red or yellowish, with many circular constrictions), then nosema is the culprit (see the illustration on nosema).

If you place the midgut on a microscope slide and crush it, nosema spores, if present, will be evident at about 40×. They are small, smooth, ovoid bodies, much smaller than pollen grains.

Nosema can be a serious disease if not checked. Because it is often confused with symptoms of other diseases, amoeba, or mite predation, diagnosis is important in order to treat for nosema properly (see "Diseases and Pests" in the References).

TREATMENT FOR NOSEMA

Good management practices and feeding the antibiotic *fumagillin* (isolated from the fungus *Apergillus fumigatus* and sold as Fumidil-B and under other brand names) as a preventive measure help control the disease and ensure healthy colonies. To control the disease and prevent it from spreading:

- Provide fresh, clean water; individually feed each colony water via Boardman feeders.
- Provide a young queen.
- Locate hives at sunny sites, sheltered from piercing winds but with good air drainage.
- Maintain adequate stores of pollen, honey, or cured sugar syrup; if stores are short, bees should be fed a heavy, medicated syrup in early fall and protein supplements.
- Keep only clean combs; sterilize or dispose of those that are soiled with fecal material.
- Provide upper hive entrance during winter to improve hive ventilation.

Combs with nosema spores should be sterilized, especially if you are going to reuse frames from dead colonies. If you do not disinfect the frames, you spread the disease. Heat equipment to 120°F (49°C) for 24 hours; combs should be free of honey and pollen and temperatures should not get above 120°F or the wax will melt. Or you can also fumigate supers.

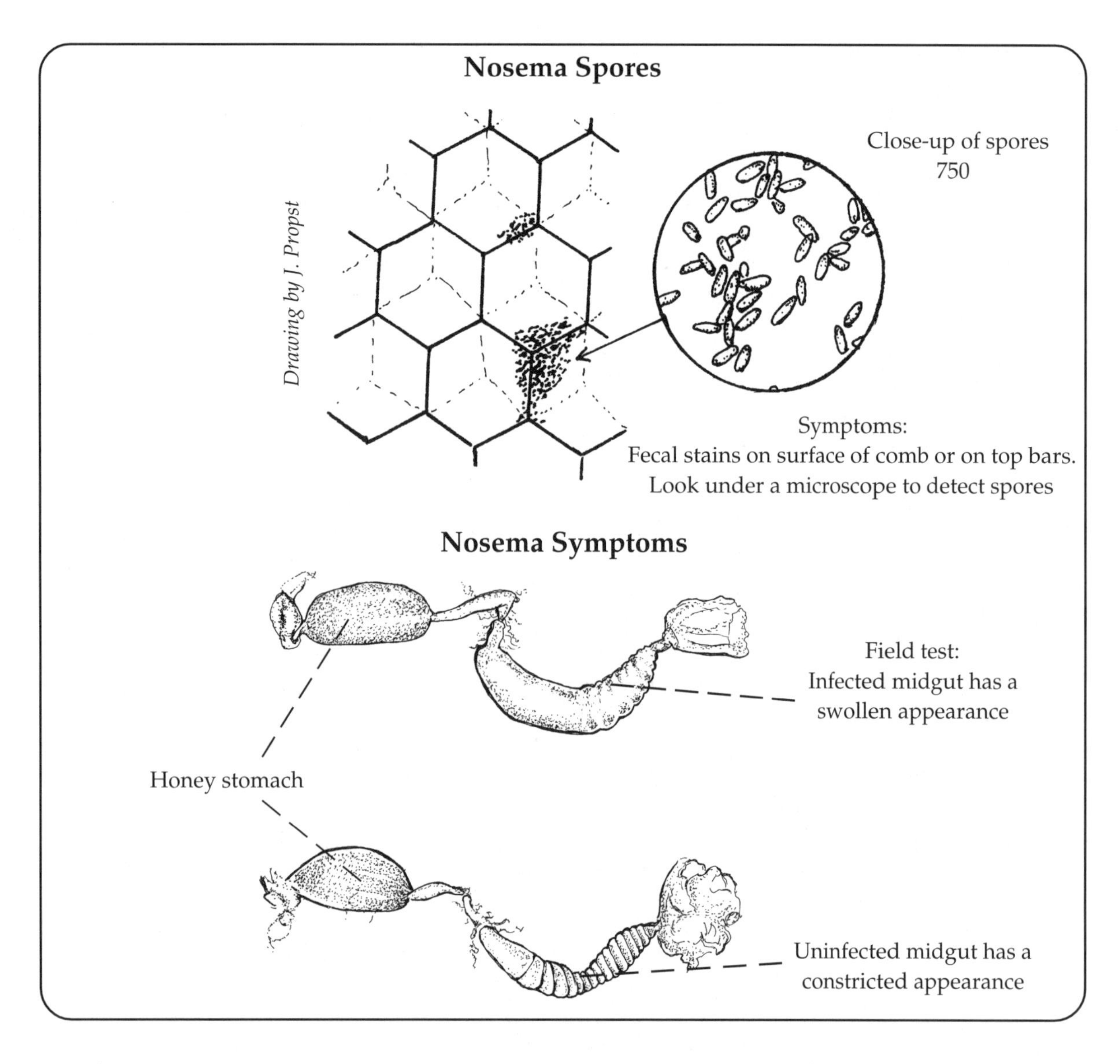

To fumigate combs:

Step 1. Place a hive body on a board or upturned outer cover.

Step 2. Soak a pad of cotton or rags in ¼ pint (118 ml) 80 percent acetic acid (available from photographic supply houses as glacial acetic acid); place pad on top bars of combs.

Step 3. Add hive bodies above the first one, placing a soaked pad on the top bars of each super in the stack. Close off the entrance, or treat the bottom board separately.

Step 4. Make the stack airtight by sealing adjacent hive bodies with masking tape and place inner and outer covers on top with a pad of acid in between.

Step 5. One week later, disassemble the stack and air out combs for two days.

Another approach is to replace old, fecal-stained, and dark combs with new foundation. A good rule of thumb is to replace frames when you can no longer see through them when held up to the sun, on a rotation cycle of every four to five years. In other words, as you go through your hives in the spring, pull out the older combs and replace them with new foundation.

The drug used to control nosema disease is bicyclohexyl-ammonium fumagillin, sold as Nosem-X or Fumidil-B. It must be fed mixed in syrup (not in powdered or patty form); it is sold by suppliers of bee equipment. This material is viable for two years, which can be extended if unopened bottles are stored in the freezer. Research has shown that spring feeding of fumagillin to established colonies can increase honey production by about 58 percent.

To dissolve the powder thoroughly, first mix it with warm water in a small jar, let stand for 5 minutes, then shake until foamy. Alternatively, you can thoroughly mix the powder with some sugar, then add it into the warm water.

The medication is fed in the formulation of 75–100 mg fumagillin per gallon of sugar syrup:

- Add 2 teaspoons to 4 ounces (½ cup) of warm water 95°F (35°C) and mix this into 2 gallons of 2:1 (sugar : water) syrup. This feeds ONE COLONY, delivering the correct amount of the medication. Proper dosage, especially for bees confined for long periods of time, is 2 gallons medicated syrup per hive. In the spring, a 1-gallon feeding is enough; use 2 gallons in the fall.
- For six packages, use the 0.5-gram bottle of fumagillin per 6 gallons of 2:1 syrup; 44 pounds of sugar in 2¾ gallons of water makes 6 gallons of 2:1 sugar syrup.
- Use the 9.5-gram bottle of fumagillin to make up 100–120 gallons of medicated syrup for feeding 100–120 package colonies in the spring or 50–60

colonies in the fall; 371 pounds of sugar in 23 gallons of water yields 50 gallons of 2:1 syrup.

Note: Always follow label directions when medicating bees.

Amoeba

Amoebae are in the kingdom Protista, which includes mostly single-celled eukaryotes (such as the protozoa, slime molds, algae, and the Sarcodina, or amoebae). The amoeba that affects honey bees is called *Malpighamoeba mellificae* Prell, which forms resistant spores called *cysts.*

A single bee can have a half-million cysts within three weeks of initial infection. Cysts are ingested by bees from infected food or water or from other contaminated material. Once ingested, the cysts migrate to the Malpighian tubules (the kidneys of bees), where they germinate within 24 hours. Within two to four weeks, new cysts are formed, pass into the intestines, and are voided with the feces.

Once the infection is under way, the bee's abdomen becomes distended, and the tubules cease functioning. The tubules take on a glossy appearance when filled with the spherical cysts, which measure 5–8 micrometers (μm) across; nosema spores are oval, measure 4–6 μm long by 2–4 μm wide, and are not in the Malpighian tubules.

Amoeba are found mostly in bees infected with nosema disease, and the cysts are often seen under the microscope in fecal material along with nosema spores. Although mostly present in workers, queens can be affected as well.

There are no clear symptoms other than dwindling colony populations, as bees die away from the colony. The effect of heavy infestation results in reduced honey yield and the impaired functioning of the tubules. Development time of the cysts is slowed in temperatures of about 68°F (20°C) or lower but increases as the temperature in the broodnest increases to 86°F (30°C). Therefore, the spring is the time when amoebae infections are most severe, peaking in May in the Northern Hemisphere.

The only control is hygienic conditions in the apiary and at the water source, and decontamination of frames or replacement of equipment; fumagillin has no effect. Some reports indicate that requeening has been a successful strategy for saving an infested colony.

Dysentery

Dysentery is not caused by a microorganism and is not a disease at all but primarily the result of poor food and long periods of confinement. In general, dysentery is caused by:

- Fermented stores
- Diluted syrup fed in fall
- Syrup with impurities such as those found in raw or brown sugar
- Dampness
- Long periods of confinement
- Too much moisture in the hive
- Poor drainage
- Honeydew in stores

SYMPTOMS

The symptoms of dysentery are similar to those of other diseases of adult bees:

- Sluggish bees
- Swollen abdomens
- Hive stained with yellow to brownish fecal material

The only way to treat dysentery is to:

- Provide a winter exit, so bees can take cleansing flights on warm winter days instead of defecating inside the hive.
- Provide good winter stores, with low water content (properly cured honey and sugar syrup).
- Feed thick syrup (2:1) in the fall, if bees need more stores going into winter.
- Medicate (as for nosema) as a preventive measure to help control the diarrhea.

Septicemia

Septicemia is caused by several different bacteria found in the hemolymph (blood) of bees, the most common of which is *Pseudomonas aspiseptica.* Although septicemia rarely if ever debilitates bee colonies, it can be recognized by these symptoms:

- Dying bees are sluggish, and hemolymph turns white instead of clear.
- Dead bees decay rapidly.
- Dead bees become dismembered when touched, as the muscle tissue degenerates.
- Dead bees have putrid odor.

Bees come into contact with the bacteria in soil, water, and infected bees by way of their breathing tubes (tracheae). It is still not clearly understood how the disease is transmitted or how to treat it, but some success has been found by requeening colonies and placing hives in locations that are sunny and dry and have good air drainage.

Other minor diseases are not mentioned here; many beekeepers never see these in their bees. For a complete look at all the diseases of bees, see "Diseases and Pests" in the References.

The Pesticide Problem

Farmers apply chemicals to protect their crop investment; most of these—fungicides and herbicides—do NOT harm honey bees. Because bees are insects, they are, however, killed by *insecticides.* The term *pesticide* is used to include all types of chemicals put on plants to control any pest, fungus, weed, or insect.

Problems related to honey bees and pesticide usage arise when insect pests threaten the crops that bees are working and growers use insecticides to protect their crops from such pests.

Bees are exposed to insecticides in several ways:

- Inadvertent but direct application on flying bees—bees may die in the field or after they return to the hive.
- Contact with recently applied insecticides—depending on the formulation, bees may die in the field or return and die in the hive.
- Consumption of contaminated water, nectar, or pollen—field bees, hive bees, and larvae will die inside the hive.
- Misapplication of material, including direct application to nontarget plants, as well as drift from treated areas and the use of inappropriate chemicals or application methods will kill field bees, hive bees, and larvae.

It is important to know the characteristics of insecticide-treated bees, because they can be easily confused with other disease symptoms. In general, you will notice:

- A sudden reduction in the number (thousands of bees) in a previously strong colony in the middle of the summer season.
- Excessive numbers of dying and dead bees, within 24 hours, in front of the hive, on the bottom board, or on top bars.
- Dying larvae crawling out of cells.
- A break in the brood-rearing cycle; disorganization of colony routine.
- Inappropriate queen supersedure.
- Within four to eight weeks, brood will either become chilled because of the lack of workers or will die from diseases or poisoned pollen.

Several types of insecticides exist, each affecting bees in a different way. The kinds of insecticides in general use today include these groups:

- Organophosphates
- Chlorinated hydrocarbons
- Carbamates
- Dinitrophenyls
- Botanicals
- Pathogens
- Pyrethroids

The symptoms for bee poisoning by these chemicals are summarized in chapter 26 of E. L. Atkins, *The Hive and the Honey Bee* (Hamilton, Ill.: Dadant and Sons, 1992), to which you should refer for more complete information.

In general, the most deadly insecticides with which you will probably come in contact are microencapsulated methyl parathion, carbaryl (Sevin), and Furadan. These are very toxic to bees and should be avoided.

Pathogen insecticides (compounds that cause diseases), unless specific for hymenopterous insects, are *not* toxic to bees, but you will see their names in conjunction with controlling many lepidopteran insects (such as the wax moth or gypsy moth). The most common pathogens are:

- Bacteria: *Bacillus thuringiensis* (Dipel, Biotrol, Thuricide).
- Virus: *Trichoplusia polyhedrosis* (Polyhedrosis virus).

What You Can Do

To reduce the chances of colony exposure to insecticides, you can take the following steps:

1. Carefully select the location of your apiary.
   - Locate and meet farmers, landowners, or land renters within a three-mile radius of your apiary.
   - Contact beekeepers in the area to learn about past problems.
   - Check plat, county, or aerial photo maps to assess apiary location in relation to areas that may be sprayed (parks, orchards, residences).
   - Become familiar with the crops grown, production methods, rotation practices, and past insect problems in your area.
   - Be aware of planting, blooming, and harvest dates of target crops.
2. Assess your chosen apiary site; weigh the potential for these chemical dangers:
   - Spray drift from nearby treated areas.
   - Frequency of sprays during the season.
   - Cyclic or unexpected outbreaks of insect pests (such as gypsy moths or mosquitoes).
   - Need for sprays during blooming period of target crop.
   - Application methods used (air or ground, low-volume, ultra-low-volume, standard, or electrostatic equipment).
3. Become familiar with:
   - The identification of crop pests in your area.
   - Pest population levels that require spray treatments (economic threshold).
   - The types of insecticides used locally, as well as their common names and formulations.
   - Registration procedures for apiary sites, so applicators can locate hives.
4. Know formulations of insecticides: Formulations with the designations WP (wettable powder), EC (emulsifiable concentrate), MC (microencapsulated), and D (dust) will kill bees on contact and, when they dry on the plant, may be picked up by bee feet or body hairs. Insecticides in these forms may also be mixed with pollen or in water puddles near sprayed areas. Addition of stickers or spreaders may significantly reduce problems caused by these formulations, by making pesticides less accessible to bees (i.e., sticking to plants).
5. Determine:
   - Local weed and wildflower blooming periods; learn to identify local honey plants.
   - Where bees are foraging at any particular time (by marking bees).
   - Where bees will forage next (sequence of blooming plants).
   - What time of day bees are on particular target crops.

6. Anticipate:
   - Changes in cropping practices.
   - Scheduled and unscheduled sprays.
   - Crop blooming periods and sequences.

Finally, when it happens, know whom to contact and what to do if bee kills are evident. Have handy the phone number and address of the local apiary inspector. Find out what legal recourse you have and how and where to take samples for analysis. In most cases, if your bees are not registered, you will have no recourse. Remember, many times your problem may lie *not* with a farmer spraying but with a neighborhood homeowner killing those "pesky" insects on the backyard rose bush or dandelions in the lawn.

Protecting Bees from Pesticides

Here are some general protective measures you can take before spraying occurs; use one or a combination of several methods:

- Make sure the applicators (local or contracted) know you and your apiary locations (supply maps).
- Check your county extension office for information on protection programs in your area.
- Make routine contacts with landowners, renters, applicators, and county agents for updates on pest problems.
- Post your name, address, and home and work phone numbers (or a neighbor's number so you can be contacted immediately) conspicuously near apiaries.
- Paint hive tops with a light color for easy aerial identification.

When spraying is imminent, here are some quick methods to use to protect your bees:

- Reduce hive entrance.
- Gorge hive with sugar syrup by pouring it directly on top of the frames (bees will stop foraging to help clean it up); pour in about one quart twice a day for one day before a spray and once a day for two or three days following a spray.
- If practical (you have only a few colonies), close the entrances with a screen to confine bees and place a screened cover on top, covered with a wet cloth or burlap. Keep this wet, especially if the weather is hot, with a sprinkler or watering can, for *at least* 24 hours. This is a dangerous step to take, because even with wet burlap, hives could overheat and the bees could die very quickly.
- Screen entrances with 8-mesh hardware cloth, and screen the top (cover top with a wet cloth).
- Activate pollen traps to collect contaminated pollen (destroy this pollen).
- Supplement feed colonies with syrup, water, and clean pollen patties during *and* after spray period.

If you have time and it is practical, the best protection method of all is to move your hives at least two miles from their previous location and target area. This method is the most expensive but also the most successful.

Once a kill has been experienced, you must immediately help those colonies that have been affected. After all, if it is early enough, they may still yield a surplus of honey or at least store enough for winter. Here are some things to do:

- Combine weakened colonies to increase populations.
- Requeen when necessary rather than waste time by letting bees rear new queens.
- Destroy contaminated stores, combs, or equipment; supply new equipment and clean combs or foundation.
- Feed syrup, pollen, or pollen substitutes to maintain the colony and stimulate brood rearing.

All beekeepers should strive to cooperate with neighboring growers for their mutual benefit. But the ultimate responsibility for a colony's protection rests with you the beekeeper, not the farmer, landowner, or applicator. You need to make the site selections as safe as possible and be alert to the expected problems while anticipating the unexpected. Utilize the numerous resources available to help you and your bees.

Such sources of information are:

- State and county extension offices (extension entomologists, agronomists, and horticulturists, as well as publications on crops, insect identification, insecticide lists, and formulations).
- State apiary inspector (for registering your apiary sites).
- Regional, state, and local beekeeping organizations.
- Libraries or city and state agencies, for maps and other references.

Brood Diseases

Brood diseases can be devastating to both novice and commercial beekeepers. Recognition of healthy and diseased brood is an important part of colony management, and awareness of how these diseases are carried may prevent a serious outbreak.

Disease carriers include:

- Beekeepers, by moving between diseased and clean colonies and not cleaning hive tools, gloves, or other beewear and equipment.
- Brood frames that are interchanged within or between apiaries.
- Old, diseased equipment or frames that are bought and interchanged or mixed with clean colonies.
- Honey, either fed directly or robbed by bees.
- Pollen sold commercially.
- Package bees or queens.
- Swarms.

Bee diseases are not mutually exclusive. A colony could have nosema and both foulbrood diseases at the same time (as well as mites)! Some conditions,

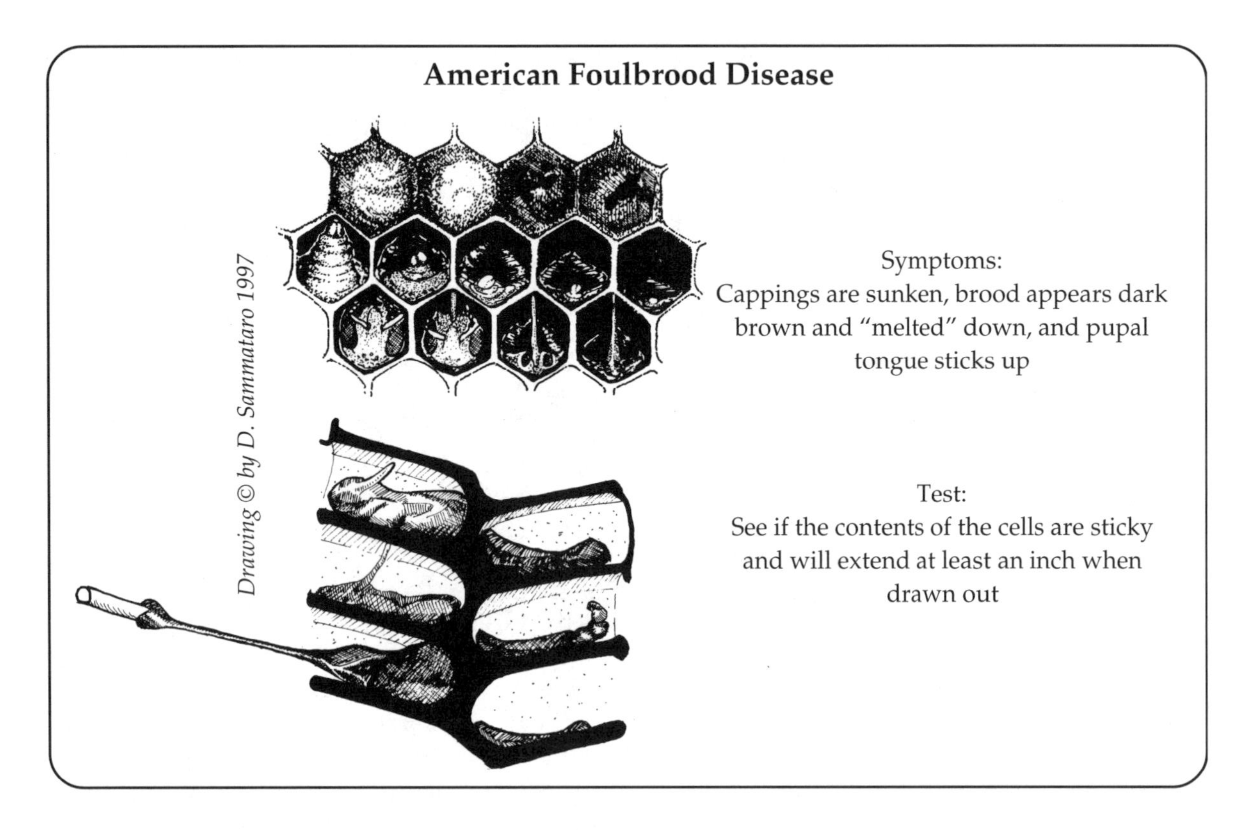

such as chilled brood, could look like diseased comb. Careful attention to the symptoms and the condition, history, and mite levels of the colony is necessary to differentiate the two.

American Foulbrood Disease

American foulbrood disease (AFB) is caused by the bacterium *Bacillus larvae,* which exists in both a spore and a vegetative stage. The disease is transmitted by the spore, and the infected brood is killed by the vegetative stage, when the spore germinates in larval guts. This is the most destructive of the brood diseases and is the reason that apiary inspection laws were first passed.

Once the vegetative stages appear in a colony, the disease is spread rapidly, and the colony weakens; in most cases, the colony will eventually die unless it is resistant to AFB. Spores can live in hive products (such as honey, wax, and propolis) for up to 69 YEARS. The LD_{50} of AFB (i.e., the amount of AFB needed to kill 50 percent of the bees) is just 35 spores fed to each one-day-old larva. Larvae affected are often older and therefore capped over and not immediately visible. Once AFB has progressed, old, diseased larvae, uncapped by the bees, are seen as black scales with their tongues stuck out.

SYMPTOMS

The symptoms of AFB are varied and could be confused with other disease or mite symptoms: make sure you learn to recognize this disease. Look for the following:

- Brood pattern is irregular rather than compact.
- Healthy larvae are glistening, white color; diseased ones lose this appearance and turn from light brown to dark brown. Larvae die upright, not twisted, in cells.
- Since the death of larvae and pupae often occurs after their cells are capped, the cappings become concave and some will be punctured by bees attempting to remove the dead brood (see the illustration on AFB).
- Surface of cappings will be moist or wet rather than dry.
- Larvae long dead develop the consistency of glue and are difficult for bees to remove.
- Eventually dead larvae dry out; the dried remains, or *scales,* adhere tightly to the bottom, back, and side walls of the cell.
- Some dead pupae, shrunken into scales, may have tongues that protrude at a right angle to the cell wall or are straight up. This may be the only recognizable characteristic.
- Unpleasant odor can permeate apiary if many colonies died over winter.

Any smelly hive, especially if winter-killed, should be suspect. If you look at the brood frames in good light, you should be able to see the protruding tongues.

AFB is transmitted from hive to hive in these ways:

- Beekeepers use disease-infested equipment, tools, or bee suits.
- Frames that contain diseased larvae and spores are moved to clean colonies.
- Bacteria are present in honey or pollen cells that were not cleared of scale. These spores are passed on to larvae by nurse bees feeding them.
- Cleaning bees spread bacteria throughout hive when attempting to remove dead brood or scales.

- Diseased robber bees enter an uninfected colony, or bees rob a diseased colony.
- Bees drift from diseased to clean colonies.
- Swarms have AFB.

If a colony is suspected of being infected with AFB, follow these steps as soon as possible:

Step 1. Reduce entrance to minimize robbing.

Step 2. Distinguish it from the rest by color or symbol, to reduce drifting.

Step 3. Begin medication (chemotherapeutic) program immediately, treating all the colonies in the apiary (see "Foulbrood Disease Chemotherapy," below).

Call your state bee inspector for advice and to confirm diagnosis. If unavailable, send a sample of the diseased brood and comb, one free of honey and about 4 or 5 square inches in area. Cut the sample out of the frame and wrap it in NEWSPAPER so it will not get moldy; do not use any other kind of wrapping. On a separate piece of paper, write your name and address and place this paper and the sample(s) in a sturdy wooden or cardboard box and mail to your state bee lab or to one of the national bee labs operated by the USDA (see "Bee Laboratories" in the References). Under separate cover, send a letter stating that you are sending samples, the problem you are having with the colony, and the following information:

- Name and address of beekeeper.
- Name and address of sender (if different from beekeeper).
- Location of samples and source.
- Number of samples sent (each labeled, numbered, and placed in a package); indicate if the samples are from the same or different apiaries.

TESTING FOR AFB

Use the "ropiness test," described in the next paragraph, on larvae that have been dead for about three weeks. Because it is difficult to determine how long a larva has been dead, randomly test between 5 and 10 cells from several frames. An accurate way of determining how long a larva has been dead is by checking for the presence or absence of its body segments or constrictions (similar to the constrictions on an earthworm). If they are absent, the larva has been dead for at least three weeks.

Insert a match, grass stem, or twig into suspect cells, stir the dead larval material, then slowly withdraw the testing stick. If a portion of the decaying larva clings to the twig and can be drawn out about 1 inch (2.5 cm) or more while adhering to the other end of the cell, its death was probably due to AFB. Be sure to burn the test stick. Scrub your smoker and hive tools with a soapy steel wool pad, and wash your hands, gloves, and bee suit thoroughly in soapy water with bleach.

TREATMENT BY BURNING HIVES

Before the availability of chemotherapy and ethylene oxide gas or gamma chambers, the only acceptable method of dealing with colonies infected with AFB was to destroy them by burning. Two methods of treating diseased hives, burning and fumigation, are discussed here.

In the first method, you burn all wooden parts, wax, bees, and honey:

Step 1. Kill all adult bees by spraying all frames at night with an insecticide (such as Resmethrin or Sevin) or with a 3–4 percent solution of soapy water (1 cup of liquid detergent per gallon of water is enough).

Step 2. To save the bees, you can shake them out of the hive into a super filled with foundation frames; feed bees on medicated syrup. This method is not advisable, because many bees will drift to other colonies, spreading the disease.

Step 3. Remove the entire hive *intact* to a field in which a pit has been dug that is at least 18 inches (45 cm) deep and contains a hot-burning fire. If you must carry the hive furniture separately, place each super on burlap or cardboard to keep dead bees and diseased honey from spreading the spores in the existing apiary.

Step 4. Burn all brood, honey, bees, and wax in the pit. You can support the hive bodies with large tree limbs placed across the pit. Make sure all is consumed and turns to ash. You can save dry wax (to be sent to a rendering plant), as long as it is securely closed against robbing bees.

Step 5. Cover ashes and pit with fresh dirt.

You can save newer hive bodies, if they are not too coated with wax and propolis (which are loaded with AFB spores). Invert, so rim edges are down, and stack hive bodies from three to four high. Then, to sterilize supers:

- Fill the inside of the stack of supers with newspaper and ignite it; when the insides of the hive bodies are scorched, extinguish the fire.
- You can also paint the insides with kerosene and light it by igniting the newspapers; this is a more thorough way to sterilize hive equipment.
- A propane torch may be used for the tops and bottom boards as well as for hive bodies. Wood should be lightly browned and all edges, corners, and seams given special attention. Scrape edges and inside thoroughly.

Before burning or using chemotherapy on bees with AFB, check with your state's bee inspector to be sure the procedure is legal and to determine the amount and kind of medication that is required.

TREATMENT BY FUMIGATION CHAMBERS

After killing the bees, place hive bodies, covers, and bottom boards in a fumigation chamber, such as an ethylene oxide gas chamber. You can use heat sterilization as well, such as a boiling paraffin bath (see Appendix D). Radiation with gamma rays will also decontaminate empty combs and equipment, but gamma radiation chambers are not commonly available. These special chambers can decontaminate empty combs and equipment by killing the disease spores and allowing you to reuse the equipment.

Many states no longer have ethylene chambers (at this writing), because the chemical is no longer available. Check with your local bee inspectors or state agriculture extension offices for licensed operators or for more information on radiation chambers in your area.

European Foulbrood

European foulbrood (EFB) is caused by the spore-forming bacterium now called *Melissococcus pluton* (formerly known as *Streptococcus pluton* and *Bacillus pluton*), although other bacteria may also infect larvae at the same time, producing similar symptoms. EFB is commonly found in weak colonies, such as those used for pollination.

The disease is usually prevalent in the spring, slowing the growth of the colony. Larvae more than 48 hours old are at greatest risk, and thus those that die are usually *not* capped but are visible in the bottom of the cells. The bacterium is found in feces, in wax debris, and on the sides of cells of infected larvae. Not as serious as American foulbrood, EFB should be treated with drugs, and the colony should be requeened or strengthened with additional bees or both.

European Foulbrood Disease

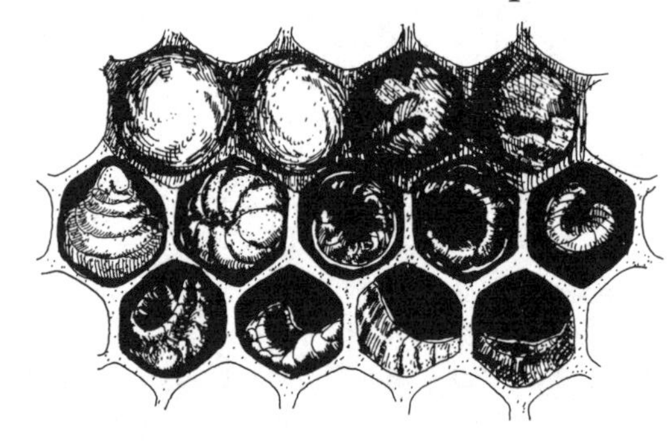

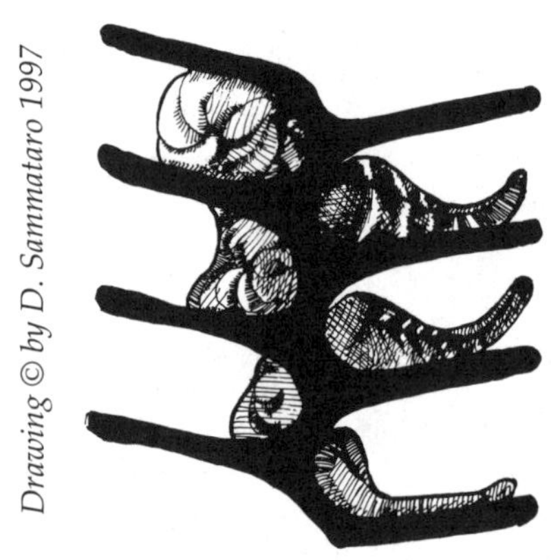

Drawing © by D. Sammataro 1997

Symptoms:
Bacteria infects younger, uncapped larvae, turning them brown

Test:
Larvae are twisted in the cells, no pupal tongue is evident, no "foul" smell is present, and the ropiness test does not apply

SYMPTOMS

The symptoms of a colony infected with EFB are different from those for AFB. Learn to recognize these:

- Larvae die in coiled, twisted, or irregular positions in their cells (see the illustration on EFB).
- Since most larvae die young, their cells are uncapped, and you can see the discolored larvae clearly.
- Larvae color may change from pearly white to light cream, then from brown to grayish brown, darkening as the dead larvae dry up. Normal, healthy larvae stay a pearly white color.
- Dry scales—the remainder of the larvae—are easily removed from their cells, unlike AFB scales, which are difficult to remove, so ropiness test will not work.
- Some larvae die in capped cells, scattered over the brood comb; cappings may be discolored, concave, and punctured.
- A sour odor may be present.
- Drone and queen larvae are also affected.

European foulbrood is transmitted from colony to colony in these ways:

- Cells in which larvae hatch may contain bacteria.
- Bacteria are present in honey or pollen or both and are passed on to larvae by nurse bees feeding them.
- As scales are removed by cleaning bees, bacteria are spread throughout the colony.
- Diseased robber bees enter a clean hive.
- Contaminated equipment is used.
- Bees from diseased hives drift to clean hives.

Bees seriously infested with varroa mites may have EFB-like symptoms; if there is any question, send a sample to the Beltsville Bee Lab via the method previously described in the section on AFB testing. Read about varroa mites and bee parasitic mite syndrome below, this chapter.

To control EFB:

- Requeen the colony, to break the brood cycle; this allows the bees to clean out dead and infected larvae.
- Use chemotherapeutic agents to treat the disease (see "Foulbrood Disease Chemotherapy," below).
- Feed with clean syrup and pollen supplement or substitute, not with pollen you purchased, as it could contain EFB spores or other disease organisms.
- Restrict drifting between colonies by relocating or redistributing hives (see the illustration on p. 38).
- Carefully inspect brood in frames before exchanging equipment.

Foulbrood Disease Chemotherapy

Drugs can be given to bees for both AFB and EFB once the disease has been diagnosed or as a preventive measure. The antibiotic drugs will not cure the disease; they prevent the spores from germinating, allowing the bees to clear out diseased brood. The antibiotic must be present while the larvae are being fed to prevent the spores from germinating inside healthy larvae. Use antibiotics only in connection with good management practices of normal colony hygiene and in maintaining healthy bees.

The antibiotic Terramycin (TM) is the only one registered in the United States (with the Environmental Protection Agency (EPA) by Pfizer) for treating AFB and EFB. If stored in a dry, dark refrigerator or in a freezer, an unopened TM packet can last several years past its expiration date. Drugs used as a preventive measure should be applied in the spring and fall, not during a honeyflow. If drugs are used during a honeyflow, the honey must not be used for human consumption. There are three methods approved for feeding medication to bees: in syrup, dry, or in an oil extender patty.

Terramycin soluble powder (or TSP), whose generic name is oxytetracycline hydrochloride (TM-HCl), is sold as an animal formula soluble powder at farm and bee supply stores. It comes in 6.4-ounce packets (2½ packets is equivalent to 1 pound), which contain 10 grams of active ingredient (or 25 grams per pound of product and therefore known as TM-25). This material is very fragile and can absorb water from the air, breaking down the components.

Terramycin is also sold bulk as TM-50D and TM-100D (in 50-pound bags). How much to feed is related to how much active ingredient is present per pound of material. To give each colony the proper dosage, follow label directions carefully, or you may not be giving enough medication to treat the disease.

However the medication is fed, you must stop feeding TM **45 days** (not 30 days, as the old label says) before you put on your honey supers.

Here are the new (revised in 1997) label directions for feeding Terramycin:

Dust or Powder. Use powdered instead of granulated sugar to make it easier to mix the antibiotic in with the sugar. The basic mixture is 1 teaspoon TM-25 in 2 tablespoons of powdered sugar; feed this

Guide for Treating Many Colonies with Terramycin-25

Number of packets (6.4 oz.) TM-25 soluble powder	Amount of sugar needed	Number of colonies fed (1 oz. per colony)
Dusting		
1 teaspoon	2 tablespoons powdered	1 (equals 200 mg)
1 packet	2 lbs. 12 oz.	50
2 packets	5 lbs. 7 oz.	100
4 packets	10 lbs. 14 oz.	200
Feed 200 mg or 2 tablespoons per colony three times at three- to five-day intervals		
Syrup (2:1 ratio of granulated sugar to water)		
1 teaspoon	1 quart syrup	1 (equals 200 mg)
1 packet	50 quarts (12.5 gallons) syrup	50
2 packets	100 quarts (25 gallons) syrup	100
Feed 1 quart three times at three- to five-day intervals; must drink all the syrup		

Extender Patty

	Shortening	**Granulated sugar**	
4 teaspoons	1/4 cup	1/4 cup rounded	1 complete dosage
1 packet	1 lb.	3 lbs.	14 (1/3 pound patty each)
3 packets	3 lbs.	9 lbs.	40

mixture of 1 ounce (2 tablespoons) to each colony three times at three- to five-day intervals. See the table on treating many colonies with TM-25. The total dosage per treated colony must equal 600 mg of oxytetracycline HCl (Terramycin).

When dusting, make sure to sprinkle the sugar mixture onto the *end* of the top bars of each colony; do not dust directly on top of brood frames containing uncapped larvae, because Terramycin is toxic to them. Repeat two more times at three- to five-day intervals to give the bees the required amount of medication. The medication is bitter and some bees will not take it; they will even propolize around it. If this is the case, try another method of feeding the antibiotic.

Stop all treatment SIX weeks before a main honeyflow. Because this form of TM (soluble powder) will absorb water from the air, make a new batch each year from an unopened or new packet. Store your mixture in an airtight container in a dark, cool place when not in use.

Syrup or Bulk Feeding. This method of feeding TM is not recommended because the drug loses viability in syrup after one week. Feed freshly made TM syrup if treating a swarm by spraying bees directly. You can feed a diseased colony with the syrup as long as the bees are VERY hungry and will take the syrup quickly. You can also spray diseased bees repeatedly, until they eat the required amount. DO NOT FEED DURING A HONEYFLOW.

Mix 1 teaspoon TM-25 for each quart (0.95 liter) of a syrup made in a 1:1 ratio of white granulated sugar to water. To dissolve the sugar, use very hot water and stir vigorously; once cooled, add the TM. Feed 1 quart three different times (three- to five-day intervals between feedings), making it fresh each time.

Extender Patty. The patty is the most stable form of TM and can last six to eight weeks. It is usually fed to bees over the winter months, and its oil base can help keep tracheal mite levels from increasing. For mixing, use 4 teaspoons of TM-25 in a patty made of ¼ cup vegetable shortening and ¼ cup, rounded, of white granulated sugar. This is the proper dosage for one colony, as it contains the needed amount of TM. Refer to the table for treating many colonies with TM-25.

Fungal Diseases

Chalkbrood

Although common in Europe for decades, chalkbrood was first reported in the United States in 1968 and has since spread throughout the country. It is caused by the fungus *Ascosphaera apis* (Maassen ex Clausen) and, although it may reduce honey production, usually will not destroy a colony. Some genetic lines, especially inbred bees, are more susceptible than others, so one control may be to requeen a diseased colony with a new strain of bees.

As a symptom, look for chalky white, mummified larvae in the cells or on the bottom board. The infected larvae are usually removed from their cells by nurse bees. Dried mummies will eventually turn dark gray to black. All these colors of mummies can be found in brood frames and on the bottom board (see the illustration on chalkbrood).

The most susceptible larvae are four days old and those larvae that are chilled, especially drone larvae. The spores of the fungus are resistant to degradation and can be viable for 15 years; spores are transmitted from bee to bee during food exchange, to or from queen bees, and by drifting bees. Contaminated combs and tools will also carry the disease. This fungus is transmitted throughout an apiary by:

- Wind
- Soil
- Nectar, pollen, and water

Chalkbrood Disease

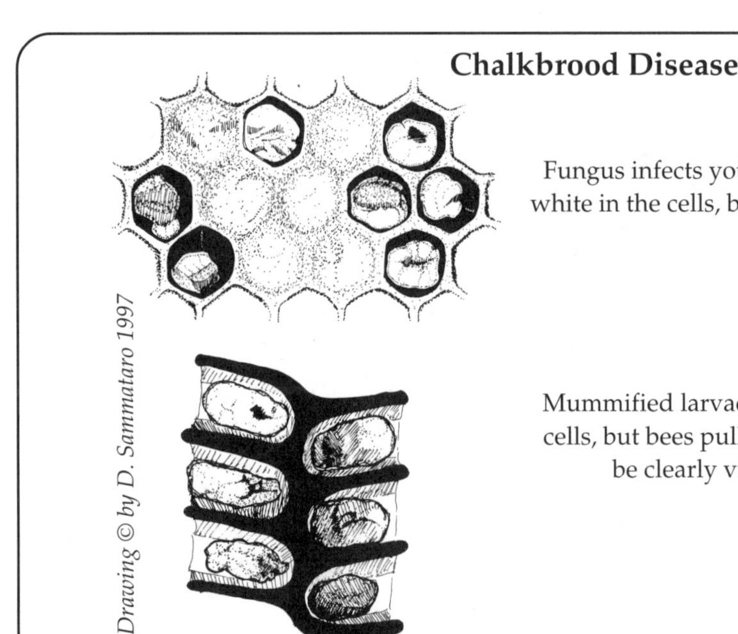

Symptoms:
Fungus infects young larvae, which appear chalky white in the cells, but turn gray and eventually black

Mummified larvae may resemble pollen packed in cells, but bees pull out these mummies, which will be clearly visible on the bottom board

- Drifting bees or diseased robber bees
- An infected queen
- Equipment

While chalkbrood is not normally a serious disease, in severe cases bee numbers can be reduced and honey crops lost. There is no chemical registered for use against this disease, but you can try the following:

- Move hives to sunny location.
- Remove infected combs and burn them.
- Add bees to strengthen the weakened, diseased colony.
- Requeen with hygienic stock.
- Feed syrup and protein supplements to keep the colony strong and healthy (see "Diseases and Pests" in the References).

Stonebrood

Another fungal disease, stonebrood, is caused by several species of *Aspergillus*, a common soil inhabitant. It is a rare disease not often seen by beekeepers and is frequently confused with chalkbrood. The only reliable way to differentiate the two is by laboratory cultivation of the fungal spores.

Viral Diseases of Bees

There are 18 known viruses that cause disease in bees. Different from bacteria, viruses are fragments of DNA or RNA (nucleic acids in a protein coat) that have become detached from the genomes (chromosomes) of bacteria. They are considered nonliving organisms because they lack all the necessary features that would allow them to reproduce on their own. They can reproduce only by altering the DNA of living host cells to manufacture more virus. Therefore, antibiotics, which kill bacterial organisms, do NOT work on viral diseases. Many virus-prone bees may have genetic predisposition to viral infection. Thus the only reliable way to control such diseases is to requeen the colony with nonsusceptible stock.

All bee viruses are probably in or on bees or in the hive in some latent form. Many of them can be activated once they find an entry into the bee's body, via the tracheae or by injury, breaking off hairs or feeding wounds caused by mites. One of the effects of mite predation is the appearance of otherwise benign viruses. As of 1997, six different viruses have been linked with mite infestation.

Sacbrood Virus

Sacbrood is a disease caused by a filterable virus and may be present in conjunction with other brood diseases. This disease is NOT one associated with bee mites but may be carried genetically by inbred bee lines.

SYMPTOMS

The symptoms for sacbrood are as follows (see also the illustration on sacbrood):

- Larvae are darkened from white to yellow; eventually they turn dark brown.
- Larvae have dark head regions.
- Black-headed larvae are bent toward cell center.
- Larvae fail to pupate and die with heads stretched out.
- Diseased larvae are easily removed in liquid-filled sacs.
- Scales (when sacs are old and dried down) are dry, brittle, and easily removed.

Sometimes bees remove these diseased larvae quickly, but if any question exists as to the identification of diseased brood, call your state bee inspector or send a sample to the state or federal bee labs (see "American Foulbrood Disease," above). Because sacbrood is a viral disease, medication is ineffective; requeening the colony may remedy the situation.

Sacbrood Disease

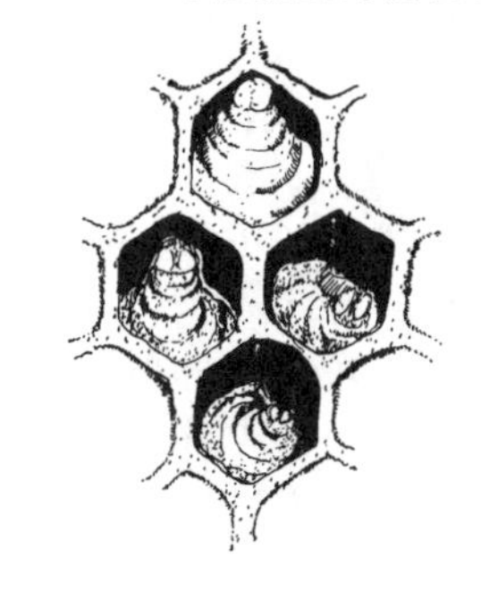

Symptoms: Virus turns larvae dark brown, with a black head

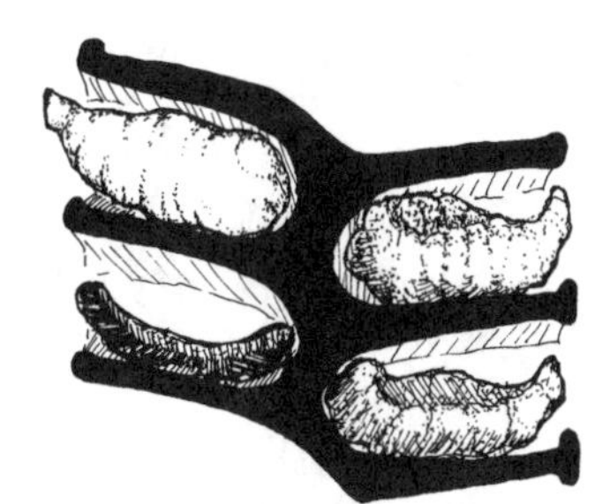

Larvae become fluid-filled sacs, which are easily removed

Drawing © by D. Sammataro 1997

The recently identified black queen cell virus may be associated with nosema disease and is found in commercial queen-rearing operations. The dead queen larvae or prepupae sealed in the queen cells turn the wax walls brown black. The larvae inside the cells are yellow, with a sacbroodlike covering. This virus may be controlled with fumagillin.

Viral Diseases Associated with Bee Mites

Acute bee paralysis (ABP) is common in a noninfective form in seemingly healthy bees, but when present in varroa-infested colonies, this virus kills both adult bees and brood. The presence of the mite intensifies this disease because feeding wounds made by the mites on the bees allow the virus access to otherwise healthy bees, enabling the virus to multiply to lethal levels. When activated in adult bees, it is transmitted to the larvae via the food from nurse bees. It is thought to be the major cause of midsummer bee mortality when varroa populations are high. Originally from the Asian honey bee (*Apis cerana*), ABP is also found in bees without varroa mites. Several strains of the virus exist.

Closely related to ABP, the Kashmir bee virus (KBV) is found in a latent form in bees and pupae. It appears to be activated from a benign to a lethal state when varroa mites feed on bees. Once introduced into bee hemolymph, it can cause mortality within three days. KBV may be associated with other bee pathogens, such as nosema, but this link is still uncertain.

Chronic paralysis virus (CPV) was one of the first viruses to be isolated. Symptoms are very similar to those of colonies suffering from tracheal mites (see below), and this virus may have been the cause of the Isle of Wight disease outbreak in the 1920s. Many of the symptoms are similar to those used to diagnose nosema, amoeba, or the presence of tracheal mites. CPV has two forms, or syndromes: Type I syndrome is recognized by trembling bees that crawl on the ground with dislocated wings (K-wing) and swollen abdomens; it can be associated with dysentery, mite infestation, and other diseases. Type II is also called hairless black syndrome because the bees lose their hair, appear shiny black or greasy, and can't fly but tremble and crawl about. The virus can enter a wound as small as a broken hair, and if bee food (syrup or pollen patties) contains contaminated hairs, the bees will get the virus. Inbred bee races that are susceptible to CPV have also been identified. If your bees come down with viral diseases, switch queen breeders and rear your own queens from different lines.

Other mite-associated viruses are cloudy wing virus, slow paralysis virus, and deformed wing virus. This last virus, usually found in varroa-infested colonies, causes the wings of young bees and pupae to become deformed, an effect often thought to be caused by mite feeding. Mite feeding does cause other bee deformities, however. Many of these viruses went undetected before the introduction of parasitic mites, especially varroa. The best way to control mite-induced diseases is to control the mites.

Mites

Two parasitic mites introduced into the United States in the late 1980s have changed beekeeping practices forever. The result has been the loss of bees (between 25 and 80 percent of bee colonies) and the decline in the number of beekeepers (up to 25 percent) nationwide, as well as the increase in costs to rear healthy bees and to lease colonies for pollination, with an accompanying rise in the price of honey. Parasitic mites currently found only in Asia and other countries have equally devastating potential. We must be careful not to introduce them elsewhere (see the information on other parasitic mites).

Tracheal Mites (Acarine Disease)

Tracheal mites, the causative agent for acarine disease, came into the United States via Mexico in 1984. From its first report, the mite has now been spread by migratory beekeepers and packages into northern states and Canada. Many colonies have died since this mite was introduced, causing losses in the hundreds of millions of dollars as a result of dead bees, honey not produced, and loss of pollination services (see "Mites" in the References). Distribution of the tracheal mite is now worldwide, except in Australia, New Zealand, and Hawaii. Originally found in Europe, its true origin is not known.

The small mite (*Acarapis woodi* [Rennie]) lives inside the thoracic tracheae (breathing organs) of adult bees. A mated female emerges from an old host bee and, by crawling up on the bee's hair, quests to find a newly emerged (callow) bee (see the illustration of the questing mite). Once the mite finds a suitable host, she enters the trachea by means of the spiracle opening and can lay about one egg per day for 8 to 12 days. After the eggs hatch, the immature mites, or larvae, live as parasites inside all castes of adult bees, feeding on bee hemolymph by piercing the walls of the tracheal tubes (see the illustration of stages of tracheal mites). New mites emerge in 11–12 days, if males, or 14–15 days, if females (see the figure on the life cycle). Mites can cause severe bee losses, some-

Other Parasitic Bee Mites (not yet in the Americas)

Euvarroa sp.

- Natural host is *Apis florea.*
- Range is Southeast Asia, Thailand, India, and Sri Lanka.
- Life cycle is similar to varroa, but they live only in drone brood.
- Smaller in size than varroa.

Tropilaelaps clareae

- Natural host is *Apis dorsata.*
- Range is Southeast Asia, Philippines, India, China, and Afghanistan.
- Appearance is elongate and large; color is reddish-brown.
- Males and females are equally large.
- Life cycle is similar to varroa, but brood is essential for survival.
- Larva is mobile and feeds actively.
- These mites do not feed on adults.
- These are a very serious pest in the tropics, but not so much of one in temperate climates.

Questing Female Tracheal Mite

Drawing © by D. Sammataro 1997

times weakening or destroying entire colonies, usually in regions where cold climates confine bees for several months. Tracheal mites cause fewer problems to colonies in warmer regions.

COLLECTING BEES TO TEST FOR TRACHEAL MITES

External signs of tracheal mites are unreliable but include dwindling populations of bees, weak bees crawling on the ground with K-wings, and abandoned hives in the spring with plenty of honey stores. To determine if you have mites, you MUST dissect bees or have them dissected by someone else. If you suspect that your apiary is infested with this mite and an inspector is unable to visit your yard, collect bees to check for mites, using this procedure:

Step 1. Sample at least 50 percent of the colonies in the apiary.

Step 2. Collect only "old" bees; they are most likely to have an infestation and are the easiest to diagnose. Old bees can usually be found on the inner cover, at the entrance, or out foraging, not near the broodnest.

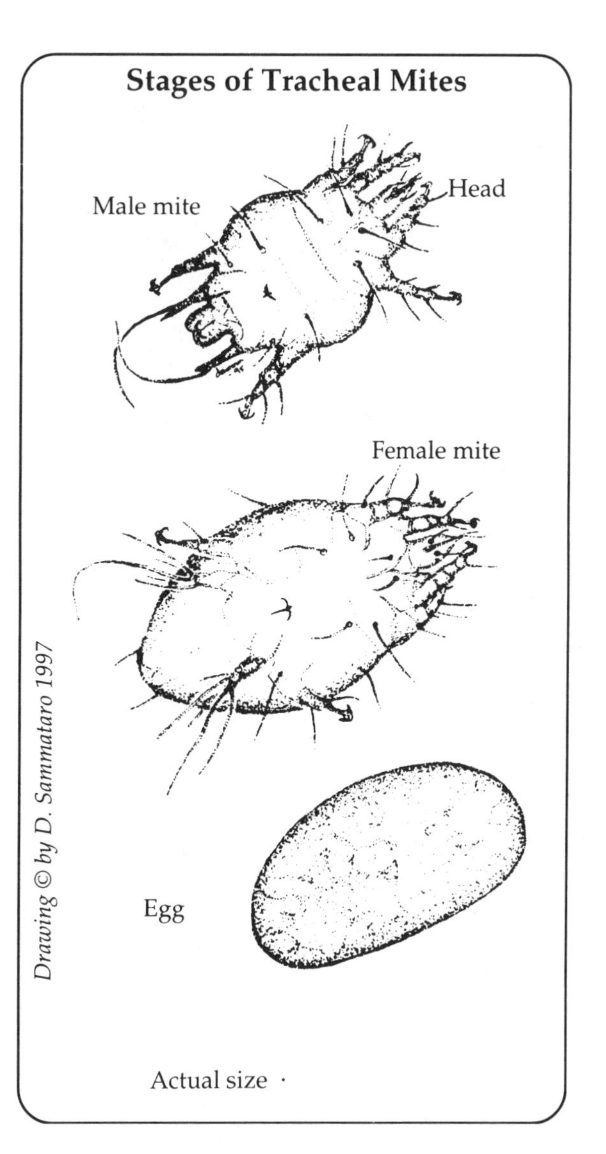

Stages of Tracheal Mites

Drawing © by D. Sammataro 1997

Step 3. Place the collected bees in a 70 percent ethyl alcohol or isopropyl (rubbing) alcohol solution, or freeze them in a glass vial, or plastic jar or bag.

Step 4. Send or deliver these specimens to the state bee inspector, the state entomologist, the Beltsville Bee lab, or a private lab, with the following information: your name, address; location of apiary tested (state, county, township); number of colonies in the apiary; and source of bees (e.g., packages from dealer X).

If sending them through the mail, use bees stored in alcohol. A positive diagnosis of the tracheal mite by gross examination of the colony or as a result of seeing bees walking around on the ground is not reliable. Some of the visible symptoms are the same as those of viral or other diseases and may not necessarily be due to the mite.

DISSECTING BEES

Patience and practice are the most important requirements for a successful dissection. A dissecting microscope (at least 40 to 60×) and a pair of fine jeweler's forceps are also needed. Practice on old summer drones first; they are easy to hold, and their tracheal tubes are larger. If requeening a colony, check your old queen, for she can infest a colony too. Finally, collect some old summer or early spring bees and dissect them. Here's how:

Step 1. Soften a frozen bee by holding it in your hand for a few minutes. If the bee was stored in ethanol, it is already soft enough, but the tissues will be darkened.

Step 2. Place the bee on its back, and pin it through the thorax, between the second and third pairs of legs, onto a piece of corkboard or a beeswax-filled jar lid, or petri dish. You may also hold the bee in your fingers, once you have become accomplished at this procedure.

LIFE CYCLE CHART

Tracheal mite (*Acarapis woodi* [R.]) *Chart © by D. Sammataro 1997*

AGE OF BEE	1 to 3 days old	3 days	8 days	12 days	
	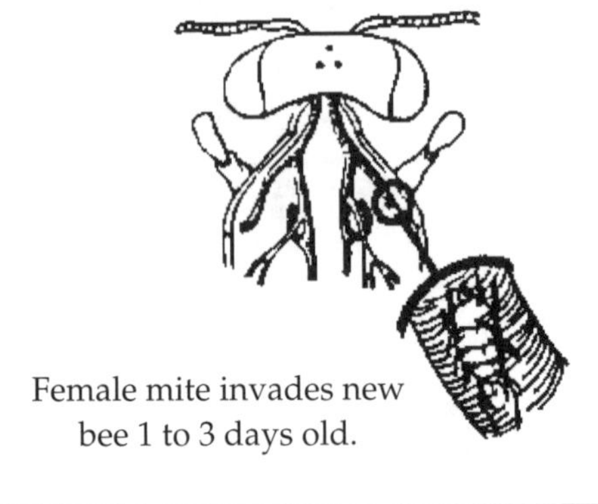			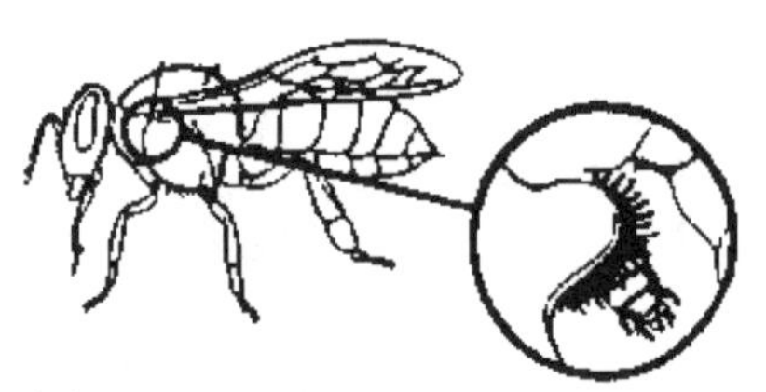	Daughter mites exit old bee, quest on bee hairs, and transfer to a new, young bee host; enter trachea to lay eggs.
	Female mite invades new bee 1 to 3 days old.	Mite feeds and lays about 1 egg per day.	Larvae hatch and feed on bee blood. Adult females hatch in 14 days, males in 12. Mating occurs in the trachea.		

Varroa mite (*Varroa jacobsoni* Oud.)

AGE OF BEE	8 days old	10 days	12 days	18 days	21 days
	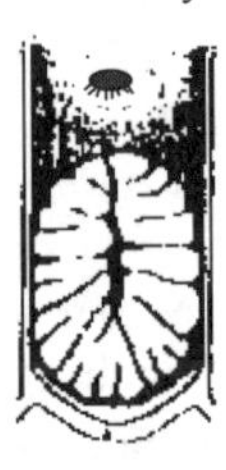		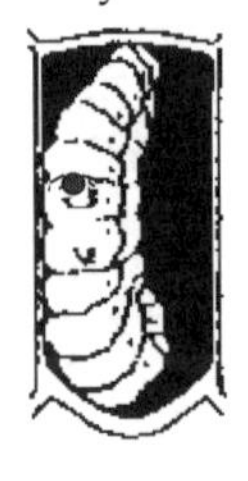		
	Female mite, attracted to the brood pheromones, invades larva before it is capped. Mite will invade drone brood first.	Female foundress mite hides in the bee brood food until cell is capped over.	When bee larva has spun its cocoon, the foundress mite feeds on its blood and begins to lay eggs.	Mite lays up to five eggs, which damage developing bee by feeding on it, allowing pathogens to enter. Mating occurs inside the cell.	Daughter mites exit as injured bee emerges; mites disperse to nurse bees and invade new larvae. Male mite usually dies in the cell.

Dissecting Bees for Tracheal Mites

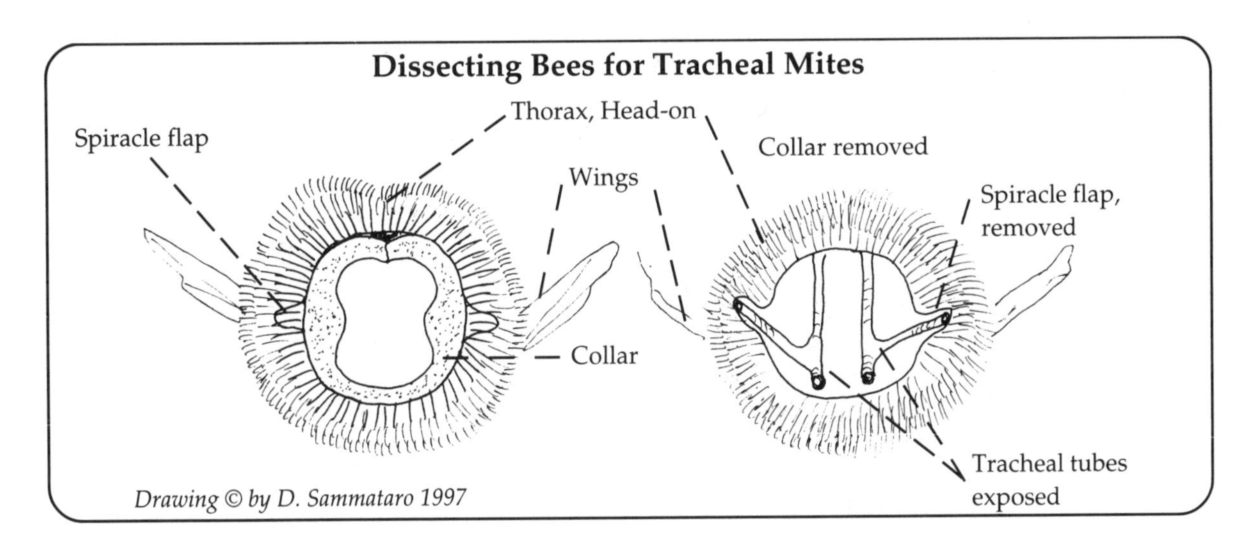

Drawing © by D. Sammataro 1997

Step 3. While looking under the microscope, take off the head, and pull and remove the collar surrounding the thoracic opening with the forceps (see the illustration on dissecting bees).

Step 4. The thoracic tracheal tubes will be exposed when this covering is removed. In a healthy bee, they look like pearly white dryer hoses.

Step 5. If mites are present, the trachea will have shadows or be spotted—the spots being all ages of mites. In severe infestations, the tube can be completely brown or black.

Step 6. Darkened tracheae will be visible to the naked eye, whereas healthy trachea will be white and shiny. You can use this method to detect heavy infestations (in the spring and fall) but not light ones, such as in the summer.

CONTROLLING TRACHEAL MITES

Chemical. Menthol, from the plant *Mentha arvensis*, is sold in crystal form (98 percent active ingredient) at many bee supply stores, and each two-story colony takes 1.8 ounces (one 50-gram packet) of these crystals. The problem with menthol is that it is temperature dependent. Menthol vapors will sometimes make bees leave the hive if the outside temperature is too hot. Conversely, the crystals will be ineffective if the outside temperature is too cold, because not enough vapor would be released. The packet should remain in the colony for at least two weeks. Remove all menthol at least one month before the surplus honeyflow, to keep honey from becoming contaminated. As for all treatments, follow label directions.

An alternate method is to use vegetable oil patties; a vegetable shortening and sugar patty kept in the colony all the time seems to protect against these mites. Some research shows that TM extender patties (those containing the antibiotic) could also be beneficial to bees by helping them overcome mite-vectored pathogens, but these patties should be placed only on overwintering colonies; see the chart on making oil or TM patties (p. 133).

Place one ¼-pound oil/sugar patty, about the size of your hand, on the top bars at the center of the broodnest. The bees at most risk are the young nurse bees, found in the broodnest. The patty should last about one month; after that, replace it with another oil patty. If some colonies appear to remove the patty much more quickly, they may be displaying hygienic behavior; try rearing queens from these colonies. Because young bees are continually emerging, it is important to have the patty present in the colony for an extended time. The best time to treat is when mite levels are climbing—fall and early spring (see the figure on the sequence of treatment times.).

Cultural Practices to Reduce Tracheal Mites. Used with menthol or oil patties, some of these techniques seem to help reduce the number of tracheal mites in a colony:

- Requeen with resistant bee stock (e.g., Buckfast)
- Reduce numbers of foragers and drones in the fall (they are heavily infested) by moving hives and destroying field force with soapy water.
- Place TM extender patties on colonies in fall and spring; keep oil-only patties on during summer.
- If the colony is highly infested, split the colony, kill older foragers, requeen, and treat with menthol crystals or oil patties.
- Keep bees healthy with plenty of pollen and honey stores, and provide food supplements if needed.

Varroa Mite

First discovered in Java and described in 1904, the varroa mite (*Varroa jacobsoni* Oudemans) was originally confined to Asia on *Apis cerana*, the Asian honey bee. As a result of moving mite-infested bees, varroa has spread worldwide since the late 1950s, except to Australia, New Zealand, and Hawaii.

In 1986, varroa was first reported in the United States and is now one of the major killers of bee colonies. Adult female mites attach themselves to adult bees and are thus inadvertently carried to other, uninfested colonies or apiaries. The movement of the African bees north from Central America, as well as the practice of shipping bees in the mail, has accelerated the mite's spread throughout the United States. The website http://www.snre.umich.edu/~sarhaus/image/animap2.gif has an animated map of varroa's spread.

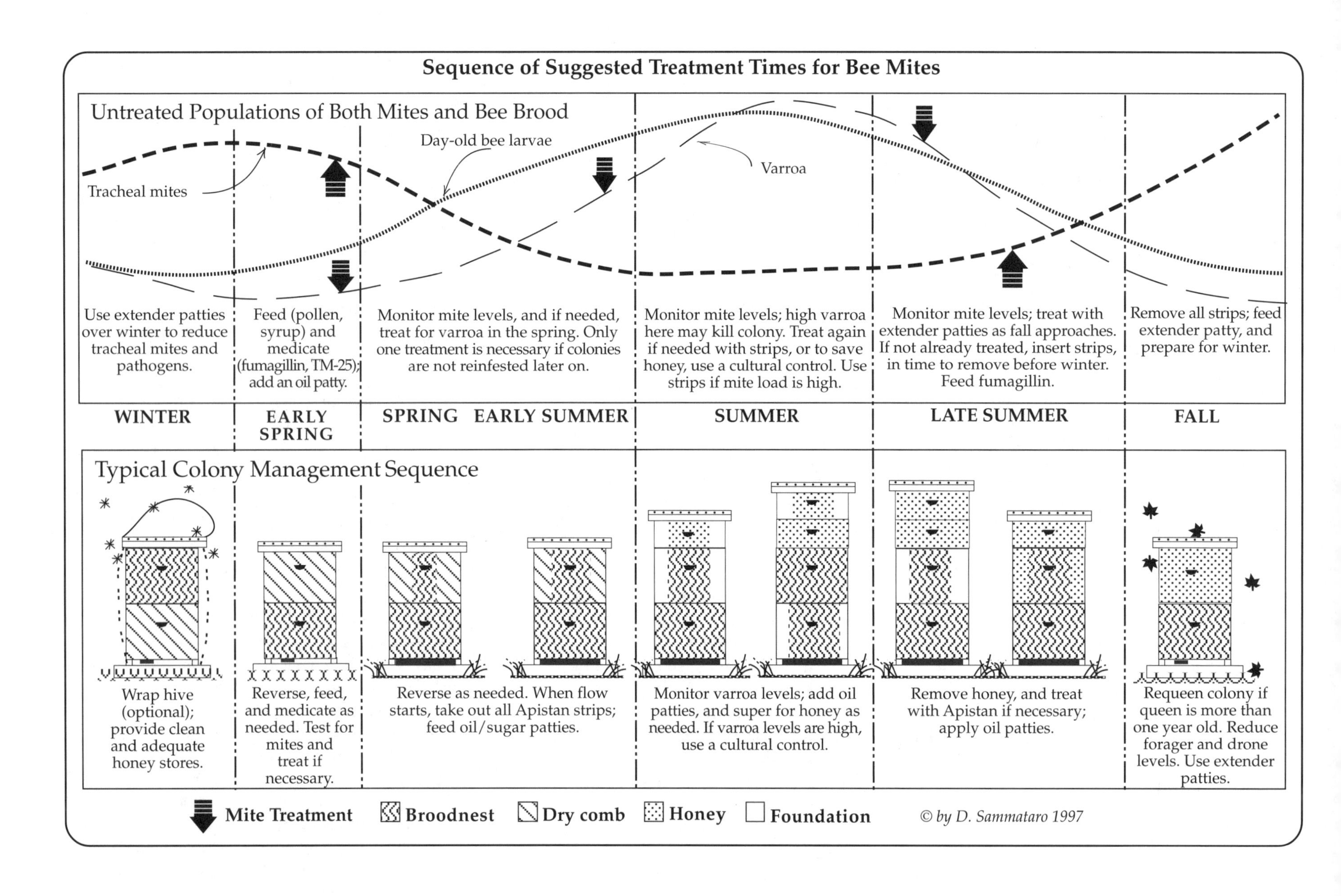

Only adult female mites are found on adult bees, where they feed on bee hemolymph by piercing the soft tissues of bees between abdominal segments or behind the head. Adult mites are about the size of a large pinhead and can be seen, after close examination, with the unaided eye (see the illustration of the varroa mite).

Varroa must complete its life cycle on bee brood. Females are attracted to the odor of the drone brood pheromone—but they will also invade worker brood if there is not enough drone—and enter prepupae as the cells are about to be capped.

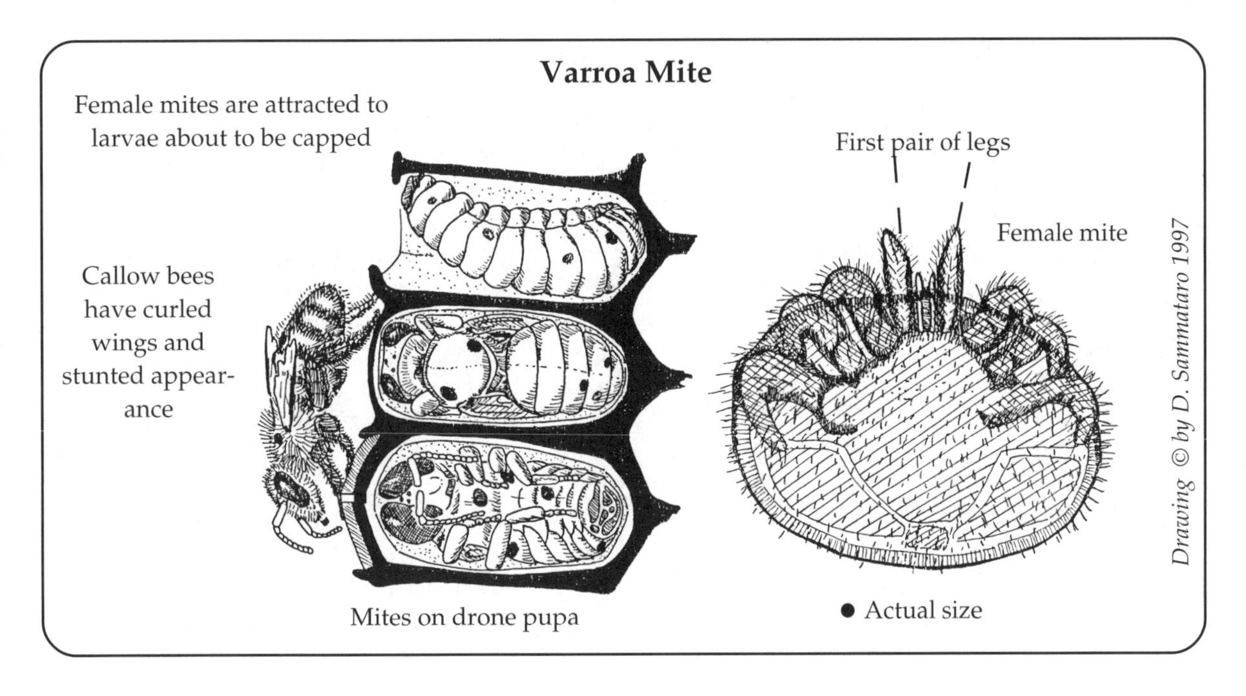

There the mites hide at the bottom of the cell and emerge from the jelly after the cell is capped. The mite will lay her eggs on the pupa, and young mites (nymphs) feed on the hemolymph of the forming bee. Daughter mites mate with their brother in the cell, after which the new females will emerge with the callow bee.

Most times the young bees, if not killed outright by the stress caused by feeding mites, are usually deformed by virus and soon die. The new females will live for a time outside on other bees (this is the phoretic state), until they invade new brood to repeat the cycle.

As of 1997, the only chemical control for varroa is Apistan strips, a plastic strip (like a flea collar) impregnated with the pesticide fluvalinate. Do not be tempted to use other materials; some chemicals can become incorporated into the honey and wax, making them unfit for human consumption or sale. Contaminants in wax will vaporize and become inhaled when burning beeswax candles.

SYMPTOMS

Symptoms of varroasis are many and can be confused with those of other diseases or situations, such as pesticide poisoning. Look for:

- Infested capped drone or worker brood; cappings can be punctured, as in foulbrood disease.
- Disfigured, stunted adult bees, with deformed legs or wings or both.
- Bees discarding larvae and pupae.
- Pale or dark reddish brown spots on otherwise white pupae.
- Spotty brood pattern and the presence of diseases.
- Uncapped cells.
- General malaise of a colony, with symptoms of multiple diseases (AFB, EFB, and sacbrood, to name a few).
- A dead colony in the early fall, right after honey has been harvested.

DETECTING VARROA MITES

There are three basic techniques you can use to detect varroa mites. It is important to be able to test your colonies periodically to determine which treatment you can use. Because Apistan strips cannot be used during the honeyflow, you may need to treat a colony using an alternate method to keep the mite levels low enough for the bees to collect honey.

Cappings Scratcher (with forklike tines).

Step 1. Pick a frame of drone brood or a large patch on several frames.

Step 2. Hold the fork parallel to the comb and insert the tines into the top third of the cappings.

Step 3. Pull the capped drone pupae straight up or lift up the handle end of the fork, leaving the tines on the comb, until the drone pupae are pulled out of their cells.

Step 4. Examine the pupae carefully. A heavy infestation is at least 2 mites per cell; a moderate level is 5 mites per 100 pupae. The mites are clearly visible: females are reddish brown and look like ticks on the white pupae. Immature mites are white or light brown.

Ether Roll.

Step 1. Collect approximately 100–300 bees in a wide-mouthed jar with a lid.

Step 2. Scrape bees (don't get the queen) into the jar; you can also modify a small car vacuum to collect bees.

Step 3. Knock bees to the bottom of the jar with a sharp blow; there should be about a 1-inch layer of bees on the bottom (see the illustration of testing for varroa).

Step 4. Remove lid, spray a 2-second burst of ether starter fluid into the jar, and replace the lid immediately. Alternatively, you can add enough 70 percent alcohol or soapy water to cover the bees.

Step 5. Agitate the jar for a minute to dislodge the mites from the bees, and then roll the jar for about

10 seconds; mites should stick to the sides of the jar. If soapy water or alcohol was used, shake the jar for about three minutes, and strain out bees using a coarse hardware cloth strainer. The mites will be in the liquid, which can later be strained through a coffee filter.

Smoke or Miticide Strips and a Sticky Board.

Step 1. Place a sticky board in the bottom of colony; you can make a board with cardboard or other stiff paper coated with Vaseline. Cut paper to fit on the bottom of the bottom board, and cover with a piece of 8-mesh hardware cloth stapled to a ¼-inch high wooden lath frame to keep bees off the board (see the illustration of testing).

Step 2. Smoke the colony with a smoker that contains 1 ounce of smoldering pipe tobacco.

Step 3. Puff bees 6–10 times; close up hive for 10–20 minutes.

Step 4. Pull out the sticky board and count mites.

You can also use miticide strips instead of smoke; insert one strip per five frames of bees for one to three days with the sticky board in place. Or, put in just the sticky board overnight to collect the natural downfall of mites. Generally, an overnight fall of over 50 mites may indicate time to treat all colonies.

These techniques will tell you, with some degree of accuracy, if you have varroa. The question is, do you treat if you see only 1 mite or if you count over 100 mites? This is a judgment call. If you see mites early in the spring, put in your strips for the recommended time. If you see mites during a honeyflow, when you cannot treat with the strips, try a few of the cultural controls to reduce the number of mites, then treat with strips later. Consult the figure in this chapter on the sequence of treatment times, and keep up with journal articles on the progress of mite control.

Testing for Varroa

Sticky Board and Apistan Strips or Tobacco Smoke

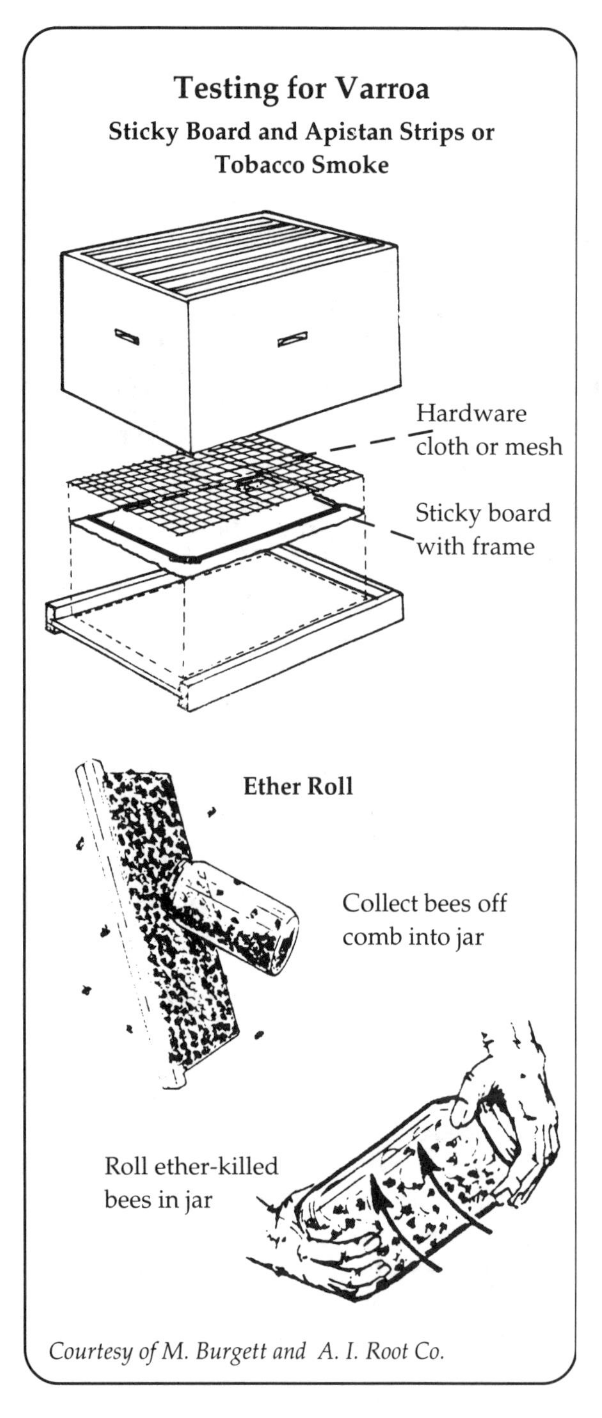

Courtesy of M. Burgett and A. I. Root Co.

TREATMENT FOR VARROA

Miticide strips. Currently, there are two treatments to control varroa: Apistan strips and CheckMite+. Insert these strips according to label directions, when you have no honey supers on the colony. Here is how to use the strips:

- Read and understand label instructions before using any strips.
- Wear *new* chemical resistant gloves when handling strips.
- Use one strip per five frames of bees; this kills only varroa mites.
- Keep strips in the colony according to the label.
- Discard used strips; do not reuse.

Test your bees to see if you have mites. Use strips once in the spring and once in the fall if your bees are in a heavily infested area. Timing of treatment is very important; refer to the figure on the sequence of treatment times (p. 140).

Tobacco smoke (or wheat flour sprinkled on adult bees) knocks mites off bees, thus lowering their numbers. Use this if you have supered for honey. Smoke the colony heavily, or dust bees lightly with flour (being careful not to get flour into uncapped brood cells). Insert sticky boards to catch the mites. This is an emergency treatment only, for when you cannot use strips. Repeat in one week, because it will not affect mites in brood cells.

Formic acid. Formic acid is now available to use for mite control. Do not use the liquid form, as it is highly caustic and can burn. Use the slow-release gel packs, available from some suppliers. It acts as a fumigant, killing both types of parasitic mites. It is also toxic to bees if not applied correctly, and it is EXTREMELY CAUSTIC TO HUMANS.

IPM Mite Management. An integrated pest management (IPM) program for mites and bees uses several control measures. They include:

- Restricting brood rearing by caging the queen and removing capped brood or requeening the colony with queen cells. This breaks the mite-brood cycle and allows the bees to clean out diseased brood.
- Trapping mites in drone brood and freezing the

frames, which reduces the mite population temporarily and can be used successfully if a frame of drone foundation is inserted in each deep super once a month and removed.
- Scraping off mites by means of a pollen trap, or trapping mites on a sticky oil strip.
- Essential oils (botanical oils of essence), which are showing promise but are still in the experimental stage. Keep current with the bee journals for more information on new controls.

Bee Parasitic Mite Syndrome

First reported by European beekeepers whose colonies were stressed by varroa mites, bee parasitic mite syndrome (BPMS) was coined by researchers at the Beltsville Bee Lab in Maryland (see "Bee Laboratories" in the References) to explain why colonies infested with both varroa and tracheal mites were not thriving. BPMS may be connected to the vectoring of virus by both mites (such as acute bee paralysis virus). The symptoms of BPMS can be present at any time during the bee season and include:

- Varroa mites present in colony (tracheal mites may be present too).
- Crawling bees on ground, with deformed wings (Deformed Wing Virus).
- Queens superseded more than normal.
- Spotty brood pattern.
- Foulbrood and sacbrood symptoms present.
- Diseased brood in all life stages of bees.
- Lowered adult bee population.
- AFB symptoms present, but no ropiness, odor, or brittle scales.
- No predominant disease bacteria found.

Although not much is known about BPMS, these treatments have been effective:

- Feed colonies with TM in syrup; feed with fumagillin.
- Treat for varroa with miticide strips.
- Treat for tracheal mites with vegetable oil patties.
- Feed pollen supplements.
- Use resistant bee stock (e.g., Buckfast for tracheal). Hygienic behavior seems to be beneficial for varroa mite control, and some bees appear to pick off the mites. Select colonies that seem to have some tolerance for mites, and breed queens from them (see Chapter 10).

Integrated pest management (IPM) is a strategy whereby one uses multiple tactics such as requeening and chemical and soft controls to "manage" mites rather than eliminate them (an impossible task), provided they do not damage the hive. It is not possible to kill all mites in a hive and all mites in a region. The presence of some mites in the hive does not detract from hive health, provided the colony is strong and mite numbers do not get out of hand.

Major Insect Enemies

Small Hive Beetle (see page 148)

Wax Moth

First reported in the United States in 1806, this pest was probably introduced with imported bees. The female greater wax moth (*Galleria mellonella* L.) is from ½ to ¾ inch (1.3 to 1.9 cm) long, is gray brown (color varies somewhat), and holds her wings tentlike over her body instead of outstretched, as would a butterfly (see the illustration of a wax moth). The wax moth is thought to have evolved with honey bees from Asia and commonly inhabits nests of all honey bee species.

This moth deposits eggs in cracks between hive parts or in any other suitable place inside the hive. After hatching, the larvae are quite active, moving up to 10 feet in optimum conditions, to infest other hives, where they tunnel into the wax combs, hiding at the midrib to keep from being discovered by house bees. The dark wax of brood combs contains the shed exoskeletons of bee larvae and some pollen, both of which are highly attractive to wax moth larvae. The larvae can grow to 1 inch long (2.5 cm) in from 18 days to 3 months, depending on the temperature. As these larvae tunnel along, silk strands mark their trails through the combs (see the illustration). Before pupating, the larvae fasten themselves to the frames or inside walls, inner covers, or bottom boards of the hive and spin a large silk cocoon, sometimes damaging the hive by chewing into the wooden parts. Left untended, wax moths can destroy weak hives within one season. Symptoms of wax moth damage are:

- Tunnels in combs.
- Silk trails, crisscrossing one another over combs.
- Small dark objects (excrement of wax moth larvae) in the silk trails in a hive.
- Silk cocoons attached to wooden parts.
- Destroyed comb; piles of debris on bottom board.

To control wax moths, use these methods:

- Maintain strong colonies (the best defense against wax moths).
- Store empty combs in cold places; cold temperatures will slow down the rate of growth and deter adult moths from laying their eggs.
- Freeze combs (like comb honey) at 20°F (−7°C) for 4½ hours; at 10°F (−12°C) for 3 hours; or at 5°F (−15°C) for 2 hours. If treating much comb honey, freeze for at least 24 hours.
- Store empty combs with moth crystals (Paradichlorobenzene, or PDB) when air temperature is above 60°F (16°C), but air out at least 24 hours before using.
- Fumigate dry combs with a mixture of 74 percent carbon dioxide (CO_2) and 21 percent nitrogen (N), at 50 percent relative humidity, at 100°F (38°C) for 4 hours; 115°F (46°C) for 80 minutes; or 120°F (49°C) for 40 minutes. Be careful, as beeswax melts at 148°F (64°C).

A cultural practice in the South is to let fire ants (*Solenopsis invicta* Bunen) kill wax moths, if combs are left near an active ant nest. But use caution, be-

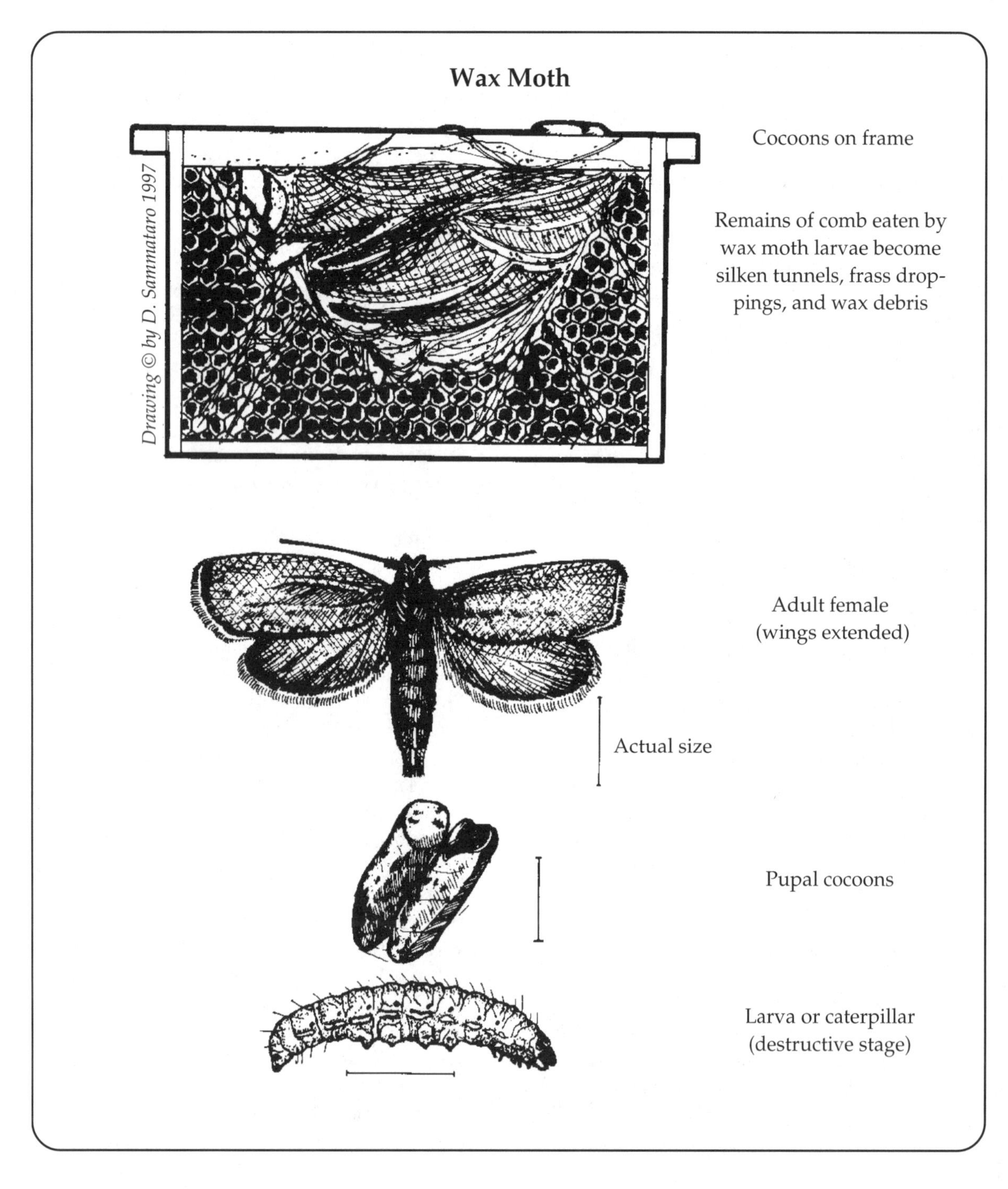

cause these ants also kill bee colonies and sting beekeepers.

Wax moths are naturally beneficial because they destroy diseased combs of feral colonies; they are also a valuable commodity, used as fish bait and pet food for reptiles and other exotic animals. They can be reared off beeswax, using baby cereal, glycerin, and honey and are often a secondary business to many beekeepers (see Appendix H).

The lesser wax moth (*Achroia grisella* Fabricius) does similar damage to wax comb. But unless the infestation is great, the damage is minor compared with that of the greater wax moth.

Africanized Bees

Recently, Africanized bees have become a serious pest in some areas of Texas, Arizona, and California. If you are living in an area that is subject to this invasion, make sure you can differentiate European honey bees (EHB) from the Africanized bees (AHB). In general, AHB are smaller than EHB, faster moving, more aggressive, and more apt to abscond and swarm (see Appendix G).

SOCIAL PARASITISM

Africanized honey bees have been reported to usurp a weak European colony, mating nuc, or colony recently stressed by beekeeper manipulation. A small cluster or swarm of AHB will land nearby, and the workers will enter the weaker colony, killing the resident queen. Once she is dead, the AHB queen will enter and resume her duties. Still being studied by researchers, such activity illustrates an important reason to keep good records of your queens, especially if you live in areas bordering those with AHBs; you will quickly be able to tell if your colonies have become Africanized.

Animal Pests

Skunks and Raccoons

Skunks (family Mustelidae), which includes weasels and badgers, and raccoons (family Procyonidae) are serious pests to bees, often visiting hives in the early evening as well as during the day. They can cause damage to both equipment and bees and dig up the beeyard looking for food to eat. By scratching at the entrance, skunks entice bees to come out of the hive, and as the bees crawl out, the skunk eats them. Skunks even teach their young that hives are a good place to get some tasty snacks, thus depleting hive populations drastically; your apiary could become decimated quickly if you do not take some countermeasures to protect colonies. Skunks also feed on bumble bee colonies. Raccoons often take and scatter anything loose in the apiary, including feeder jars and frames of brood or honey that have been left out. Some raccoons can lift off the covers of hives. Depending on the severity of winter in your area, these pests could be nearly a year-round problem.

In areas that have a lot of coyote and coatimondi (usually found in southwestern U.S. deserts), it has been reported that these creatures can get into bee-hives in search of liquid and food. In this case, heavy rocks placed on top of the hives deter such pests.

Signs of their visits are:

- Defensive bees.
- Grass near hive entrance is torn up.
- Scratch marks on the hive front or on earth at the hive entrance.
- Weak colonies for no other apparent reason.
- Scat and droppings near hives.
- Area near entrance is muddy after a rain.

Discouraging and eliminating these pests may be accomplished by:

- Using hive stands, at least 18 inches high, to keep bees out of reach (see "Hive Stands" in Chapter 4). This is the best and easiest way to eliminate skunk predation.
- Sprinkling rock salt crystals on the ground around the hive. Although this method may deter these pests until it rains, it will also kill vegetation around the hives.
- Placing paradichlorobenzene (moth ball crystals) in jars with holey lids approximately one yard from the hives; the crystals need to be replaced and kept out of the rain.
- Trapping skunks, which may be illegal in your area and will cause the skunks to discharge, not making you popular with your neighbors. You can live-trap raccoons and move them to another area; use cat food or marshmallows as bait.
- Killing skunks and raccoons in their lairs, if you can find them (this may be illegal).
- Using poison baits; this method is not recommended because it is not selective enough and can harm other animals. Before killing any animal or using poison bait traps, contact your state game and wildlife departments, and comply with regulations for controlling fur bearers.
- Placing a strip of carpet tacking, nail side up, on the landing board; this does not always discourage the skunks, who many times pull it out.
- Extending a piece of hardware cloth in front of the entrance, which will allow bees to sting the skunk's belly; this is a good temporary measure. Make sure it is securely fastened to the bottom board, or the skunk will tear it off.

Bears

Bears (*Ursidae*) eat brood and honey and do extensive damage to equipment, especially in Canada, where large bear populations exist. However, bears are now found in almost all states in the continental United States and they are capable of destroying apiaries. Signs of bear damage are overturned hives; smashed hive bodies; frames scattered over the apiary; and entire supers that have been removed from the apiary and scattered 30–50 yards away.

An electric fence around the apiary is probably the only effective control against this large animal. Raised platforms where hives are kept are extremely inefficient and difficult to work. Locating apiaries away from bear routes may help, because these animals keep to knolls, forest edges, and stream banks. Do not leave combs or hive debris around an apiary; such material will attract not only bears but other pests as well. Paint hives to blend into the background.

Alternative ways to reduce bear damage include moving bees to a new location and seeking the assistance of local conservation departments.

Mice

Mice (*Mus*) are the most damaging animals to bee hives, next to vandals. They enter hives in the fall and winter and, although they appear not to harm the bees, can cause extensive comb and woodenware damage. They may destroy weak colonies by feeding on pollen, honey, wax moth larvae and cocoons, bee brood, and bees. Their droppings and urine are another irritation that often disrupts cluster behavior, especially if the colony is weak. Mice are often found in the winter packing.

Because mice carry viral diseases (Hanta) and vermin (fleas) that may affect humans, keeping them out of bee hives is important to your health. If you find mice in equipment stored in your bee house, make sure your room is well ventilated before sweeping it up, and clean the hive bodies outdoors. Signs of mouse damage are chewed combs or wood; droppings on the bottom board; holes chewed in entrance reducers, thus enlarging the opening to enable mice to enter; and nesting materials (grass, paper, straw, cloth, or such) in hives, usually among the comb.

Colonies at forest edges and in fields of tall grasses are especially at risk. The following measures may help to control damage from mice:

- Place hives on stands (although mice can climb).
- Use entrance reducers; some beekeepers line them with metal sheeting to keep mice from chewing the wood.
- In the fall, close the entrance with 4-mesh hardware cloth or metal mouse guards. Mice can squeeze through a space that measures ⅜ × 3 inches (1 × 8 cm). Some beekeepers drill several ⅜-inch holes in deep supers to help with winter ventilation.
- Keep weeds down around hives.
- Place poison grain on bottom boards or around the base of the hive. This measure is not recommended because its effect is not selective. The only safe use of poison bait and traps is in an enclosed space where extra equipment is stored, such as your honey house.

Vandals (*Homo sapiens*)

There has been an increase in the number of hives stolen or otherwise vandalized in recent years, which makes vandals an important vertebrate pest of bees. The increasing demands and increased value of equipment, honey, bees, and hives for pollination services have contributed to the prevalence of thieves. Furthermore, colonies are also vandalized by the curious who think that they will be able to obtain free honey simply by opening up a colony. Those bent on mischief can overturn or otherwise damage hives.

Vandals can be discouraged by placing apiaries near year-round dwellings. If it is not possible to place them near one's own residence, land can often be rented from a homeowner for a few pounds of honey each year. Branding your hive bodies and frames is good protection. If your hives are stolen, for example, and the bee inspector finds your brand on hives in some other yard, the person responsible for the act is more likely to be apprehended and your equipment returned to you (see Chapter 3 for other ideas on personalizing your hive furniture).

Instead of painting your hives white and placing them in open, highly visible areas, try going to paint stores and getting cans of premixed colors that other consumers have returned. Mixing these together often results in a nice, mud-colored paint that makes your hive boxes disappear into the background. Tree hedges or judiciously placed shrubs to create screens around the yard also help to discourage would-be thieves. So does fencing with a locked gate!

Minor Insect Enemies

Although bees are often preyed on by other insects and spiders, such predators usually do not have any appreciable effect on a colony's well-being. In some areas, however, any of these predators might be a serious problem. Spiders (Araneae) also prey on adult bees; some species even wait for bees to arrive at a flower before attacking them. The most common types of spiders that would catch a bee are the orb weaver, grass, jumping, and house spiders.

Although not a serious problem in the temperate climates, in the subtropical areas ants (Formicidae) are a serious pest, and hives must be placed on top of greased posts or in oil-filled cans to keep out marauding ants. The more harmful ones in North America include the Argentine ants (*Iridomyrmex humilis*), a newer arrival from South America; fire ants; and carpenter ants (*Camponotus* spp.). Ants can be controlled by keeping the apiary free of weeds, debris, and rotting wood and by placing hives on stands and painting the legs with oil. For more serious infestations, poisons may be needed to control ants; consult your local extension agents on what to use.

Other ants, earwigs, and cockroaches may use various hive parts, especially the inner cover, as a shelter or nest. Earwigs (Dermaptera), found on top of the inner cover, may be annoying to bees. Keep vegetation mowed around hives. Nematodes, small worms, may also live on bees but are not really serious. Termites (Isoptera) can damage hive parts, especially those on the ground. As before, keep your apiary mowed and hives on stands and many of these problems will disappear.

To control these insect pests, store equipment in cold or freezing temperatures. NEVER use insecticides or pest strips in stored equipment or a storage area: these will also kill bees if absorbed by the wax combs.

Insect predators of bees are numerous but do not typically pose a serious threat to colonies, and no control measures are needed. Your bees may be caught by:

- True bugs (Hemiptera), including assassin bugs (Reduviidae) and ambush bugs (Phymatidae), which eat insects.
- Robber flies (Asilidae).
- Mantids (order Mantodea).
- Hornets and wasps (Vespidae). These may be a problem in the fall or if colonies have died from pesticides or mite predation. Wasps, hornets, and yellow jackets will clean out hives of dead bees, feeding on the dead insects, brood, pollen, honey, and even wax moths.
- Dragonflies (Anisoptera) and damselflies (Zygoptera).

Other insects live on the stored products in a colony or on the insects that eat the stored products, and they can be a problem if there are many dead colonies. These insects are:

- Moths: dried fruit moth (*Vitula edmandsae*) and Indianmeal moth (*Plodia interpunctella*).
- Beetles (Coleoptera), which may live inside a hive, eating debris and litter found there. The most common ones are dermestid beetles (Dermestidae), weevils (Curculionidae), sap beetles (Nitidulidae), and scarab beetles (Scarabaeidae);

Poisonous Plants

Name	Toxic Part
Abies alba, silver fir	some aphid's honeydew
Aconitum spp., monkshood	honey/pollen
Aesculus californica, California Buckeye	honey/pollen
Andromeda spp., andromeda	honey/pollen?
Arbutus unedo, strawberry tree	nectar
Astragalus spp., locoweed, tragacanth	nectar
A. miser v. *serotinus*, timber milk vetch	nectar
Camellia reticulata, netvein camellia	nectar
Coriaria arborea, New Zealand tutu	honeydew
C. japonica	honeydew
Corynocarpus laevigatus, New Zealand laurel or karaka	nectar
Cuscuta spp., dodder	nectar
Cyrilla racemiflora, southern leatherwood, titi	nectar
Datura stramonium, jimsonweed	honey/pollen
D. metel, Egyptian henbane	honey
Digitalis purpurea, foxglove	pollen
Euphorbia segueirana, spurge	honey/pollen
Gelsemium sempervirens, yellow jessamine	nectar/pollen?
Hyoscyamus niger, black henbane	nectar/pollen
Kalmia latifolia, mountain laurel	nectar
Ledum palustre, wild rosemary	honey
Macadamia integrifolia, macadamia	cyanide gas from bloom
Nerium oleander, oleander	honey
Papaver somniferum, opium poppy	pollen
Ranunculus, buttercup	nectar/pollen
Rhododendron spp., azalea and *R. anthopogon*, *R. ponticum*	nectar/pollen
Sapindus spp., soapberry	nectar?
Senecio jacobaea, tansy ragwort	nectar
Sophora microphylla, yellow kowhai	nectar
Stachys arvensis, nettle betony, staggerweed	nectar?
Taxus spp., yew	pollen
Tripetaleia paniculata, an azalea	nectar
Tulipa spp., tulips	stigma nectar?
Veratrum spp., Western false hellebore	nectar
Zygadensis (= Zygadenus) venenosus, death camas	nectar/pollen

the last two eat stored pollen. Some are predators, such as ground beetles (Carabidae), or parasites, like the blister beetles (Meloidae), which eat or parasitize live bees.

Certain flies (Diptera) bother bees at times but are mostly considered a minor nuisance unless their natural prey is unavailable. Some flies are predators, but others are opportunists, found in colonies that died of other causes. Others can parasitize bees, but these are found mainly in tropical climates.

The following have been noted in the literature as being pests of bees:

- Humpbacked flies (Phoridae), blow flies (Calliphoridae), thick-headed flies (Conopidae), flesh flies (Sarcophagidae), and tachinid flies (Tachinidae).
- The bee louse (*Braula coeca*) eats food at the bee's mouth. This fly, which looks like a varroa mite, except that it has six instead of eight legs, may reach damaging levels in some regions. It can be controlled by using miticide strips as you would treat for varroa.

Miscellaneous Minor Pests

Although many birds are insectivorous, few, if any, eat bees in large quantities in North America. Bee-eaters, common in Asia, Africa, and Europe, can decimate apiaries and can eat many virgin queens on mating flights. In North America, flycatchers and kingbirds feed on bees, and woodpeckers can damage old, abandoned equipment. But you should make no attempt to control birds by poisoning or shooting them, which is illegal.

Other minor pests that could be of major concern in some areas include frogs, toads, lizards, squirrels, opossums, rats, and shrews. Livestock will knock over hives if they are not otherwise protected in pastures; and remember, horses and bees do not mix.

There are many other mites (Acari), such as the

pollen mite, which feed on stored pollen. And various other mites feed on one another, hive debris, or fungi that live inside a hive. They are generally harmless to bees and may even serve a beneficial function as garbage collectors.

Poisonous Plants

Sundew (Droseraceae), Venus flytrap (*Dionaea muscipula*), and pitcher plants (Sarraceniaceae) are insect-eating plants that attract their victims by secreting a sweet sap or odor or both. These plants grow in wet areas and are not usually attractive to bees; the number of bees lost to them in minimal.

The list on page 147 summarizes information on toxic plants. Certain environmental conditions, such as abnormally cold or dry weather, may cause otherwise nontoxic plants, such as linden (*Tilia*) trees, to yield toxic nectar or pollen or both. Furthermore, wild bees, such as bumble bees, can collect these substances that are toxic to honey bees without adverse effects.

Small Hive Beetle (*Aethina tumida*)

The small hive beetle (SHB) was first identified in Florida in the spring of 1998. Native to tropical or subtropical areas of Africa, it is not known how it came to the U.S. The beetle is not considered a serious pest in South Africa, but in Florida beekeepers have seen the quick collapse of strong colonies. As of February 2001, the beetle has been found year-round in apiaries in Florida, Georgia, and North and South Carolina. They have also been found along most Eastern states, to Maine, and in some Mid-Atlantic states but do not seem to be a serious problem.

Description

The adult beetle is small (about 1/3 the size of a bee), reddish brown or black in color, and covered with fine hair. The larvae are small, cream colored, and similar in appearance to young wax moth larvae. Beetle larvae can be identified because they have three sets of legs just behind the head. Wax moth larvae have three sets of legs behind the head plus a series of paired prolegs, which run the length of the body.

Life Cycle

Female beetles lay large egg masses on or near the combs, which hatch in a few days into larvae. Beetle larvae consume pollen and wax, honey, bee eggs, and larvae. After 10 to 16 days the larvae crawl out and drop to the ground, where they pupate in the soil. They require sandy soil for this stage in their life. Adults emerge from the soil in approximately 3 to 4 weeks, and the females can lay eggs about one week after emergence. Beetles are good flyers and easily disperse to new colonies, where they deposit eggs to begin a new generation. In northern states, beetles completely shut down reproduction during winter and overwinter in the bee cluster.

Damage

The SHB has the potential to be a pest of significant economic importance in areas where it overwinters. Whether or not it can successfully establish itself in temperate regions or in areas without sandy soil is not yet known. In addition to consuming the resources of the colony, the adult beetles defecate in the honey, causing it to ferment and run out of the combs. Full honey supers stored in the honey house or on hives above bee escapes and weak hives with honey but few bees seem most vulnerable to attack. When SHB infestations are heavy, even in strong colonies queens will stop laying eggs and the bees may abscond.

Detection

Beetles can be seen running across the combs to find hiding places or under top covers or on bottom boards. If an infestation is heavy, both adults and masses of larvae may be seen on the combs and bottom board. Corrugated cardboard with the paper removed from one side, placed on the bottom board at the rear of the hive, has been successfully used in detecting adult beetles. Fermented honey (smelling like decaying oranges) exuding from full honey supers in storage or on active colonies is a sign that SHB are present.

Control

Contact your state apiary inspector if you find or suspect SHB. In 1999, coumaphos received a section 18 (emergency use) registration of CheckMite+ strips to control SHB. Because this pest is not yet found throughout the U.S., however, beekeepers are strongly urged to freeze or burn any infested hive and bees. Freezing at 10°F (-12.2C) for 24 hours will kill all life stages of SHB. In addition:

- maintain only strong, healthy colonies
- keep apiaries clean of *all* equipment not in use
- extract honey as soon as it is removed from colonies
- destroy these beetles as soon as they are detected

Hive Treatment

Wear gloves when handling these strips (see coumaphos, under varroa mite control). Prepare a piece of corrugated cardboard about 6 inches by 6 inches (remove paper from one side). Cut the miticide strip in half crosswise and staple both pieces to the corrugated side of the cardboard. Then place the cardboard in the center of the bottom board with the strips facing down for at least 3 days but no more than 7 days.

Pollination and Bee Plants

Pollination

In this section we discuss how flowers, from apples to rutabagas, are pollinated, set seed, and bear fruit. We also outline the importance of adequate pollination from bees and other pollinators. Growers and beekeepers may require that wild or native bee pollinators help honey bees with the pollination tasks. Many beekeepers are supplementing the pollination workforce by rearing native bees such as sweat bees, bumble bees, and leafcutter bees, to name a few (see "Non-Apis Bee Pollinators" in the References).

Some Definitions

Flowers are the reproductive parts of a plant, where seeds are formed and from which fruits and vegetables develop. For the cycle to begin, a pollen grain comes in contact with the *stigma* (or female part) of the same flower species.

Fertilization takes place when pollen from the *anther* (male organ) unites with a female ovule, which forms the seed and fruit. Each species of plant has its own unique shape and form of pollen grains, which enables paleobotanists, studying ancient plants, to identify pollen from the mud at the bottom of 10,000-year-old bogs. The transfer of pollen from male to female sex organs is called *pollination.* All plants must be pollinated before seed (or fruit) will set. Pollen is transferred from the anthers to the stigma by wind, water, gravity, mammals, birds, humans, and insects.

A transfer that takes place on the same blossom or on another blossom on the same plant is called *self-pollination.* Beans, for example, are self-pollinating. Though many kinds of beans and other plants do not need insect visitors, they do benefit from the extra pollen carried by these insects and may even set better or more fruit. This is true for soybeans and lima beans.

But if the pollen goes from a Red Delicious apple tree to a Granny Smith apple tree, this is called cross-pollination. Apples and many fruits have a further complication. Many varieties are self-sterile, which means that the pollen from the Red Delicious, for example, will NOT pollinate itself or flowers from other Red Delicious trees. It must have pollen from another variety of apple to set fruit. The placement of apple varieties, the size of the blocks of *pollinizers,* and length of rows may be factors in getting good fruit set in the orchard. It is important not to have too big a block of any one variety in any one area of the orchard.

Many plants are wind pollinated, including all the grasses (and their cultivated cousins corn, oats, wheat, and rice), tomatoes, ragweed, and evergreen trees. Such pollen is light, is produced in enormous quantities, and is the cause of allergic reactions or hay fever in many people.

The Mechanics of Fertilization

When a bee visits an apple flower, she picks up pollen grains from the anther and moves on to another apple bloom, where her body may brush up against the stigma (see illustration of the fertilization of a flower). Grains of pollen stick to the moist tip of the stigma, after which incredible things begin to happen. The pollen grains start to grow a root, called a *pollen tube,* down the pistil, to deliver two sperm cells to the female ovule (embryo seed). This tube has to grow down the entire length of the stigmatic tissue to reach the ovary. Once it has found an unfertilized ovule, the two sperm cells are released. When one sperm cell fuses with an egg nucleus, it becomes the seed. The other sperm cell goes to the center of the ovule to unite with the polar nuclei; this develops into tissue called the endosperm, which nourishes the developing embryo. The endosperm becomes the seed leaves, or cotyledons, of the new plant.

After fertilization, the ovules secrete hormones that stimulate the wall of the ovary to thicken into the surrounding fruit tissue. From this complex, double fertilization, almost all flowering plants on earth are pollinated. Even "seedless" varieties of some crops need to be pollinated: they usually start to form a seed, but it is aborted early in its development.

Bees as Pollinators

The most efficient pollinators—highly motile, small, and plentiful—are the insects. Major insect pollinators include beetles, flies, butterflies, moths, and bees. Bees are probably the principal pollinating agents of plants whose flowers have colors within a bee's visual range of blue, yellow, green, and ultraviolet.

Although honey bees are the insect of commerce, for some plants, such as alfalfa, they are not very efficient. They do not like to work the flowers because of their unique tripping mechanism, which hits the bee's body while she works (see the illustration of tripping). The alfalfa leafcutter bee is now used exclusively in some parts of the country.

But for the most part, bees are the best pollinators. It takes several trips by many bees, for example, to

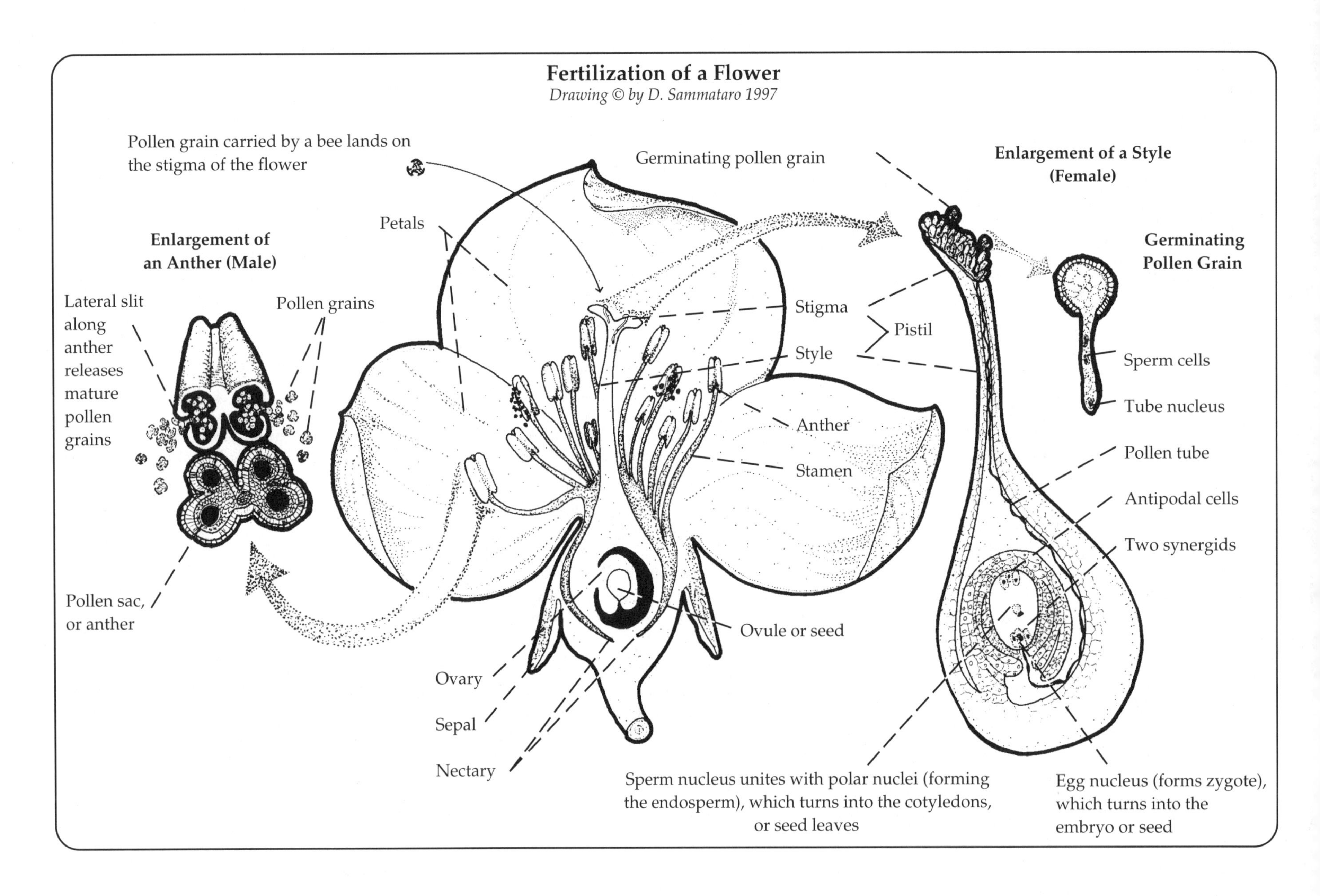

pollinate one apple or one cucumber adequately, because there are many seeds in each fruit, and each one needs a pollen grain. For cucumbers, complete pollination takes over 20 visits by bees, and the female blossom lasts only one day. Pollen must be moved from the male flower to the female flower before it closes up for the night; bees are the most cost-effective way of doing this.

Honey bees are desirable pollinating agents for these reasons:

- Colonies of bees can be moved to crops needing pollination.

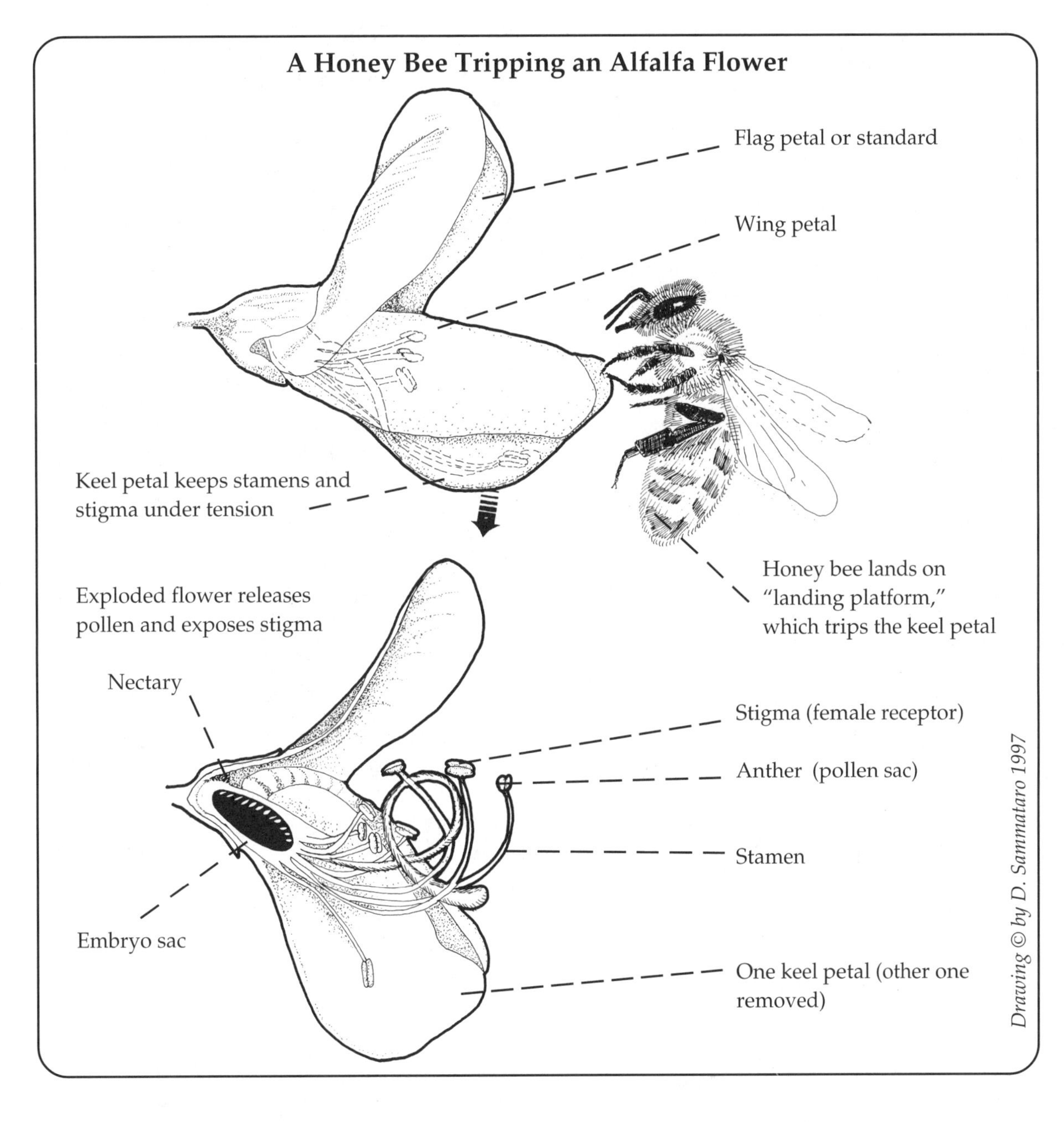

- Each colony contains large populations of foragers to work crops within a narrow pollinating window.
- Bees will usually work only one type of flower on each trip (flower fidelity), not mixing pollen types. For example, when the honey bee flies out to gather nectar and pollen from an apple blossom, this is the only type of flower she will visit on this flight.
- Crops can be sprayed with certain attraction pheromones, which assures that the bees will work only the target crop.

Recommendations for Growers

Many growers and orchardists have planted large blocks of crops that require migratory honey bee pollinators. A good rule of thumb is to place one or two colonies per acre of crop, and more would be better. In early spring, the more bees in the orchard, the closer they will be to the target crops. If the weather turns bad, more bees in the field will help ensure adequate pollination. Also, colonies in the spring may be weaker than colonies in the summer; to compensate for this, make sure there are four or more frames of brood covered with bees per colony.

If growers are not diligent, bees can be killed by pesticides sprayed on the crops or on nontarget plants and weeds or as it drifts over hives or into the water supply. Work closely with the grower to spray when bees are NOT in the field. Make it a clear part of the lease contract (see below). There are many lists published on pesticides toxic to bees (see "Pesticides" in the References and "The Pesticide Problem" in Chapter 13).

In many cases, native pollinators have been killed by destructive farming practices or pesticide use. Many non–honey bee pollinators are valuable to growers and need to be cultivated. You, as a beekeeper, can also raise alternate pollinators as a sideline (see "Non-Apis Bee Pollinators" in the References). These may be more important to future growers as the number of feral bee colonies succumb to mite infestation. But these non-apis bees and honey bees both need flowers for an entire season to stay alive and well fed. By placing your colonies near

Common Bee Plants

Alfalfa	Horsemints
Almond	Joe-pye weed
Apples and fruit trees	Knapweed
Asters	Knotweed or smart-weed
Berries	Lespedeza
Bird'sfoot trefoil	Lima bean
Black gum or tupelo	Lindens or basswood
Blueberry	Locusts
Blue curls	Loosestrife
Brazilian pepper tree	Mangrove
Buckeye or horse chestnut	Maples
Buckthorn	Mesquite
Buckwheat	Milkweed
Cactus	Mints
Catnips	Mustards and canola
Citrus	Palms
Cleome or spider plants	Privet
Clovers and sweet clovers	Redbud
Cotton	Russian olive
Cranberry	Sage
Dandelion	Serviceberry
Dead nettle	Sourwood
Elm	Soybean
Eucalyptus	Squash and melons
Figwort	Sumac
Fireweed	Sunflowers
Goldenrod	Sweet corn (pollen)
Hawthorn	Thistles
Heath and heathers	Thyme
Hollies	Tulip poplar
Honeysuckle	Vetch
	Vitex
	Willows

For more information, see "Honey Plants" in the References

uncultivated areas (or by planting certain forage crops), you will have better success in keeping and establishing all kinds of bees (see the figure on common bee plants).

Leasing Bees

Many beekeepers lease their hives to fruit and vegetable growers whose crops benefit from or require bees for pollination. The need for bee pollination is increasing, due in part to declining bee populations (especially feral or wild honey bee nests), caused by urbanization of natural foraging land, pesticide use, mites, and pollution.

Some factors to consider when leasing or renting bees are:

- Number of hives: if other factors are favorable, count on one colony per acre of fruit crops, more for other crops.
- Weather: optimum flying conditions for bees include temperatures between 60° and 90°F (15.6 and 32.3°C), winds of less than 15 miles per hour, and fair, sunny days.
- Colony strength: each colony should have at least four frames of brood and bees and a laying queen.
- Timing: set out bees just as the crop comes into bloom; if set out too early, bees may work other blooming plants and may not switch to the target crop.
- Leasing fees: although there is no flat fee for leasing bees, factors that may affect the price include time of year, pesticide hazard, loss of queen, bees and honey, and the difficulty of getting in to and out of the field.

Some beekeepers remove frames of pollen from colonies to stimulate bees to collect more pollen. Others use pollen traps for the same reason or install queenless colonies that have queen cells. All these techniques require close attention to the condition of the bees and the crop as well as to the weather.

Pollination Contracts

To be fair, the beekeeper and the grower should draw up a pollination contract or agreement. This document will help prevent misunderstandings while detailing the expectations of all parties. Key points include:

- Date of placement of bees into the crop and of their removal (relative to bloom time and condition).
- Location of crop.
- Number and strength of colonies.
- Pattern of colony placement.
- Rental fee and date on which it is to be paid.
- Grower agrees not to apply bee-toxic pesticides while bees are in the crop or will give the beekeeper 48 hours notice.
- Grower will warn beekeeper of other spraying in the area.
- Grower will reimburse beekeeper for any additional movement of colonies in, out of, or around the crop.
- Grower will provide right of entry to beekeeper for management of bee colonies.

Sample contracts can be obtained from *Bee Culture* magazine (A. I. Root Co., Medina, OH 44256-0706); the Department of Entomology, Cooperative Extension Service (CES), Pennsylvania State University, 501 ASI Bldg., University Park, PA 16802; or the Department of Entomology and Nematology, CES, Institute for Food and Agriculture, University of Florida, Gainesville, FL 32611 (Sheet ENY-110).

Commercial Crops Benefiting from Honey Bee Pollinators

Many commercial crops benefit directly by insect pollination. For more information, see "Pollen and Pollination" in the References.

Here is the list of commercially grown plants that benefit from but do not require bee visitation:

- Asparagus
- Apricots
- Broad beans
- Caraway
- Cherimoya
- Chestnut
- Chives
- Citrus
  - Grapefruit
  - Lemon
  - Mandarin
  - Orange
- Clove
- Clovers, minor
- Coconut
- Coffee
- Cotton
- Cowpeas
- Cut-flower seeds
- Drug plants
- Feijoa
- Flax
- Guava
- Herbs (spices)
- Kapok
- Lespedeza
- Lima beans
- Loquat
- Mangosteen
- Nectarines
- Oil palm
- Okra
- Onion and leek
- Opium poppy
- Papaya
- Pears
- Peppers
- Pyrethrum
- Safflower
- Scarlet runner beans
- Strawberry
- Tephrosia
- Tomatoes
- Vanilla
- Vegetable seeds
  - Anise
  - Chervil
  - Endive

Commercial Crops Requiring Honey Bee Pollinators

Here is a list of crops that require insect pollination. Many commercial growers rent honey bee colonies to pollinate these crops. Pollination is a $15 billion industry in the U.S.

- Alfalfa
- Allspice
- Almonds
- Alsike clover
- Apples
- Avocado
- Berseem
- Blackberries
- Blueberries
- Buckwheat
- Cacao
- Carambola
- Cardamom
- Cashew
- Celeriac
- Chayote
- Cherries
- Chinese gooseberry or kiwi
- Cicer milkvetch
- Cinnamon
- Citron
- Citrus
  - Pummelo
  - Tangelo
  - Tangerine
- Clovers, minor
- Cole crops
  - Broccoli
  - Brussels sprouts
  - Cabbage
  - Cauliflower
  - Collards
  - Kale
  - Tendergreens
- Coriander
- Cranberries
- Crimson clover
- Crownvetch
- Cucumbers
- Currants
- Cut-flower seeds
- Dewberry
- Dill
- Drug plants
- Eggplants
- Fennel
- Garlic
- Gooseberries
- Herbs (spices)
- Huckleberry
- Jujube
- Kenaf
- Kohlrabi
- Kola nut
- Lavender
- Leek
- Litchi
- Longan
- Lotus
- Macadamia
- Mango
- Muskmelons
  - Cantaloupe
  - Casaba
  - Crenshaw
  - Honeyball
  - Honeydew
  - Persian melon
- Mustard
- Niger
- Nutmeg
- Onion
- Parsley
- Parsnip
- Passion fruit
- Peaches and nectarines
- Pears
- Persimmon
- Pimenta
- Plums and prunes
- Pumpkin and squash
- Quinine
- Radish
- Rape
- Raspberries
- Red clover
- Rutabagas
- Sainfoin
- Sapote
- Sunflower
- Sweet clovers
- Sweetvetch
- Tea
- Trefoils
- Tung
- Turnips
- Vegetable seeds
  - Artichoke
  - Asparagus
  - Caraway
  - Carrots
  - Celery
  - Chicory
  - Chives
- Vetch (hairy)
- Watermelon
- Welsh onion
- White clover

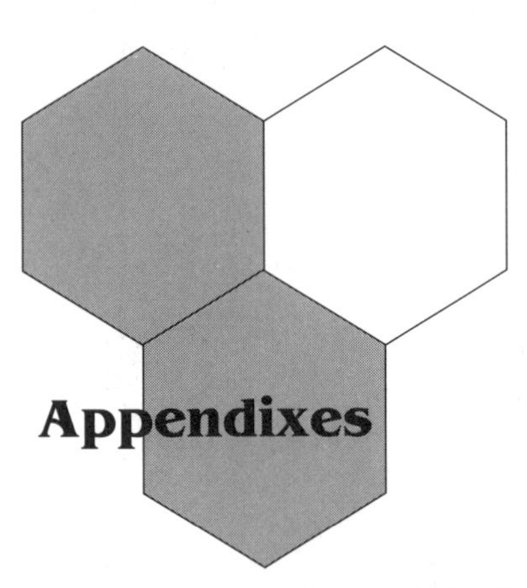

Appendixes

Appendix A
Anatomy of the Honey Bee

The anatomy of the honey bee is similar to that of other insects, except for the specialization of certain organs and structures needed by bees to carry out functions peculiar to their social life. Parts common to other insects include the three basic insect parts—head, thorax, and abdomen; the hard, waxy protein (chitin) covering; the free respiratory system (no lungs); the ventral or bottom spinal cord; and the open circulatory system (no veins or arteries).

The internal organs of the worker, queen, and drone are basically the same, consisting of the pharynx, foregut, midgut, and hindgut. Attached to the anterior end of the hindgut are the Malpighian tubules, which act as a liver, cleaning the bee's blood as it flows through the body cavity.

Some of the more specialized structures and functions not seen externally include the honey sac (worker), the ovary (queen), the male genitalia (drone), and the various glands that produce the pheromones.

Honey Sac

The esophagus of the bee begins at the back of the mouth and continues through the thorax, terminating in the anterior part of the abdomen, where it expands into the crop, or honey sac. Collected nectar, honeydew, and water are temporarily stored in this sac by workers as they return with it to the colony. The sac walls are readily expandable because of internal invaginations, thus the honey sac can carry a heavy load of liquid. A valve at the posterior part of the sac called the *stomach mouth,* or *proventriculus,* controls whether the contents of the honey sac pass into the remaining parts of the alimentary canal. On returning to the hive, the contents of the honey sac are transferred to young hive bees, who add the appropriate enzymes to alter the liquid, work the nectar with the proboscis for some time to remove moisture, and place the drop in the cells for further drying and curing.

Antennae and Eyes

Most of the tactile (touch) and olfactory (smell) receptors of bees are located on the antennal segments. These receptors guide bees both inside and outside the hive, enabling them to differentiate between hive, floral, and pheromone odors. Once detected, odor and tactile stimulations are transmitted down the nerve cord from the brain, ending in the affected area. Workers are about 10 to 100 times more sensitive than humans to flower or beeswax fragrances and can follow an odor stream upwind, which is important when locating a queen in a swarm. The surface of the antennae includes plates, pits, pegs, hairs, and pores that respond to different odors. Other uses of the antennae include carbon dioxide detection as well as the relaying of humidity and temperature differences.

The two types of eyes, the ocelli and the compound eyes, enable bees to see quite well. The two compound eyes, made up of thousands of hexagonal facets, help direct the light to the sensory cells and then to the brain. Groups of these facets have special functions, such as recognizing polarized light or certain patterns, and the range of color vision. Thus, a bee sees its world as a mosaic of movement and color.

Internal Anatomy of the Queen and Drone Honey Bee

Examine the anatomy of the queen and drone. The queen has larger ovaries than the worker, a sting, but no wax glands or pollen-carrying structures on her hind legs. The drone's male organ, or endophallus, is internal until mating time, when it is completely everted. When the drone copulates with a queen, the endophallus is pulled out and broken off, and the drone dies. See the figures showing the external anatomy and the table on morphological differences between honey bee castes and the drones.

Morphological Differences between Honey Bee Castes and Drones

Item	Queen	Drone	Worker
Larval diet	Royal jelly	Drone jelly?	Worker jelly
Cell orientation	Vertical	Horizontal	Horizontal
Development time	16 days	24 days	21 days
Eyes	3,920 to 4,920 facets	13,000 facets	400 to 6,300 facets
Brain	Small	Large	Large
Mandibular gland products[a]	9-oxodec-*trans*-2-enoic acid;[b] 9-hydroxydec-*trans*-2-enoic acid[c]	Absent	10-hydroxy-(*E*)-2-decenoic acid
Mandibles	Unmodified for cutting only	Small and notched	Modified for comb building; flat center ridge
Tongue	Short	Short	Short
Honey stomach	Small	Small	Well-developed
Wings	Appear reduced in length	Extend over most of abdomen	Extend over most of abdomen
Legs	Middle tibia (no spine)	Middle tibia (no spine)	Middle tibia (spine); hind leg (pollen-collecting apparatus)
Wax glands	Absent	Absent	Present (4 pairs)
Nasanoff gland	Absent	Absent	Present
Ovaries	Well developed; 300 ovarioles	Absent	Undeveloped; between 2 and 12 ovarioles
Spermatheca	Present	Absent	Rudimentary
Sting	Lightly barbed, curved, waxy	Absent	Strongly barbed, straight, no wax
Antennae	1,600 chemoreceptors	300,000 chemoreceptors	2,400 chemoreceptors
Reproductive role	Egg laying	Mating	Maternal care
Number per colony	1 or 2	About 100 to 2,500	About 30,000 to 60,000
Food	Fed royal jelly	Begs or eats honey	Honey, pollen
Life span	1–8 years	Up to 60 days	Up to 4–6 months

[a]Brood food comes from the mandibular (white liquid) and the hypopharyngeal (clear) glands of young, nurse bees.
[b](*E*)-9-oxo-2-decenoic acid (9-ODA).
[c](*E*)-9-hydroxy-2-decenoic acid (9-HDA).

Appendix B
Pheromones

Honey bee behavior both inside and outside the hive is regulated to a large extent by chemical substances called pheromones. Pheromones are secreted by an animal and trigger certain behavioral responses or physiological activities in other members of the same species.

Queen Pheromones

Located in the mandibles of a queen's head are the mandibular glands. These glands produce and secrete the pheromones called queen substance, which include the chemicals (*E*)-9-oxo-2-decenoic acid (9-ODA) and (*E*)-9-hydroxy-2-decenoic acid (9-HDA). They are sometimes labeled other names, but these seem to be the current ones. These components elicit various responses in worker and drone honey bees.

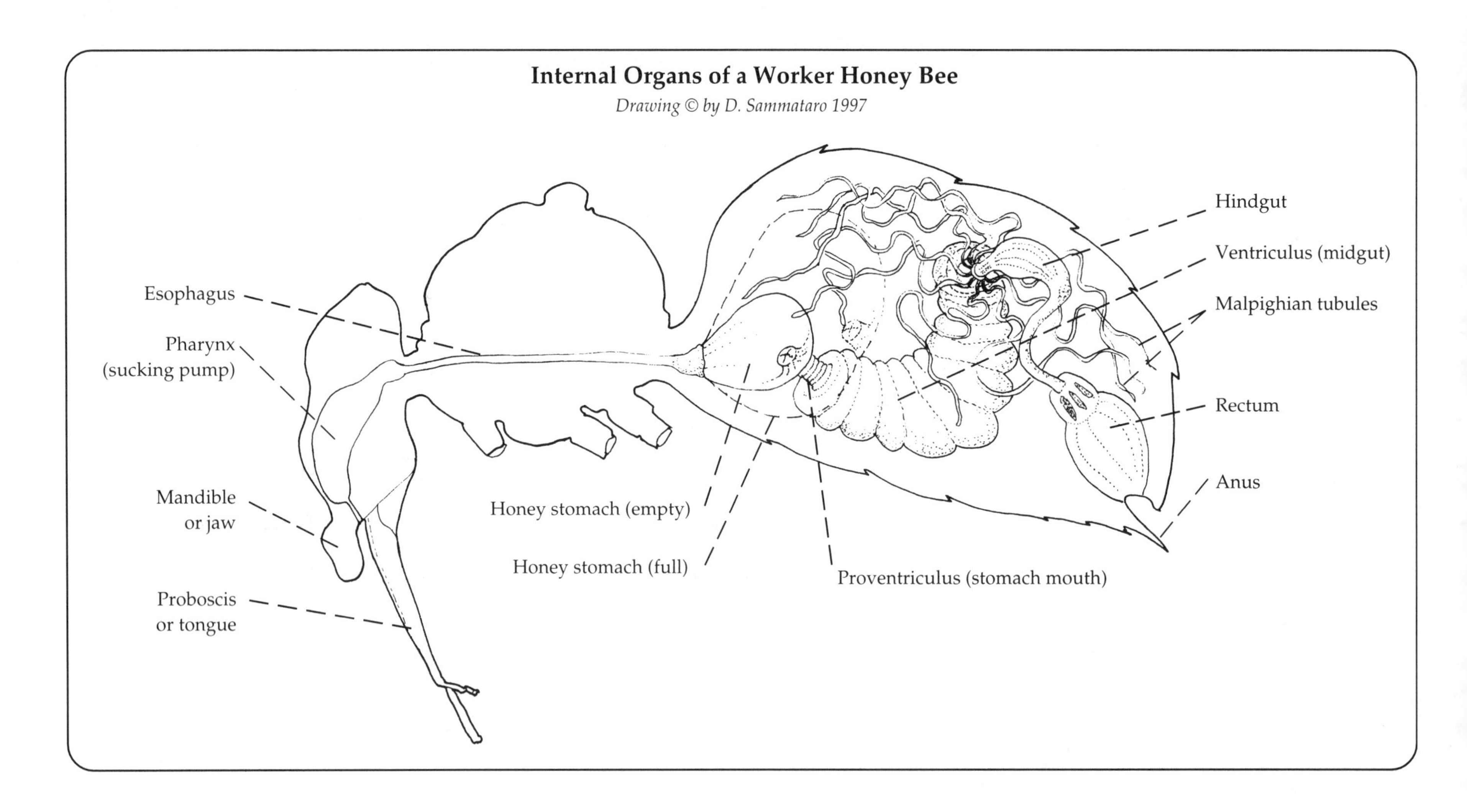

Inside the hive these substances have been shown to inhibit ovary development in workers and deter them from constructing queen cells. Their absence invokes the opposite response—queen cell construction is undertaken. (If bees are unsuccessful in rearing a replacement queen, ovary development takes place in some worker bees, and they become laying workers.)

A swarm—either flying out of the hive, to a homesite, or in a cluster—is aware of its queen's presence by means of these and other substances. Queen substance also guides drones toward queens who are on mating flights. Researchers are still investigating other pheromones released by the queen.

Worker Pheromones

There are several different chemical pheromones produced by workers. Two of these are alarm pheromones. One identified alarm odor (isopentyl acetate, also called isoamyl acetate) is released from a membrane at the base of the sting. The dispersal of this chemical is enhanced by the fanning action of the bee that secretes it. It smells like banana oil and stimulates bees to sting or fly at intruders. Other alarm compounds from the sting have recently been isolated and are currently being studied. Another alarm odor (2-heptanone) is released from the mandibular glands of workers. Items anointed with this odor are attacked by bees.

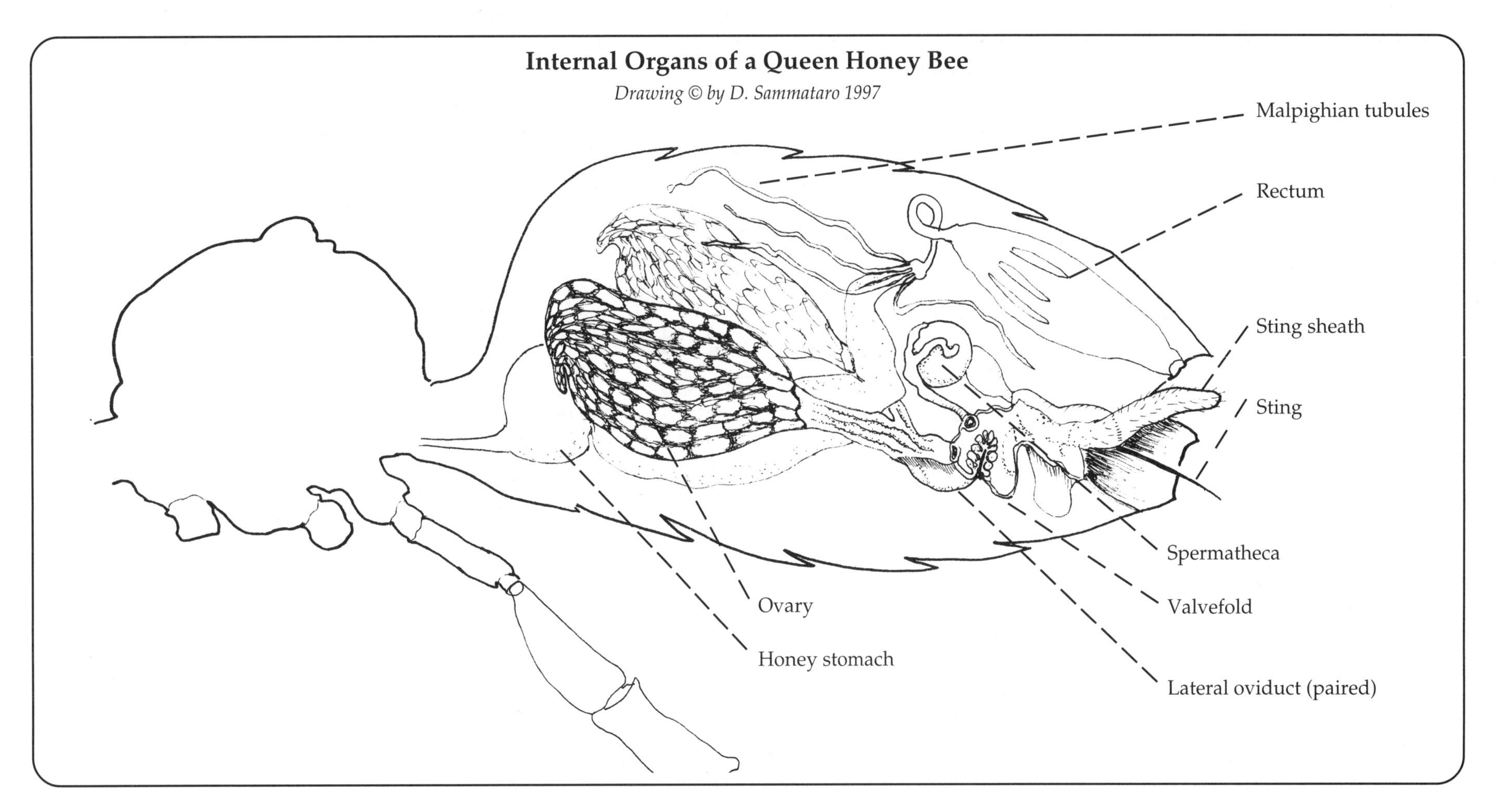

Internal Organs of a Queen Honey Bee

Drawing © by D. Sammataro 1997

A third important worker pheromone is the scenting or orientation odor. It consists of many chemicals; the three major ones are geraniol, nerolic acid, and (*E,E*)-farnesol. This chemical cocktail is produced by the Nasanoff gland located near the dorsal tip of the abdomen. On smelling these chemicals, bees move toward the source (as in a swarm). They are important in orienting bees to the entrance of a new nest or orienting bees that have been disrupted by the actions of a beekeeper.

There are many other pheromones produced by honey bees, not all of which have been identified or are understood. Researchers are continuing to study drone and brood pheromones. Check the References for more complete information on these topics.

Sting Anatomy

The sting structure of bees is a modified egg-laying device, or ovipositor, developed into a defensive structure. Made up of two barbed lancets, these are supported by the quadrate plates and extensive musculature in the last abdominal segment (called VIIT for the seventh tergite; see the figure showing the sting anatomy of a worker, p. 162). When the bee stings, the two lancets move into the skin with a sawing action and stay in place by means of the barbs.

The poison gland secretes venom into an attached poison sac, which holds the venom until it is pumped through the sting. This gland was originally labeled the acid gland and is so found in the older bee literature. Another smaller gland, the Dufour (or alkali) gland, is also present but has an unknown function.

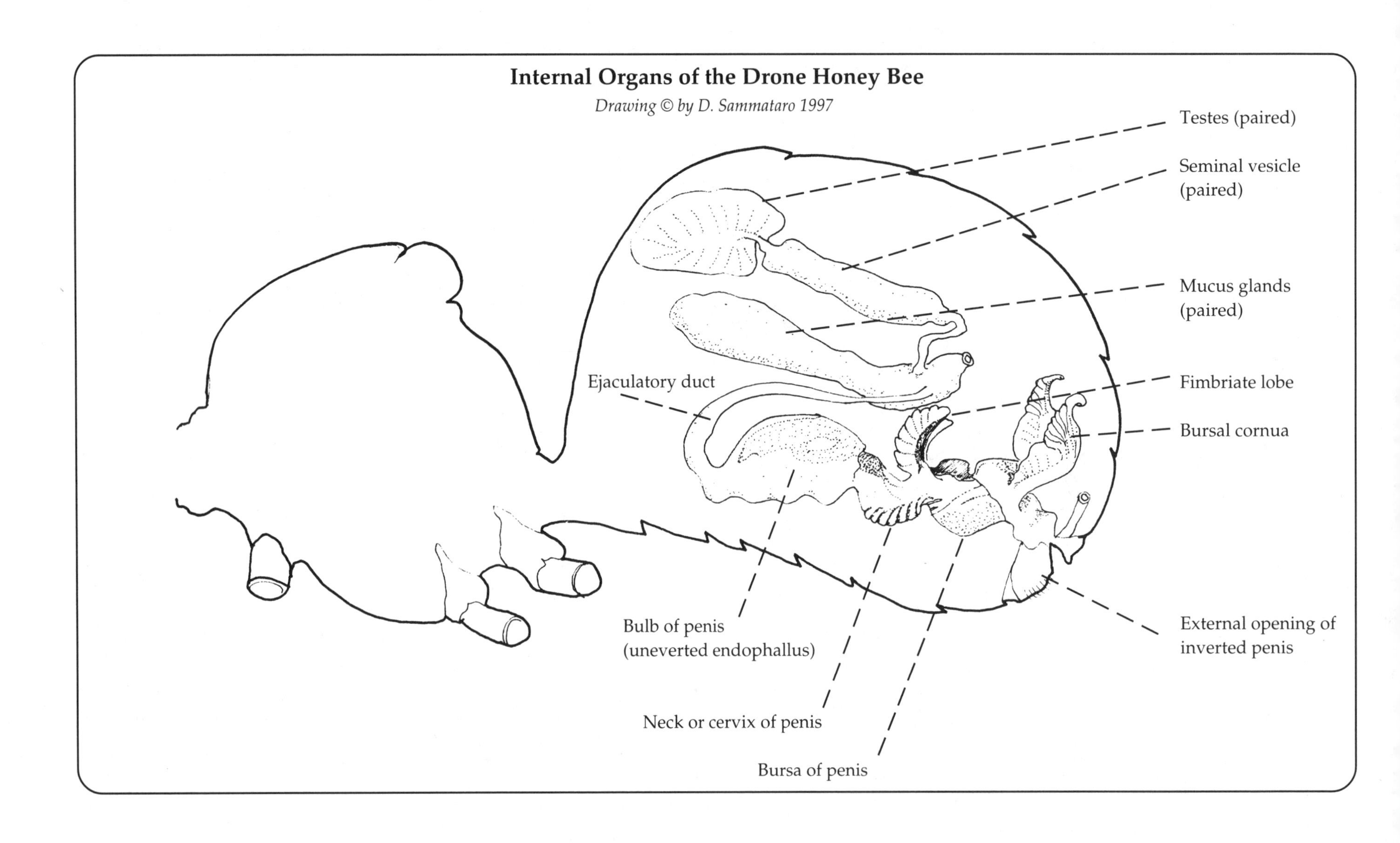

Appendix C
Bee Sting Reaction Physiology

Local Reaction

What happens in your body when you are stung by a bee? As with any bacterial invader, the body's natural defenses are called on to help. Basically, bee venom is a foreign protein (called an *antigen*) that stimulates the production of the body's defense proteins (called *antibodies*). Antibodies belong to a family of proteins known as gamma globulin and are also called immunoglobulins. The bee sting antigens appear to stimulate a specific class of immunoglobulins known as immunoglobulin E (IgE in the figure on bee sting reaction physiology).

Because the bee venom antigen reacts with specific antibodies (in this case, IgE), people not otherwise exposed to honey bee proteins must be stung at least once before any type of reaction will occur. After the

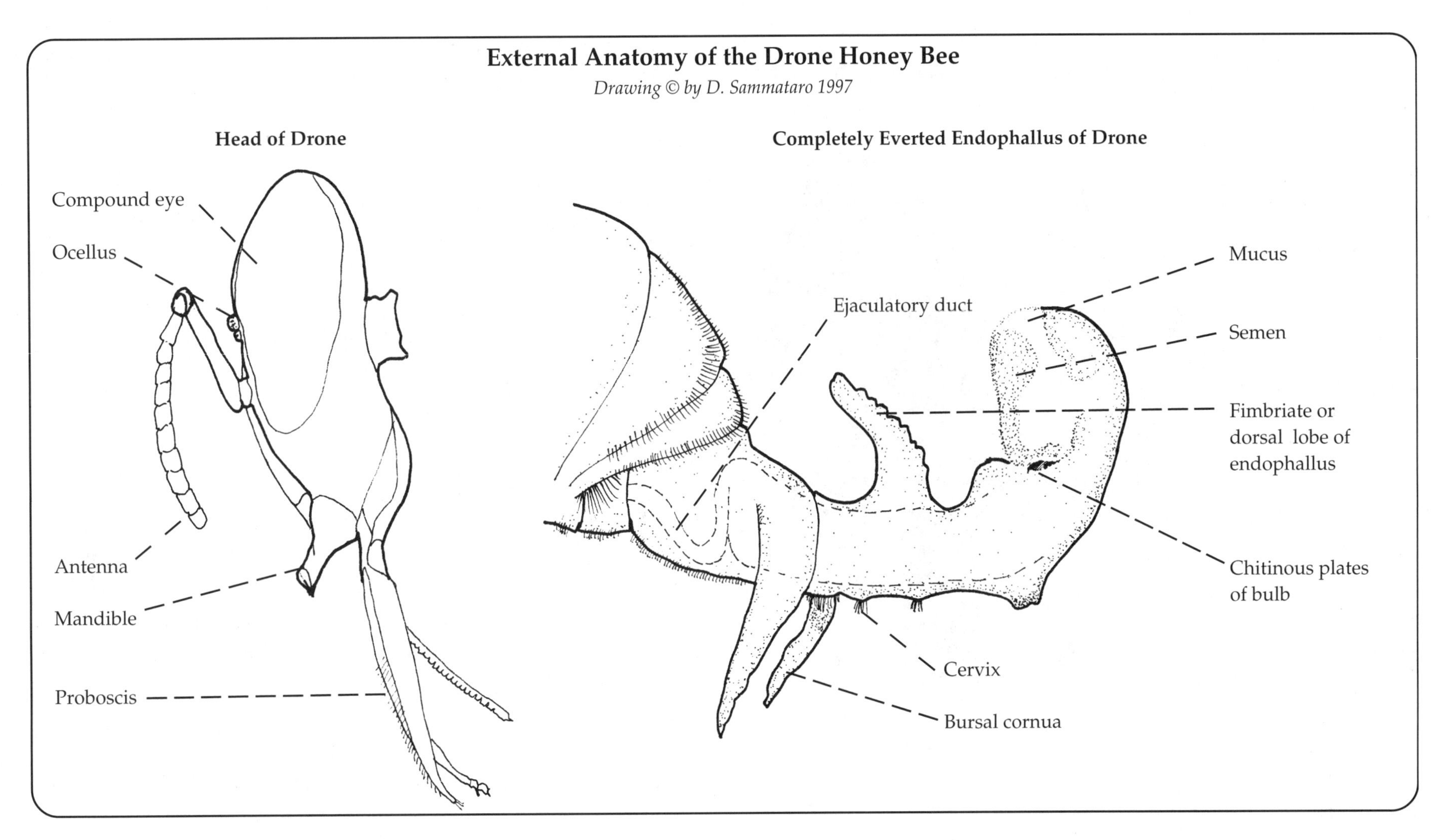

initial inoculation, the body seems to "remember" that particular antigen and will likely react faster to subsequent stings, with further antibody production.

In a local reaction, the antigen of the bee venom appears to react with the IgE bodies, which are attached to tissue cells (called *mast cells*). Mast cells contain numerous vesicles filled with histamine and other substances promoting inflammation. The action of the antigen reaction with the IgE / mast cell complex causes the histamine-filled vesicles to empty. Once released into the body, histamine has several effects, which include the expansion of blood vessels (vasodilation), the increased permeability of capillary cell walls to proteins and fluids, and the constriction of the respiratory passages. The first two actions may be responsible for the inflammation, swelling, and itching associated with bee stings. Most beekeepers are reported to have this kind of local reaction. With repeated stingings, the body becomes immune to the bee venom, and the venom will probably cause little, if any, discomfort.

If you are stung many times at once by honey bees or by Africanized honey bees, cover your eyes and mouth (targeted by bees) and run away as fast and as far as possible from the stinging insects. The lethal dose reported by researchers is 10 bee stings per pound of body weight, which means an average male weighing 160 pounds can receive over 1500 stings and still live.

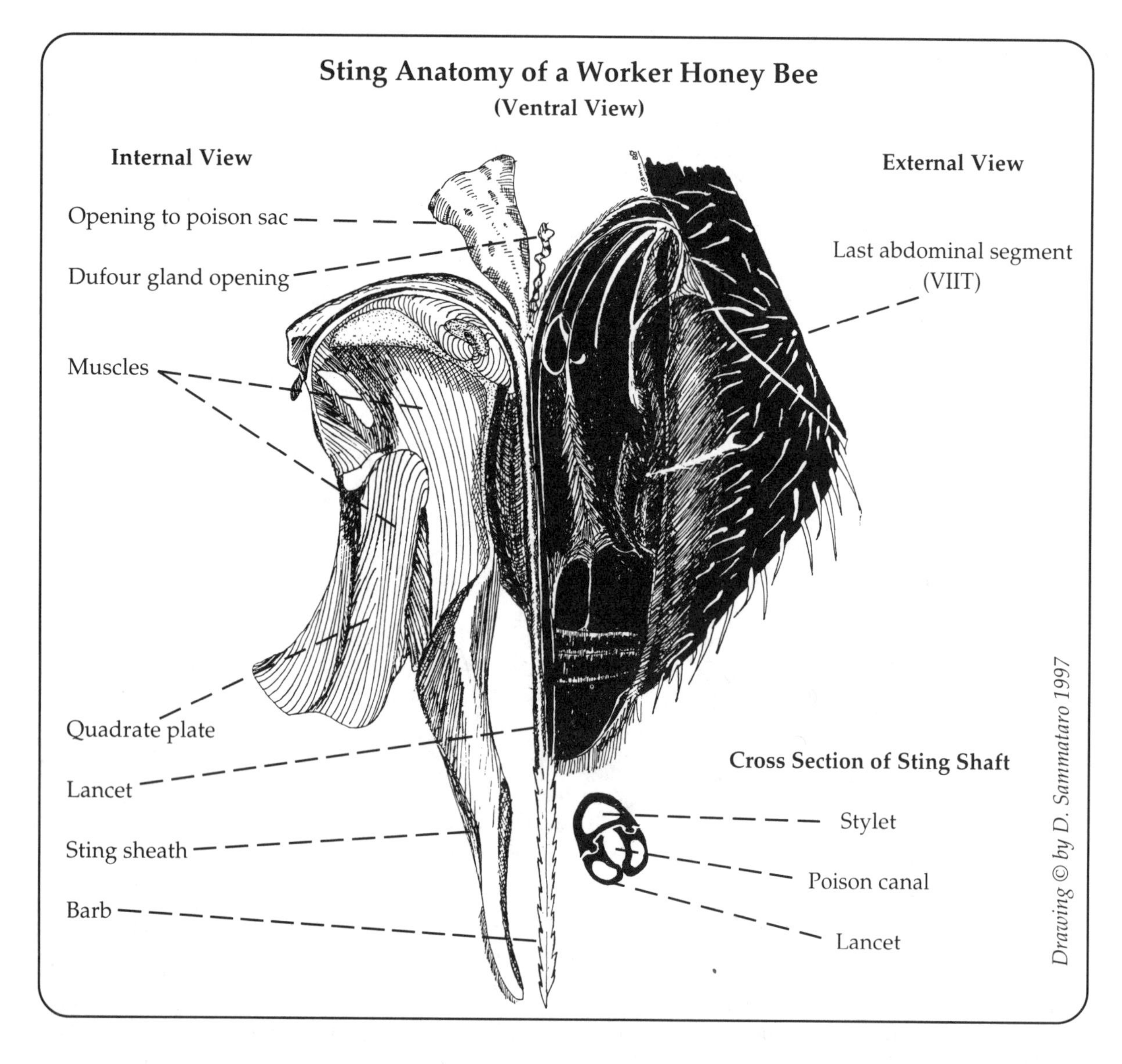

Systemic Reaction

In a systemic reaction, the same mechanisms come into play as in the local reaction, with one big difference: the antigen /IgE /mast cell complex reaction can cause death. This allergic reaction, called *hypersensitivity*, appears to be a result of the large amounts of histamine being released from the mast cells. Because the body remembers the bee venom antigen, the subsequent inoculations usually cause a faster reaction, which means more histamine is released each time a person is stung. Usually, a systemic reaction builds up gradually, with the victim showing greater distress (such as difficulty in breathing) after each sting.

In some people, the second bee sting may be enough to kill them. An antihistamine and adrenaline (epinephrine) should be immediately administered to counteract the effects of the released histamine and give relief for breathlessness.

Desensitization or Immunity

People who develop hypersensitivity to bee stings can become desensitized. Most beekeepers become less sensitive or immune to bee stings after repeated exposure. Desensitization can also be undertaken by an allergist. In either case, the immune processes (desensitization) are probably the same. Frequent injections of the venom appear to induce the body to manufacture a "blocking" antibody, IgG. The IgG competes with the IgE in its reaction activities to bee venom antigens. Because the IgG antibodies are not fixed to mast cells but float freely, they seem to be better able to combine with the bee sting venom antigens. Less histamine is therefore released, and the discomfort or allergic response is prevented. What an allergist does is to control the amount of venom that the victim receives, allowing the body to form enough of these blocking antibodies to combat the allergic reaction.

Some beekeepers can, over long periods of time, become allergic even to beeswax, honey, bee debris and bodies, and propolis. The percentage of beekeepers who do become allergic to bee stings, although hard to assess because individual body chemistry and allergy history are so varied, is small. Most people will probably already know if they are allergic, but those who do not can contact local allergy clinics or hospital outpatient facilities for testing.

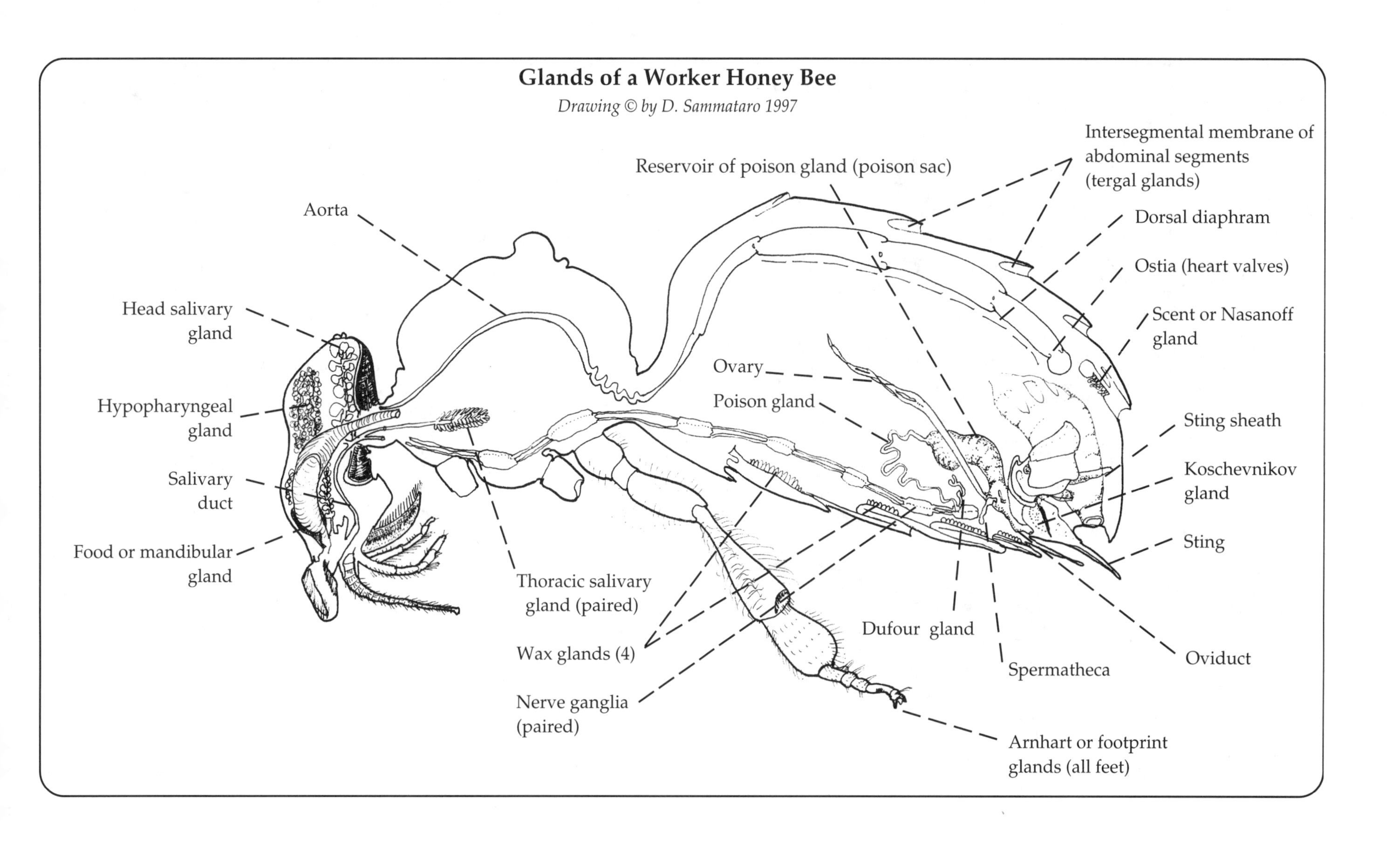
Glands of a Worker Honey Bee
Drawing © by D. Sammataro 1997
Intersegmental membrane of abdominal segments (tergal glands)
Reservoir of poison gland (poison sac)
Aorta
Dorsal diaphram
Ostia (heart valves)
Head salivary gland
Scent or Nasanoff gland
Ovary
Poison gland
Hypopharyngeal gland
Sting sheath
Koschevnikov gland
Salivary duct
Sting
Food or mandibular gland
Thoracic salivary gland (paired)
Dufour gland
Wax glands (4)
Oviduct
Spermatheca
Nerve ganglia (paired)
Arnhart or footprint glands (all feet)

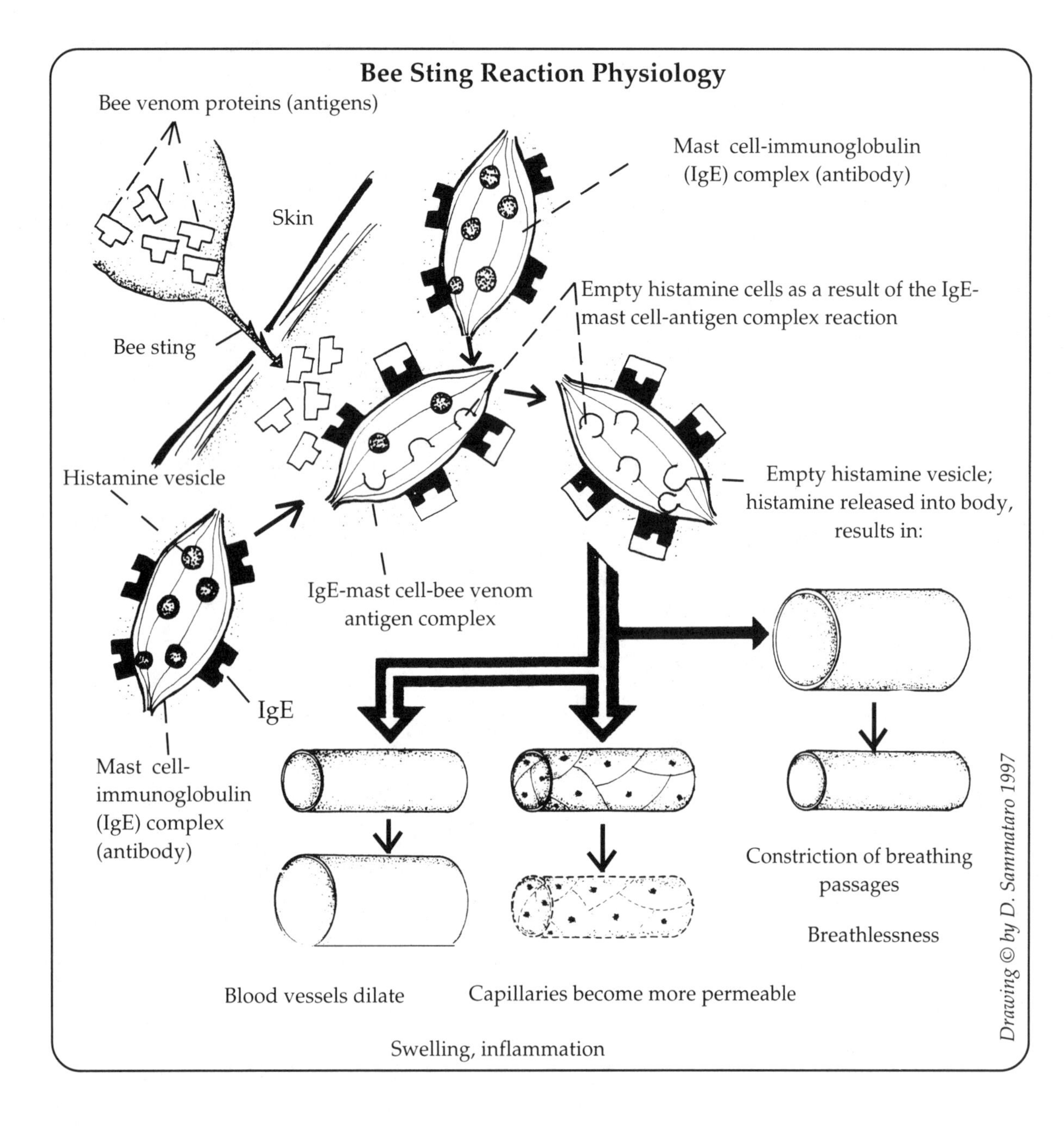

Appendix D
Preserving Woodenware

There are many compounds available at most hardware stores that are used to preserve wood. Not all of them are created equal, however, and many cannot be used in a beehive without adversely affecting the bees or contaminating the hive products. For example, Creosote, pentachlorophenol, chromated copper arsenate (CCA), tributyl tin oxide (TBTO), ACA, and ACZA must NOT be used on bee hives.

Acid copper chromate, copper naphthenate, zinc naphthenate, and copper quinolinolate can be used safely in all parts of hives. Ensure that the treated wood is thoroughly dried before use. When it is dry, treated wood can be painted with latex paint. Some visible bleeding of preservative may occur through white and light pastel colors when painting over oil-based treatments.

Preservatives may be applied by brush, dip treatment, hot and cold baths, and commercial pressure-treating processes. Note that these preservatives should be used on tops, bottoms, and supers only, NOT on frames (which do not need preservatives).

Paraffin Dipping

Paraffin dipping has been used by commercial beekeepers in many countries. The treatment is very cheap per unit of woodenware, and it cleans propolis and wax from old boxes being dipped for the first time. It can also be used to sterilize equipment contaminated by disease (such as American foulbrood).

The wax dipper vat should have these features:

- A metal vat to contain the paraffin wax.
- An enclosure to provide a "double skin" around the vat.
- An electric heat source, for safety.

Have a vat deep enough to hold two Langstroth

three-quarter (6⅝-inch) depth boxes and still leave some free room. All joints are welded from both sides to reduce the risk of bursting when in use. Hot dripping wax is dangerous.

Paraffin wax is available from oil companies; the grade with a melting point about 140°F (60°C) is best. Heat wax to 320°F (160°C), a process that takes up to two hours. Use a thermometer attached to a stick; DO NOT rely on guesswork.

Immerse boxes for about two minutes, using a heavy weight to keep them below the surface. Diseased boxes must be treated for 10–15 minutes. The supers are lifted from the wax onto a draining board. Boxes previously painted with oil-based paints blister and will need to be scraped.

Boxes can be used straight after dipping without being painted. Paraffin wax is a good preservative on its own, but you can paint the boxes while still hot with a water-based paint. They get two quick coats of paint, which is actually pulled into the wood by the drying wax.

The secret is to give the boxes long enough in the hot wax so the wood is heated right through and good penetration of the wax can be obtained. All the woodenware must be as dry as possible, as the wax will seal in any water in the wood and could cause it to rot from the inside out. The boxes are very hot when they come out of the dipper, so use thick leather or fireproof gloves or tongs.

Finally, a few safety precautions:

- Locate the wax dipper away from anything flammable.
- Wear eye protection in case of splashing wax.
- Have a metal cover handy to put over the vat in case the wax boils over.

For more information, see *Wood Preservation of the Farm* (B. C. Canadian Ministry of Agriculture, Fisheries and Food, 1993), and A. Matheson, Easily-constructed paraffin wax dipper, *New Zealand Beekeeper* 41(4):11–12 (1980).

Appendix E
Metric Conversions

Metric Units

1 gram (g) × 1000 = 1 kilogram (kg)
1 gram ÷ 1000 = 1 milligram (mg or 10^{-3})
1 meter × 1000 = 1 kilometer (km)
1 meter ÷ 1000 = 1 millimeter (mm or 10^{-3})
1 liter ÷ 1000 = 1 milliliter (ml or cc or 10^{-3})
1 liter ÷ 1,000,000 = 1 microliter (μl or 10^{-6})

Weight

kg × 2.2046 = pounds (lbs.)
pounds × 0.4536 = kg
g × 0.035 = ounces (oz.) (1 g = 0.035 ounce)
ounces × 28.35 = g (1 ounce = 28.35 g)
1 mg = 0.000035 ounce

Length

1 centimeter (cm) = 10 mm
1 mm = 0.039 inch (in.)
1m = 3.28 feet (ft.)
1 km = 1000 m = 0.62 mile (mi.)
cm × 0.3937 = inches
m × 1.094 = yards (1 yard = 3 feet)
inches × 2.54 = cm
miles × 1.6093 = km
km × 0.6214 = miles

Volume

1 ml = 0.034 ounce
1 fluid ounce = 2 tablespoons = 29.6 ml
1 liter = 0.27 gallon (gal.)
Liters × 1.057 = quarts
1 gallon = 3.79 liters = 231 cubic inches
1 liter = 61.02 cubic inches or 0.353 cubic feet

Temperature

°F = (°C×1.8) + 32
°C = (°F − 32) × 0.555

Source: T. J. Glover, *Pocket Reference,* 2d ed. Littleton, Colo.: Sequoia Publishing, 1997.

Appendix F
Fun Facts about Bees

Flight Speed

- 6–9 miles per hour loaded, 8 mph empty
- 15 mph maximum wingbeat at 300 Hz (500 Hz mosquitoes, 30 Hz grasshoppers)

Flight Range

- 2 mile radius, or 8000 acres
- Maximum radius of 5–8 miles
- Total flight during life 500 miles (about 800 km) in 5–30 days

Nectar Collecting

- 50–80 percent of flying bees collect nectar
- 100–1500 visits per load
- 1–29 trips per day
- 0.36–0.50 milliliters per load (0.50 ml is about one eyedropper drop)
- 5–150 minutes per trip
- Full load is 85 percent of bee's body weight, or 30 mg nectar
- 150 pounds of honey, in mileage, equals 13 round-trips to the moon
- Bees consume 35–60 kg honey per season

Sugar Requirements

- Resting drone: 1–3 mg sugar/hour
- Flying drone: 14 mg/hour
- Resting worker: 0.7 mg/hour
- Flying worker: 11.5 mg/hour

Pollen Collecting

- Collecting stimulated by presence of uncapped brood
- 15–30 percent of flying bees are collecting pollen
- 8–100 visits per load, or 50 to 350 flowers
- 1–50 trips per day
- 6–200 minutes per trip
- Full load is 35 percent of bee's body weight, or 7.5 to 15 mg
- 250,000 to 6,000,000 pollen grains on one bee depending on pollen source
- 250 grams pollen collected in one day = 17,000 flights

Bee Stings

- Each bee sting contains 150 μg of venom
- There are 190,000 bee stings per ounce of venom
- 6.4 grams of venom equals 424,000 stings
- Lethal dose of venom is 10 stings per pound of body weight

Nurse Bee Visits to Larvae

- Each cell takes 25–30 bees 41 minutes to prepare
- 1300 bees inspect and visit each larva, feeding them 2 percent of the time
  - –Up to 7200 visits per larva
  - –650 bees cap cells
  - –60 bees clean cells

Estimating Weight of Colony

- A deep frame fully filled with honey weighs about 10 pounds (4.536 kg) × number of frames in hive = total weight of hive box
- A full medium frame (U.S.) = 7 pounds
- A full shallow frame (U.S.) = 5 pounds

Queen

- Queen fed every 20–30 minutes at peak brood rearing
- In normal hive, in active season, there will be:
  - –300 to 1000 drones
  - –25,000 older foragers
  - –25,000 young hive bees
  - –9,000 uncapped larvae
  - –6,000 eggs incubating
  - –20,000 capped brood incubating

Estimating the Number of Bees

Count the number of bees returning to the hive for 1 minute (B/M). To estimate number of deep frames of bees in colony, the equation is B/M × 30 minutes × .0005 = numbers of deep frames of bees.

(The 0.0005 assumes one deep frame, with both sides contains 2000 bees (1 ÷ 2000); the 30 minutes assumes the amount of time needed for any one bee to make a return trip.)

If hive is weak, you could use one deep = 1750 bees (0.00057) or ½ pounds bees/frame.

Appendix G
Africanized Bees
Behavioral Differences between Africanized (AHB) and European (EHB) Honey Bees

Behavior	AHB	EHB
Physical Characterisitics		
Worker length	12.73 mm	13.89 mm
Cell size[a]	4.6–5 mm	5.2–5.4 mm
Wingbeat[b]	290 Hz	243–270 Hz
Wing position	Wings elevated off abdomen	On abdomen
Fresh weight[c]	2 g	2.5 g
Nest size, volume	22 liters	45 liters
Comb area (1000 cm^2)	8–11	23.4
Honey area	0.9	2.8
Queen's egg laying		
Maximum/day	4000	2500
Total/year	105,000	58,000
Behavioral Differences		
Swarming	3–10 times per year	1–4
Absconding	30 percent of colonies abscond	Not common
Supersedure	>40 percent in tropics	Unknown
Propolis	More (?)	Less (?)
Mating time	Afternoon	Early afternoon
Defensiveness/stinging	Bounce off veil, sting in vicinity	Sting in apiary
Time before first sting	14 seconds	229 seconds
No. stings/minute	35	1.4
Pursuit distance	160 meters	22 meters
Recovery time	28 minutes	3 minutes
Entrance	Attack other bees at entrance	Not so prevalent
	Propolize entrance	Not as much (?)
	Returning bees fly into entrance	Land on landing board
Foraging	More pollen	Normal amount
Unassisted spread per year	200–500 km	4 km
Maximum km to new nest site	About 75 km	5 km
Running on comb	Excessive, runs off comb, festoons at lower edge; parades around inside box or inner cover	Normal; not usual to run much
Dequeening	Can be very difficult; if requeening with EHB, about 40 percent loss	Normally, 10 percent loss
Life expectancy	Shorter than EHB; 12–18 days as foragers	Longer than AHB

Appendix G Continued

Behavior	AHB	EHB
Development Stage		
Egg to adult worker	18.5 days	21 days
Egg to adult queen	15 days	16 days
Worker's first flight	3 days old	10–14 days old
Queen's mating flight	5–6 days old	7–10 days old
Drone mating flight	7.5 days old	13 days old
Brood production	Almost 2× as much when resources good	Slower
Drone production	Stimulated	Suppressed

Sources: D. Caron, *History of Africanized Bees* (in press); E. Crane, *Bees and Beekeeping: Science, Practice, and World Resources* (Ithaca, N.Y.: Cornell Univ. Press, 1990).

[a]Cell size: measure 10 cells three times; if average ≥ 49 mm = AHB, and average ≥ 52 mm = EHB.

[b]Wingbeat: measure free-flying, undisturbed bees.

[c]Fresh weight of 30 callow bees.

Appendix H
Rearing Wax Moths

Wax moth larvae (waxies) are valued by anglers as bait for ice fishing in the winter as well as for fishing in the summer; many claim waxies are the best bait. Wax worms are also an alternative food source for protein-hungry people. They are easy to raise, become fairly large, and are nutritious and tasty. If you want to try raising waxies, use the larvae of the greater wax moth (*Galleria mellonella* L.), whose larvae grow bigger and faster than those of other species.

Wax Moth Diets

Here are two different diets:

- Boil together for several minutes ⅓ cup sugar (or honey), ⅓ cup glycerol or glycerin, and ⅓ cup water. Cool, then add ¼ teaspoon of a vitamin supplement, such as you would use to feed chickens, but do NOT add antibiotics. Then stir in 5 cups dried, fine-textured cereal, such as a dry baby cereal. Mix together and store in the refrigerator until ready to use.
- Warm together ⅞ cup (200 ml) glycerin and ⅞ cup (200 ml) honey. Add to two boxes of a complete, fortified, mixed-grain baby cereal and mix together. Cover and let stand overnight, or warm gently in the microwave so you can work the cereal thoroughly into a fine texture; store the diet in a sealed container in the refrigerator until ready to use. This will be a dry mix.

Beginning your Culture

Select a glass (or plastic) jar with a wide top. A peanut butter jar, a 2-quart fruit jar, or wide-mouth quart Mason jars work fine. Fill container 2 inches deep with the prepared diet.

Cover the jar with one sheet of a plain paper towel, a 20-mesh wire screen disk (window screen), and a jar lid with a large hole cut in the center (or you could use a Mason jar rim). The paper towel is to keep foreign material from entering and the young wax moths inside; but the wax moths will eat through the toweling if it is not protected by the wire.

Now add eggs, larvae, pupae, or adults to your culture. Whichever you add, they should all be at the *same stage* (i.e., don't mix eggs and pupae). Add 10–50 larvae, pupae, or adults or about 100 eggs. Do not crowd too many into one jar, as you will get smaller worms. It is best to start with adults that will lay eggs, because larvae that you collect from stored equipment may have parasites or diseases.

Now place your culture jars in a warm 80–93°F (27–34°C) room out of the sun for four weeks. Moths can tolerate a range of 77–99°F temperatures (25–37°C), but remember, the warmer the room, the faster they will grow. Date the jar when you start your culture so you can estimate when your waxies will be ready.

Harvesting and Increasing Your Colonies

If you are using your worms just for bait, harvest them before they spin their cocoons. Put them in cold storage for four or five hours at 32–33°F (0–1°C). To assure an even temperature of 32°F, put ice cubes and some water in a large bowl. Now put the waxies in a smaller jar, and place the jar in the ice bath. Keep them in the bath until all the ice has melted.

Afterward, they will not spin their cocoons, and you can store the larvae at 40–45°F (4–7°C) until ready to sell as bait. Place the waxies in a separate container, and keep them moist (not wet) with paper toweling, wood shavings, or other absorbent material.

If you plan on harvesting larvae to eat, collect them as they begin to spin their cocoons, because they void all fecal matter just before spinning. Try not to disturb the diet when collecting them. If you wish to hold the larvae for a long period, you can store them at 59°F (15°C) and a relative humidity of 60 percent, for up to one year, which will keep them from pupating.

Continuing the Culture

To keep your wax moth colony going, you will need to have some adult moths lay eggs. Either place 25–50 adults, or about 50 cocoons, in a separate, covered jar. You need no food in these jars, for neither cocoons nor adults feed. Now add a piece of paper, folded like an accordion or fan. By adding 25–50 cocoons, you will get a good mix of males and females. Within 10 days, the adult moths emerge. Each female moth will mate in the jar and begin to lay over 100 eggs in the folds of the paper.

Once you see the eggs (they are small, round, tan-colored balls, usually laid in strips), place a 2-inch piece of egg-filled paper on top of the diet in a separate jar. Don't wait too long, or your larvae could hatch out, wander out into your house, or starve to death.

Helpful Tips

If your worms suddenly start dying, are attacked by fungi or a virus, or change to a dark color, they could be diseased. Burn that batch and any other batch that may have been contaminated, wash the jars in hot, soapy water with bleach, and rinse by pouring boiling water in them. Then start with a fresh batch.

Tasty Insect Treats

It has been suggested that the growing larvae can be delicately flavored by incorporating small amounts of various spices into their diet. Wax worms are great as snack items. Fried in hot oil, they pop like popcorn and, if lightly salted, are reported to have as good or better flavor than potato chips or corn puffs.

For more information, see R. L. Taylor and B. J. Carter, *Entertaining with Insects,* 2d ed. (Yorba Linda, Calif.: Salutek Publishing, 1992), or write George Keeney, director, Insectary, Department of Entomology, The Ohio State University, 1735 Neil Ave., Columbus, OH 43210-1220.

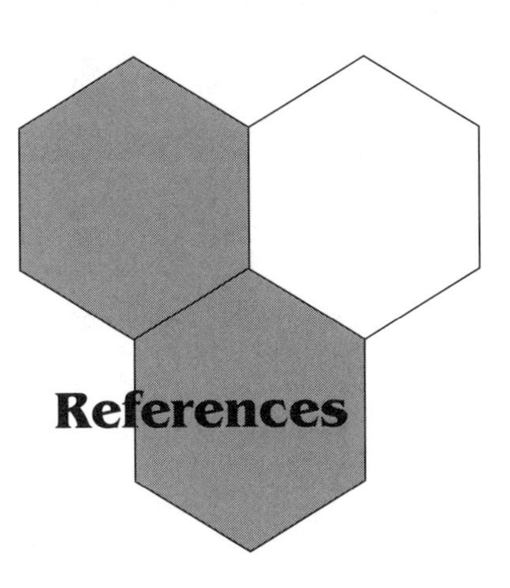

References

Abbreviations

ABJ = *American Bee Journal*
Ag. = Agriculture
Ann. = Annals
Ann. Rev. Ent. = *Annual Review of Entomology*
Apic. Res. = *Apicultural Research*
ARS = Agricultural Research Service
Assn. = Association
Bull. = Bulletin
CES = Cooperative Extension Service
Circ. = Circular
Dept. = Department
Dept. Ent. = Department of Entomology
Ent. = Entomology
Exp. Appl. Acarol. = *Experimental and Applied Acarology*
Exp. Sta. = Experimental Station
Ext. Serv. = Extension Service
GPO = Government Printing Office
IBRA = International Bee Research Association
J. = Journal
JAR = *Journal of Apicultural Research*
J. Econ. Ent. = *Journal of Economic Entomology*
J. Kans. Ent. Soc. = *Journal of the Kansas Entomological Society*
n.d. = no date
POB = Post Office Box
Pub. = Publication
Tech. Bull. = Technical Bulletin
Univ. = University
USDA = United States Department of Agriculture

Bee Journals/Publications

American Bee Journal. Dadant and Sons, Hamilton, IL 62341. ☎ 217-847-3324.

American Beekeeping Federation. Publishes the *ABF Newsletter,* POB 1038, Jesup, GA 31545-1038.

L'Apicolture Moderno. via Ormea 99, 10126, Turin, Italy.

Apidologie. Éditions Scientifiques Elsevier, 29, rue Buffon, F-75005 Paris.

Apitalia. Apimondia, 101 Corso Vittorio Emanuele, Rome, Italy 00186.

The Australasian Beekeeper. Pender Bee Supplies, PMB 19, Maitland, NSW 2320, Australia.

Australian Bee Journal. Graves, ed., 23 McBride Rd., Upper Beaconsfield, Victoria, 3808, Australia.

Bee Biz. M. Allan, ed., 41 George St., Eastleigh, Hampshire, S050 9BT, United Kingdom.

Bee Craft. S. White, ed., 15 West Way, Copthorne Bank, Crawley, Sussex RH10 3QS, United Kingdom.

Bee Culture. A. I. Root Co., Medina, OH 44256-0706. ☎ 1-800-289-7668. Web site: <http: / /www.airoot.com>.

Beekeeper's Quarterly. Northern Bee Books, Scott Bottom Farm, Mytholmroyd, Heden Bridge, West York HX7 5JS, United Kingdom.

Beekeeping and Development. Troy, Monmouth NP5 4AB, United Kingdom.

BeeScience. L. Connor, POB 817, Cheshire, CT 06410.

British Bee Journal. 46 Queen St., Geddington NR Kettering, Northamptonshire NN14 1AZ, United Kingdom.

Canadian Beekeeping. Box 678, Tottenham, Ontario, Canada L0G 1W0.

Entomological Society of America. *Journal of Economic Entomology, Annals of the Entomological Society of America,* and *American Entomologist.* 9301 Annapolis Rd., Lanham, MD 20706-3115.

Honey Bee Science. Institute of Honey Bee Science, Tamagawa University, Machida-shi, Tokyo, 194 Japan.

Indian Bee Journal. All India Beekeepers' Assn., 1325 Sadashiv Peth, Bharah Natya Mandir Rd., Poona 411 030, India.

Insectes Sociaux. International Journal for the Study of Social Arthropods. M. Breed, Environmental, Population, and Organismal Biology, Campus Box 334, University of Colorado, Boulder, CO 80309.

International Bee Research Association. 18 North Road, Cardiff CF1 3DY, United Kingdom. Publishes *Bee World, Apicultural Abstracts,* and *Journal of Apicultural Research.*

Irish Beekeeping. Seamns Reddy, 8 Tower View Park, Kildare, Ireland.

Korean Journal of Apiculture. The Apicultural Society of Korea, Dept. of Applied Ent., College of Ag., Seoul National University, Suwon 441-744, Korea.

National Honey Market News. USDA Ag. Marketing Service, Fruit and Vegetable Division, 2015 South First St., Room 4, Yakima, WA 98903

Die Neue Bienenzucht. Hamburger Str. 109, D-2360 Bad Segeberg, Germany.

The New Zealand Beekeeper. Farming House, 211-213 Market Street South, POB 307, Hastings, New Zealand.

Scottish Bee Journal. R. N. H. Skilling, FRSA, 34 Rennie St., Kilmarnock, Ayrshire, Scotland.

The Scottish Beekeeper. A. E. McArthur, ed., Melbourne House Regent 84, Dalmuir G81 3QU, Clydebank, Scotland.

South African Bee Journal. POB 41, Modderfontein, 1645, Republic of South Africa.

Speedy Bee. Box 998, Jesup, GA 31545-0998

Bee Laboratories

Arizona: Carl Hayden Bee Research Lab, 2000 E. Allen Rd., Tucson, AZ 85721.

Canada: Agriculture Canada, Research Branch, Box 29, Beaverlodge, Canada TOH 0C0.

Canada: Biosystematics Research Centre, B149, K. W. Neatby Bldg., Ottawa, Canada K1A OC6.

Louisiana: Honey Bee Breeding and Genetics and Physiology Lab, 1157 Ben Hur Rd., Baton Rouge, LA 70820-5502. ☎ 504-767-9280.

Maryland: Bee Research, Division of Beneficial Insects Lab, Room 209A, Bldg. 476, BARC-East, Beltsville, MD 20705-2350. ☎ 301-504-8205

Texas: Africanized Bee Lab, 2413 East Hwy. 83, Weslaco, TX 78595. ☎ 210-969-4870.

Utah: Bee Biology and Systematics, National Resources Biology Bldg., Utah State University, Logan, UT 84322-5310. ☎ 801-750-2520.

Note: Owing to fiscal restraints, some USDA labs may close in some areas.

Bee Libraries and Museums

Cornell University, Everett F. Phillips Library (M. Quinby, L. L. Langstroth, and Dyce Collections), Ithaca, NY 14850.

Crane, E. Directory of the world's beekeeping museums. *Bee World* 60:9–23 (1979).

Dadant and Sons, Hamilton, IL 62341.

Florida Dept. of Agriculture and Consumer Science, Division of Plant Industry Library, POB 1269, Gainesville, FL 32001.

International Bee Research Association, 18 North Road, Cardiff CF1 3DY, United Kingdom.

Michigan State University, Special Collections Division, East Lansing, MI 48824.

Smithsonian Institution Libraries, Entomology Branch, Washington, DC 20560.

University of California at Davis, Shields Library, Dept. of Special Collections (J. S. Harbison and J. E. Eckert Collections) Davis, CA 95616.

University of Connecticut, Waterbury Branch (P. J. Hewitt Collection), 32 Hillside Ave., Waterbury, CT 06710.

University of Guelph, McLaughlin Library (B. N. Gates Collection), Guelph, Ontario, Canada N1G 2W1.

University of Massachusetts, Morrill Library (C. C. Crampton Collection), Amherst, MA 01003.

University of Minnesota, Entomology Library (F. F. Jaeger Collection), St. Paul, MN 55108.

University of Wisconsin, Steenbock Memorial Library (C. C. Miller Collection), Madison, WI 53706.

USDA National Agricultural Library, Beneficial Insects Branch, Technical Information Systems, Beltsville, MD 20705.

Bee Organizations, North America

American Apitherapy Society, Box 54, Hartland Four Corners, VT 05049.

American Assn. of Professional Apiculturists, M. Ellis, 301 Centennial Mall South, POB 94756, Lincoln, NE 68509.

American Bee Breeders Assn., F. Rossman, POB 905, Moultrie, GA 31776-0905.

American Beekeeping Federation, T. Fore, POB 1038, Jesup, GA 31545.

American Honey Producers Assn., J. Stroope, Rt. 3, Box 258, Alvin, TX 77511.

American Mead Assn., POB 4666, Grand Junction, CO 81502.

Apiary Inspectors of America, B. Smith Jr., 50 Harry S. Truman Parkway, Annapolis, MD 21401.

California Bee Breeders Assn., Fred Selby, 2624 Cramer Lane, Chico, CA 95926.

Canadian Assn. of Professional Apiculturalists, Dr. G. Otis, Dept. of Environmental Biology, University of Guelph, Guelph, Ontario, Canada N1G 2W1.

Canadian Honey Council, L. Gane, Box 1566, Nipawin, Saskatchewan, Canada SOE 1E0.

Eastern Apicultural Society, L. Suprenant, Sec., Box 330A, County Home Road, Essex, NY 12936.

Mid-U.S. Honey Producers Marketing Assn., G. Reynolds, Box 363, Concordia, KS 66901.

National Honey Board, B. Smith, Executive Director, 390 Lashley St., Longmont, CO 80501-1421.

National Honey Packers and Dealers Assn., R. Sullivan, 5 Ravine Dr., POB 776, Matawan, NJ 07747.

State beekeeping organizations: check with your state agriculture department or the April issue of *Bee Culture* magazine.

Western Apicultural Society, N. Stewart, 2110 X Street, Sacramento, CA 95818.

Bee Sting Reaction

Barr, S. E. Allergy to Hymenoptera stings: Review of world literature, 1953–1970. *Ann. Allergy* 29:49–66 (1971).

Busse, W. W. Allergic reactions to insect stings. *In* H. F. Conn, ed., *Conn's Current Therapy,* pp. 651–653. Philadelphia: Saunders, 1981.

Busse, W. W., O. E. Reed, et al. Immunotherapy in bee sting anaphylaxis. *J. American Medical Assn.* 231(11): 1154–1156 (1975).

Camazine, S. Hymenopteran stings: Reactions, mechanisms, and medical treatment. *Bull. Ent. Society of America* 17–21 (1988).

Frankland, A. W. Treatment of bee sting reactions. *Bee World* 44(1):9–12 (1963).

Frankland, A. W. Bee sting allergy, *Bee World* 57(4): 145–150 (1976).

Riches, H. R. C. Hypersensitivity to bee venom. IBRA reprint M109. *Bee world* 63:7–22 (1982).

Schmidt, J. O. Allergy to venomous insects. *In* J. E. Graham, ed., *The Hive and the Honey Bee.* Hamilton, Ill.: Dadant and Sons, 1992.

Schmidt, J. O. Toxicology of venoms from the honey bee genus *Apis. Toxicon* 33:917–927 (1995).

Visscher, P. K., R. S. Vetter, and S. Camazine. Removing bee stings. *Lancet* 301–302 (1996).

Yu, Y., A. L. DeWeck, et al. Anti-IgE autoantibodies and bee sting allergy. *Allergy* 50:119–125 (1995).

Beeswax

Beeswax from the Apiary. Advisory leaflet no. 347. Edinburgh, Scotland: H.M.S.O. Press, 1971.

Berthold, R., Jr. *Beeswax Crafting.* Cheshire, Conn.: Wicwas Press, 1993.

Bisson, C. S., G. H. Vansell, and W. B. Dye. *Investigations on the Physical and Chemical Properties of Beeswax.* Bull. no. 716. Washington, D.C.: USDA, 1940.

Brown, R. *Beeswax.* Burrowbridge, Somerset, United Kingdom: BBNO, 1981.

Casteel, D. B. *The Manipulation of the Wax Scales of the Honey Bee.* Bull. no. 161. Washington, D.C.: USDA, 1912.

Coggshall, W. L., and R. A. Morse. *Beeswax: Production, Harvesting, Processing, and Products.* Cheshire, Conn.: Wicwas Press, 1984.

Constable, D., N. Battershil, et al. *Beeswax Craft.* Tumbrige, United Kingdom: Search Press, 1996.

Hepburn, H. R. *Honeybees and Wax: An Experimental Natural History.* New York: Springer-Verlag, 1986.

Root, H. H. *Beeswax.* Brooklyn: Chemical Publishing, 1951.

Taylor, R. *Beeswax Molding and Candle Making: A Guide to the Refinement.* New York: Interlaken, n.d.

Tulloch, A. P. Beeswax: Composition and Analysis. *Bee World* 61(2):47–62 (1980).

Books on Bees and Beekeeping

Adams, J. F. *Beekeeping: The Gentle Craft.* Garden City, N.Y.: Doubleday, 1972.

Aebi, O., and H. Aebi. *The Art and Adventure of Beekeeping.* Santa Cruz, Calif.: Unity Press, 1979.
Aebi, O., and H. Aebi. *Mastering the Art of Beekeeping.* Santa Cruz, Calif.: Unity Press, 1979.
Alford, D. B. *The Life of the Bumblebee.* London: Davis-Poynter, 1978.
Barth, W., ed. *Five Hundred Answers to Bee Questions.* Medina, Ohio: A. I. Root, 1955.
Beekeeping Questions and Answers. Hamilton, Ill.: Dadant and Sons, 1978.
Bonney, R. E. *Hive Management: A Seasonal Guide for Beekeepers.* Pownal, Vt.: Garden Way, 1990.
Bulter, D. G. *The Honeybee.* Oxford: Clarendon, 1949.
Bulter, D. G. *World of the Honeybee.* New York: Macmillan, 1955.
Campion, A. *Bees at the Bottom of the Garden.* London: A. and C. Black, 1984.
Collisen, C. E. *Fundamentals of Beekeeping.* University Park: Pennsylvania State Univ. College of Ag. Ext. Serv., 1984.
Crane, E. *The Archaeology of Beekeeping.* Ithaca, N.Y.: Cornell Univ. Press, 1983.
Crane, E. *Bees and Beekeeping: Science, Practice, and World Resources.* Ithaca, N.Y.: Cornell Univ. Press, 1990.
Crompton, J. A. *Hive of Bees.* New York: Doubleday, 1958.
Dadant, C. P. *First Lessons in Beekeeping.* 1952. Reprint, Hamilton, Ill.: Dadant and Sons, 1978.
Dade, H. A. *Anatomy and Dissection of the Honeybee.* Maidstone, Kent, United Kingdom: Bee Research Assn., 1961.
Delaplane, K. S. *Honey Bees and Beekeeping.* Athens: Univ. of Georgia, CES, 1993.
Dodd, V. *Beemasters of the Past.* West Yorkshire, United Kingdom: Northern Bee Books/Hebden, 1983.
Donovan, R. E. *Hunting Wild Bees.* Tulsa, Okla.: Winchester Press, 1980.
Dyce, E. J. *Beekeeping Terms.* Ithaca, N.Y.: Cornell Univ., Dept. Ent., 1960.
Eckert, J. E., and F. R. Shaw. *Beekeeping.* New York: Macmillan, 1960.
Erickson, Eric H., Jr., S. D. Carlson, and M. B. Garment. *A Scanning Electron Microscope Atlas of the Honey Bee.* Ames: Iowa State Univ. Press, 1986.
Five Hundred Answers to Bee Questions. 1955. Reprint, Medina, Ohio: A. I. Root, 1973.
Free, J. B. *Bees and Mankind.* London: E. Arnold, 1982.
Free, J. B. *Honeybee Biology.* Ilford, United Kingdom: Central Assn. of Beekeepers, 1982.
Gojmerac, W. L. *All About Bees, Beekeeping, and Honey.* New York: Drake Publications, 1976.
Gojmerac, W. L. *Bees, Beekeeping, Honey, and Pollination.* Westport, Conn.: Avi Publications, 1980.
Gould, J. L., and C. Gould. *The Honey Bee.* New York: Scientific American Library, W. H. Freeman, 1988; 2d ed. 1995.
Graham, J. M., ed. *The Hive and the Honey Bee.* Hamilton, Ill.: Dadant and Sons, 1992.
Grout, R. A., ed. *The Hive and the Honeybee.* Hamilton, Ill.: Dadant and Sons, 1975.
Hambleton, H. E. *Honey Bee.* Pub. no. 4494. Washington, D. C.: Smithsonian, 1962.
Hooper, T. *Guide to Bees and Honey.* Emmaus, Pa.: Rodale Press, 1983.
Hoyt, M. *The World of Bees.* New York: Bonanza, Coward, McCann, 1965.
Hubbell, S. *A Book of Bees—and How to Keep Them.* New York: Random House, 1988.
Jaycox, E. *Beekeeping Tips and Topics.* Urbana: Univ. of Illinois Press, 1982.
Kelley, W. T. *How to Keep Bees and Sell Honey.* Clarkson, Ky.: Walter T. Kelley, 1987.
Lindauer, M. *Communication among Social Bees.* Cambridge: Harvard Univ. Press, 1961.
Longwood, W. *The Queen Must Die.* New York: Norton, 1985.
Loring, M. *Bees and the Law.* Hamilton, Ill.: Dadant and Sons, 1981.
Mace, H. *Complete Handbook of Beekeeping.* London: Ward Lock, 1976.
Manley, R. O. B. *Honey Farming.* Halifax, United Kingdom: Northern Bee Books, 1985.
Menzel, R., ed. *Neurobiology and Behavior of Honeybees.* New York: Springer-Verlag, 1987.
Meyer, O. *The Beekeepers Handbook: Manual of Bee Management.* New York: Sterling, 1983.
Meyer, O. *Microscopy on a Shoestring for Beekeepers and Naturalists.* Cheshire, Conn.: Beekeeping Education Service, 1984.
Moffett, J. *Some Beekeepers and Associations.* Cushing, Okla.: J. O. Moffett, 1979.
More, D. *The Bee Book.* New York: Universe Books, 1976.
Morse, R. A. *Bees and Beekeeping.* Ithaca, N.Y.: Cornell Univ. Press, 1975; New York: Linden Books, 1978.
Morse, R. A. *Bees, Beekeeping, Honey, and Pollination.* Westport, Conn.: Avi Publications, 1980.
Morse, R. A. *A Year in the Bee Yard.* New York: Charles Scribner's Sons, 1983.
Morse, R. A. *Complete Guide to Beekeeping.* New York: Dutton, 1996.
Morse, R. A., and P. K. Flottum, eds. *The ABC and XYZ of Bee Culture.* 40th ed. Medina, Ohio: A. I. Root, 1990.
Morse, R. A., and T. Hopper, eds. *Illustrated Encyclopedia of Beekeeping.* Dorset, United Kingdom: Alpha Books, 1985.
Naile, F. *America's Masters of Bee Culture: The Life of L. L. Langstroth.* Ithaca, N.Y.: Cornell Univ. Press, 1976.
Needham, G. R., et al., eds. *Africanized Honey Bee and Bee Mites.* Chichester, United Kingdom: E. Horwood; New York: Halsted Press, 1988.
Onstott, K. *Beekeeping as a Hobby.* New York: Harper and Brothers, 1941.
Potter, A. *The Killer Bees.* New York: Grosset and Dunlap, 1977.
Powell, J. *The World of a Beehive.* Boston: Faber and Faber, 1979.
Ruttner, F. *Biogeography and Taxonomy of Honeybees.* New York: Springer-Verlag, 1988.
Scott, A. A. *Murmur of Bees.* Oxford: Oxford University Press, 1980.
Scott, W. *Backyard Beekeeping.* Dorchester, England: Prism Press, 1977.
Smith, F. G. *Beekeeping in the Tropics.* Bristol United Kingdom: Western Printing Service, 1960.
Taylor, R. *Beekeepers Record and Journal.* Interlaken, N.Y.: Linden Books, 1984.
Taylor, R. *The How-to-Do-It Book of Beekeeping.* Naples, N. Y.: Walnut Press, 1995.
Tew, J. E. *National Beekeeping Publications List.* Wooster, Ohio: CES, 1992.
Tibbets, A. B. *The First Book of Bees.* New York: F. Watts, 1952.
Tompkins, E., and R. M. Griffith. *Practical Beekeeping.* Charlotte, Vt.: Garden Way, 1977.
Trump, R. F. *Bees and Their Keepers.* Ames: Iowa State Univ. Press, 1987.
Vernon, F. *Beekeeping.* Teach Yourself Books. London: Hodder and Stroughton, 1976.

Vivian, J. *Keeping Bees.* Charlotte, Vt.: Williamson Publishing, 1986.
Von Frisch, K. *The Dancing Bees.* London: Methuen, 1954.
Waine, A. C. *Background to Beekeeping.* Hampshire, United Kingdom: Bee Books Old and New, 1979.
Wedmore, E. B. *A Manuel of Beekeeping.* Hampshire, United Kingdom: Bee Books Old and New, 1979.
Weiss, E. A. *The Queen and I.* New York: Harper and Row, 1978.
Whitcomb, H. J. *Bees Are My Business.* New York: C. P. Putnam's Sons, 1955.
Winston, M. L. *The Biology of the Honey Bee.* Cambridge: Harvard Univ. Press, 1987.

Books on Bees - Historic

Alley, H. *The Bee-Keeper's Handy Book.* Boston: A. Wenham, 1883.
Atkins, W., and K. Hawkins. *How to Succeed with Bees.* Watertown, Wis.: G. B. Lewis, 1924.
Bonsels, W. *Adventures of Maya the Bee.* New York: T. Seltzer, 1922.
Coleman, M. L. *Bees in the Garden and Honey in the Larder.* New York: Doubleday/Doran, 1939.
Comstock, A. B. *How to Keep Bees.* New York: Doubleday/Page, 1905.
Cook, A. J. *The Beekeeper's Guide.* East Lansing, Mich.: Ag. College, 1888.
Cowan, T. W. *Wax Craft.* London: Sampson, Low, Marston, 1908.
Dadant, C. P. *First Lessons in Beekeeping.* Hamilton, Ill.: *ABJ,* 1917.
Dadant, C. P. *Dadant System of Beekeeping.* Hamilton, Ill.: *ABJ,* 1920.
Dadant, C. P. *Huber's Observations on Bees.* Hamilton, Ill.: *ABJ,* 1926.
Edwards, T. *The Lore of the Honey-Bee.* London: Methuen, 1923.
Flower, A. B. *Beekeeping Up to Date.* London: Cassell, 1925.
Gilman, A. *Practical Bee Breeding.* New York: J. P. Putnam's Sons, 1929.
Harrison, C. *The Book of the Honey Bee.* London: John Lane/Bodley Head, 1903.
Hawkins, K. *Beekeeping in the South.* Hamilton, Ill.: *ABJ,* 1920.
Herrod-Hempsall, W. *Beekeeping New and Old.* Vols. 1 and 2. London: *British Bee J.,* 1930.
Langstroth, L. L. *Langstroth on the Honey-Bee.* Rev. by Dadant. Hamilton, Ill.: Dadant, 1888.
Latham, A. *Allen Latham's Bee Book.* Hapeville, Ga.: Hale, 1949.
Lyon, D. E. *How to Keep Bees for Profit.* New York: Brosset and Dunlap, 1910.
Maeterlinck, M. *Life of the Bee.* New York: Dodd-Mead, 1924.
Manley, R. O. B. *Honey Farming.* London: Faber and Faber, 1946.
Miller, C. C. *Fifty Years among the Bees.* Medina, Ohio: A. I. Root, 1911.
Miner, T. B. *American Bee Keeper's Manual.* New York: C. M. Saxton, 1850.
Naile, F. *Life of Langstroth.* Ithaca, N.Y.: Cornell Univ. Press, 1942.
Pellett, F. C. *History of American Beekeeping.* Ames, Iowa: Collegiate Press, 1938.
Pellett, F. C. *The Romance of the Hive.* New York: Abingdon, 1931.
Phillips, E. F. *Beekeeping.* Norwood, Mass.: Norwood Press, 1915.
Phillips, E. F. *Beekeeping.* New York: Macmillan, 1943.
Root, A. J. *The ABC of Bee Culture.* 1st ed. Medina, Ohio: A. I. Root, 1877.
Sechrist, E. L. *Honey Getting.* Hamilton, Ill.: Dadant and Sons, 1944.
Sechrist, E. L. *The Bee Master.* Roscoe, Calif.: Earth-master, 1947.
Sechrist, E. L. *Amateur Beekeeping.* New York: Devin-Adair, 1955.
Snelgrove, L. E. *Swarming, Its Control and Prevention.* Weston-Super-Mare, United Kingdom: Snelgrove, 1946.
Spinner, J. *The House Apiary.* Exeter, United Kingdom: Cornwell, 1952.
Stuart, F. S. *City of Bees.* New York: McGraw Hill, 1947.
Sturges, A. M. *Practical Beekeeping.* Philadelphia: David McKay, 1924.
Teale, E. W. *The Golden Throng.* New York: Dodd-Mead, 1945.
Webb, A. *Beekeeping for Profit and Pleasure.* New York: Macmillan, 1944.
Wedmore, E. B. *A Manual of Beekeeping for English-Speaking Beekeepers.* London: Longmans, Green, 1932.

Courses

Fairview College, Box 3000, Fairview, Alberta, Canada T0H 1L0.
Technical and Further Education (TAFE) Information Centre, 47 York Street, Sydney, New South Wales 2000, Australia. Phone: 61-2-9299-5011. Fax: 61-2-9299-5298. Email: TAFEIC@tafensw.edu.au. For information on their beekeeping course: <http://www.tafensw.edu.au/cgi-bin/rdbweb/handbook/XGETCOURSE?VCOURSE No=535>. For a course list (the above course is under "Beekeeping"): <http://www.tafensw.edu.au/handbook/tafeinfo/welcome.html>.
Telford Rural Polytechnic, Nick McKenzie, Freepost 73901, Apiculture Secretary, PB 6, Balclutha, New Zealand.

Diseases and Pests

Akratanakul, P. *Honeybee Diseases and Enemies in Asia: A Practical Guide.* Rome: FAO, 1987.
Bailey, L. *Honey Bee Pathology.* New York: Academic Press, 1981.
Bailey, L., and B. V. Ball. *Honey Bee Pathology.* 2d ed. London: Academic Press, 1991.
Burke, P. W. *Bee Diseases and Pests of the Apiary.* Pub. no. 429. Ontario, Canada: Ministry of Ag. and Food, n.d.
Directions for Sending Diseased Brood and Bees for Diagnosis. Pub. no. CA-33-30. Beltsville, Md.: USDA-ARS, Ent. Research Division, 1967.
Hansen, H. *Honey Bee Brood Diseases.* Ithaca, N.Y.: Wicwas Press, 1977.
Identifying Bee Diseases in the Apiary. Ag. Information Bull. no. 313. Washington, D. C.: GPO, 1967.
Matheson, A. World bee health report. *Bee World* 77:45–51 (1996).
Morse, R. A., and R. Nowogrodzki, eds. *Honey Bee Pests,*

Predators, and Diseases. 2d ed. Ithaca, N.Y.: Cornell Univ. Press, 1990.
Nosema and Amoeba. Advisory leaflet no. 473. Edinburgh, Scotland: H.M.S.O. Press, 1972.
Shimanuki, H., and D. Knox. *Diagnosis of Honey Bee Diseases.* USDA Ag. handbook AH-690. Washington, D. C.: USDA, 1991.

Equipment

Anderson, E. J. An improved solar wax extractor. *Pennsylvania State Univ. Ag. Report* 225 (1960).
Bell, R. My homemade hive scale. *ABJ* 2:97 (1979).
Briscoe, D. A. Using the sun to melt wax. *New Zealand Journal of Beekeepers* (November): 10–13 (1970).
Build Your Own Beehive. Plan P-4. Charlotte, Vt.: Garden Way Plans, 1983.
Build Your Own Honey Extractor. Plan P-1. Charlotte, Vt.: Garden Way Plans, 1982.
Busker, L. H. The WATDIT hive stand. *Bee Culture* 10:521–525 (1970).
Crane, E., and A. J. Graham. *Beehives of the Ancient World.* Reprint no. M117. London: IBRA, 1985.
Detroy, B. F., E. H. Erickson, and K. Diehnett. Plastic hive covers for outdoor winterizing of honey bees. *ABJ* 12:583–587 (1982).
Detroy, B. F., and E. R. Harp. *Trapping Pollen from Honey Bee Colonies.* Pub. no. AZ-PRR-163. Madison: USDA-ARS and Univ. of Wisconsin Extension, 1977.
Dyce, E. J. *Wood Preservatives and Their Application.* Ithaca, N.Y.: Cornell University, CES, 1950.
Gilberd, D. J. Make your own bale weigher. *New Zealand Farmer* 9:57 (1986).
Gojmerac, W. L. *Building and Operating an Observation Beehive.* Pub. no. A2491. Madison: Univ. of Wisconsin, Ext. Serv. 1978.
Goodman, R. D. A solar beeswax extractor. *Australian Beekeeper* (November):113–115 (1980).
Harding, J. P. A simple method of weighing a hive. *Bee World* 43(2):40–41 (1962).
Harp, E. R. *A Simplified Pollen Trap for Use on Colonies of Honey Bees.* Pub. no. 33-111. Beltsville, Md.: USDA-ARS, Ent. Research Division, 1966.
How to Construct and Maintain an Observation Bee Hive. Leaflet no. 2853. Berkeley, Calif.: CES, 1976.
Introducing hive monitor weighing base, a revolutionary advance in hive management. *Bee Culture* 105(4):138 (1977).
Jaycox, E. R. *Making and Using a Solar Wax Melter.* No. H-680. Urbana-Champaign: Univ. of Illinois, 1973.
Kalnins, M. A., and B. F. Detroy. Effect of wood preservative treatment of beehives on honeybees and hive products. *J. Ag. and Food Chemistry* 32:1176–1180 (1984).
Lesher, C., and R. A. Morse. The efficiency of solar wax extractors. *ABJ* 12:820–821 (1982).
Miller, S. R. *Let's Build a Bee Hive.* Phoenix, Ariz.: Miller (2028-A W. Sherman St.), 1976.
Nobbs, Rev. E. *Make Your Own Skep and Revive a Lost Art.* VBBA no. 8. Derby, United Kingdom: Village Bee Breeders' Assn. (VBBA), 1969.
Plans and Dimensions for a 10-Frame Bee Hive. Pub. no. CA 33-24. Beltsville, Md.: USDA-ARS.
Seeley, T. D., and R. A. Morse. *Bait Hives for Honey Bees.* Ithaca, N.Y.: Cornell Univ., CES, 1982.
Shaw, F. R. An improved device for weighing colonies. *ABJ* 96(8):322 (1956).
Showler, K. *The Observation Hive.* Steventon, United Kingdom: Bee Books Old and New, 1978.
Solar Beeswax Extractor. Pennsylvania State Univ. Ag. Ext. Serv. plan no. 790-301. 1969.
Stanger, W., and R. A. Parson. *Beehive: California Plan.* Pub. no. 217. Berkeley: Univ. of California, CES, 1974.
Al-Tikrity, W. S., R. C. Hillmann, A. W. Benton, and W. W. Clarke, Jr. Three methods for weighting honey-bee colonies in the laboratory and field. *ABJ* 111(4):143–145 (1971).

Feeding Bees

Barker, R. J. Influence of diet on sugars found by thin layer chromatography in thoraces of honey bees, *Apis mellifera* L. *J. Experimental Zoology* 188:157–164 (1974).
Barker, R. J. Laboratory comparisons of high fructose corn syrup, grape syrup, honey, and sucrose syrup as maintenance food for caged honey bees. *Apidologie* 9:111–116 (1978).
Erickson, E. H., and E. W. Herbert Jr. Soybean products replace expeller processed soyflour for pollen supplements and substitutes. *ABJ* 120:122–126 (1980).
Haydak, M. H. Honey bee nutrition. *Ann. Rev. Ent.* 15:143–155 (1970).
Johansson, T. S. K. *Feeding Honeybees Pollen and Pollen Substitutes.* London: IBRA, 1977.
Johansson, T. S. K., and M. P. Johansson. *Some Important Operations in Bee Management.* London: IBRA, 1978.
Lehner, Y. Nutritional considerations in choosing protein and carbohydrate sources for use in pollen substitutes for honeybees. *JAR* 22(4):242–248 (1983).
Little, L. H. *How to Feed Bees.* Circ. no. 517. Knoxville: Tennessee Dept. of Ag. and Univ. of Tennessee, Ag. Ext. Serv., n.d.
Loper, G. M., and R. L. Berdel. The effects of nine pollen diets on broodrearing of honey bees. *Apidologie* 11(4):51–59 (1980).
Standifer, L. N., M. H. Haydak, J. P. Mills, and M. D. Levin. Influence of pollen in artificial diets on food consumption and brood production in honey bee colonies. *ABJ* 113(3):94–95 (1973).

Honey and Honey Products

Arnon, S. S. Honey and other environmental risk factors for infant botulism. *J. Pediatrics* 94:331 (1979).
Bauer, L., A. Kohlich, et al. Food allergy to honey: Pollen or bee products? Part 1. *J. Allergy and Clinical Immunology* 97:65–73 (1996).
Beck, B. F. *Honey and Health.* New York: McBride, 1938.
Beck, B. F., and D. Smedley. *Honey and Your Health.* New York: McBride, 1944.
Crane, E., ed. *Honey: A Comprehensive Survey.* London: Heinemann and IBRA, 1975.
Crane, E. *A Book of Honey.* Oxford: Oxford Univ. Press, 1980.
Demuth, G. S. *Comb Honey.* USDA farmers' bulletin no. 503. Washington, D. C.: GPO, 1912.
Dyce, E. J. *Finely Crystallized or Granulated Honey: General Information.* Ithaca, N.Y.: Cornell Univ., Dept. Ent., 1961.
Dyce, E. J. *Marketing Honey.* Ithaca, N.Y.: Cornell Univ. Press, 1961.
Dyce, E. J. *Some Basic Requirements for an Efficient Honey House.* Ithaca, N.Y.: Cornell Univ., Dept. Ent., 1961.

Dyce, E. J., and R. A. Morse. *Methods of Removing Honey from Colonies.* Ithaca, N.Y.: Cornell Univ., Dept. Ent., 1968.
Field, O. *Honey by the Ton.* London: Barn Owl Books, 1983.
Gojmerac, W. L. *What You Should Know about Honey.* Madison, Wis.: Eureka Valley Enterprises, 1981.
Gojmerac, W. L. *Honey: Guide to Efficient Production.* Bull. no. A-2083. Madison, Wis.: Univ. of Wisconsin, Ext. Serv., 1982.
Helbling, A., C. Peter, et al. Allergy to honey in relation to pollen and honey bee allergy. *Allergy* 47:41–49 (1992).
Honey Market News. Fruit and Vegetable Division, Ag. Market Service, Washington, D.C.: USDA Ag. Market Services, ongoing.
Horton, W., and I. S. Thursby. *Florida Honey and Beekeeping.* Bull. no. 66. Tallahassee: Florida Dept. of Ag., 1958.
Huhtanen, C. N. Incidents and origin of *Clostridium botulinum* in honey. *J. Food Protection* 44:812 (1981).
Hunt, C. L. *Honey and Its Uses in the Home.* USDA farmers' bulletin no. 653. Washington, D. C.: GPO, 1924.
Jarvis, D. C. *Folk Medicine.* New York: Holt, 1958.
Killion, C. E. *Honey in the Comb.* 1951. Reprint, Paris, Ill.: Killion and Sons, 1981.
Morse, R. *Notes and Annotated Bibliography on the Manufacture of Honey Jelly.* Ithaca, N.Y.: Cornell Univ., Dept. Ent., 1957.
Morse, R. *Comb Honey Production.* Ithaca, N.Y.: Wicwas Press, 1978.
National Honey Board. B. Smith, Executive Director, 390 Lashley St., Longmont, CO 80501-1421. Web site: www.nhb.org.
Pascoe, T. Beekeeping and Back Pain. *Bee Biz* 2:10–11 (1996).
Penner, L. R. *The Honey Book.* New York: Hastings House, 1980.
Sechrist, E. L. *Honey Getting.* Hamilton, Ill.: Dadant, 1944.
Selling Honey. Medina, Ohio: A. I. Root, 1994.
Smith, M. R., W. F. McCaughey, and A. R. Kemmerer. Biological effects of honey. *JAR* 8:99–110 (1969).
Style, S. *Honey: From Hive to Honeypot.* San Francisco: Chronicle books, 1993.
Sugiyama, H. Number of *Clostridium botulinum* spores in honey. *J. Food Protection* 41:848 (1978).
Taylor, R. *How to Raise Beautiful Comb Honey.* Interlaken, N.Y.: Linden Books, 1977.
Taylor, R. *The New Comb Honey Book.* Interlaken, N.Y.: Linden Books, 1981.
Thein, F. C., R. Leung, et al. Asthma and anaphylaxis induced by royal jelly. *Clinical and Experimental Allergy* 26:216–222 (1996).
Tonsley, C. *Honey for Health.* New York: Award Books, 1970.

Honey Cookbooks

Bass, L. L. *Honey and Spice.* Ashland, Ore.: Coriander Press, 1983.
Beck, B. F., and D. Smedley. *Honey and Your Health.* New York: Dodd, Mead, 1944.
Berto, H. *Cooking with Honey.* New York: Gramercy Publishing, 1972.
Brogdon, P. C. *Preserve with Honey?* Chattanooga, Tenn.: CES, n.d.
Charlton, J., and J. Newdick. *A Taste of Honey.* Edison, N.J.: Chartwell Books, 1995.
Davenport, M. *Cooking with Honey!* Tigard, Ore.: Paddlewheel Press, 1992.
Drops of Gold. Ohio Dept. of Ag. Columbus, Ohio: National Graphics, n.d.
Ellison, V. H. *The Pooh Cook Book.* New York: E. P. Dutton, 1969.
Geiskopf, S. *Putting It Up with Honey.* Ashland, Ore.: Quicksilver Production, 1979.
Gems of Gold, with Honey. Whittier, Calif.: Honey Advisory Board, n.d.
Gross, S. *The Honey Book.* St. Charles, Ill.: Kitchen Harvest Press, 1974.
Elkon, J. *The Honey Cookbook.* New York: Knopf, 1975.
Hanson, L. G., and L. A. Davis. *The Basic Beekeeping and Honey Book.* New York: D. McKay, 1977.
Harmon, A., and E. Miner Jr. *Kitchen Creations with Honey.* Frederick, Md.: Fredericktown Printing Services, 1979.
Honey Cook Book. Medina, Ohio: A. I. Root, 1991.
The Honey Kitchen. Hamilton, Ill.: Dadant and Sons, 1980.
Honey Sales. Pub. no. PL 89-755. Washington, D. C.: Food and Drug Administration, n.d.
Honey Sampler. Chicago: American Honey Advisory Board, n.d.
Kees, B. *Cook with Honey.* Brattleboro, Vt.: Stephen Green Press, 1973.
Kiser, M., and V. Nessie. *Honey Naturally.* Sonoma: California Honey Advisory Board, 1983.
Kraynek, S. L. D. *A Honey Cook Book.* Medina, Ohio: A. I. Root, 1991.
Lonic, L. *The Healthy Taste of Honey.* Virginia Beach, Va.: Donning, 1981.
LoPinto, M. *Eat Honey and Live Longer.* New York: Twayne, 1957.
Norman, J. *Honey.* New York: Bantam, 1990.
Old Favorite Honey Recipes. Madison, Wis.: American Honey Institute, 1945.
Parkhill, J. *No Diet Way to Health, Beauty, and Happiness.* Berryville, Ark.: Country Bazaar Publishing, 1979.
Parkhill, J. *Honey—God's Gift.* Berryville, Ark.: County Bazaar Pubishing, 1981.
Parkhill, J. *Nature's Golden Treasure Honey Cookbook.* Berryville, Ark.: Country Bazaar Publishing, 1981.
Parkhill, J. *Wonderful World of Honey.* Berryville, Ark.: Country Bazaar Publishing, 1981.
Perlman, D. *The Magic of Honey.* New York: Galahad, 1971.
Tonsley, C. *Honey for Health.* Geddington, United Kingdom: British Bee Publications, 1980.
White, J. *Honey in the Kitchen.* London: Morrison and Givy, 1978.

Honey Plants

Arnold, L. E. *Some Honey Plants of Florida.* Bull. no. 548. Gainesville: Univ. of Florida, Ag. Exp. Sta., 1954.
Ayers, G. S., and J. R. Harman. Bee forage of N. A. and the potential for planting for bees. *In* G. Graham, ed., *The Hive and the Honey Bee,* pp. 437–535. Hamilton, Ill.: Dadant and Sons, 1992.
Blake, S. T., and C. Roff. *The Honey Flora of Queensland.* Brisbane, Australia: Dept. Ag. and Stock, 1959.

Burgett, D. M., B. A. Stringer, and L. D. Johnston. *Nectar and Pollen Plants of Oregon and the Pacific Northwest.* Blogett, Ore.: Honeystone Press, 1989.
Crane, E. *Surveying the World's Honey Plants.* Ilford, United Kingdom: Central Assn. of Beekeepers, 1983.
Crane, E., P. Walker, and R. Day. *Directory of Important World Honey Sources.* London: IBRA, 1984.
Crompton, C. W., and W. A. Wojtas. *Pollen Grains of Canadian Honey Plants.* Pub. no. 1892/E. Ottawa, Ontario: Ag. Canada, 1993.
Hooper, T., and M. Taylor. *The Beekeepers' Garden.* London: Alphabooks, 1988.
Howes, F. N. *Plants and Beekeeping.* London: Faber and Faber, 1979.
Kirk, W. D. J. *A Colour Guide to Pollen Loads of the Honey Bee.* Cardiff, United Kingdom: IBRA, 1994.
Lovell, H. B. *Honey Plants Manual.* Medina, Ohio: A. I. Root, 1956.
Lovell, J. H. *Honey Plants of North America.* Medina, Ohio: A. I. Root, 1926.
Matheson, A., ed. *Forage for Bees in an Agricultural Landscape.* Cardiff, United Kingdom: IBRA, 1994.
Milum, V. G. *Illinois Honey and Pollen Plants.* Urbana: Univ. of Illinois, Dept. Ent., 1943.
Morton, J. F. *Honeybee Plants of South Florida.* Reprinted from vol. 77 of the *Proceedings of the Florida State Horticultural Society.* Miami: Florida State Horticultural Society, 1964.
Nectar and pollen plants. *In Beekeeping in the United States.* Handbook no. 335. Washington, D. C.: USDA-ARS, 1967.
Pellett, F. C. *Useful Honey Plants.* Hamilton, Ill.: *ABJ,* n.d.
Ramsay, J. *Plants for Beekeeping in Canada.* London: IBRA, 1987.
Sanford, M. T. *A Florida Beekeeping Almanac.* Extension Entomology, Columbus: Ohio State Univ., 1981.
Sanford, M. T. *Florida Bee Botany.* Circ. no. 686. Gainesville, Fla.: Univ. of Florida, CES, 1986.
Sawyer, R. *Pollen Identification for Beekeepers.* Cardiff, United Kingdom: Univ. College of Cardiff Press, 1981.
Sawyer, R. *Honey Identification.* Cardiff, United Kingdom: Cardiff Academic Press, 1988.
Scullen, H. A., and G. H. Vansell. *Nectar and Pollen Plants of Oregon.* Station Bull. no. 412. Corvallis: Oregon State Ag. Ext. Serv. and USDA, 1942.
Taylor, R. *Beekeeping for Gardeners.* Interlaken, N.Y.: Linden Books, 1981.
Vansell, G. H. *Nectar and Pollen Plants of California.* Bull. no. 517. Rev. ed. Berkeley, Calif.: College of Ag., Ag. Exp. Sta., 1941.
Williams, H. E. *Nectar and Pollen Sources for Honey Bees.* Pub. no. 1181. Knoxville: Univ. of Tennessee, Ag. Ext. Serv.; Washington, D.C.: USDA-CES, 1988.
Wilson, W. T., J. O. Moffett, and H. D. Harrington. *Nectar and Pollen Plants of Colorado.* Bull. no. 503-5. Ft. Collins: Colorado State Univ., Exp. Sta., 1958.

International Beekeeping

Adjare, S. *Beekeeping in Africa.* Bull. no. 68/6. Rome: FAO, Ag. Serv., 1990.
Attfield, H. H. D. *A Beekeeping Guide.* Bull. no. 9. Mt. Rainier, Md.: VITA, n.d.
Beekeeping in Kenya. Canadian Apiculture Team. Animal Production Division, Nairobi: Ministry of Ag., 1974.
Clauss, B. *A Beekeeping Handbook.* South Africa: Ghanzi District Council, n.d.
Connor, L. J., et al., eds. *Asian Apiculture.* Cheshire, Conn.: Wicwas Press, 1993.
Crane, E. *Bibliography of Tropical Apiculture.* London: IBRA, 1978.
Crane, E., and A. J. Graham. Bee hives of the ancient world. *Bee World* 66:25–41 (1985).
Dutton, R. W., et al., eds. *Honeybees in Oman.* Oman: Sultanate of Oman, 1982.
Farmer's Guide. Glasgow, Scotland: Univ. Press, Jamaica Ag. Society, 1962.
Franssen, C. J. H. *Beekeeping in Java.* Reprint no. 3-1546. London: IBRA, n.d.
Free, J. B. Biology and behavior of the honeybee. *A. florea* and possibilities for beekeeping. *Bee World* 62(2):45–58 (1981).
Lindauer, M. Communication among the honey bees of India. *Bee World* 38:3–24, 34–39 (1957).
Morse, R. A., and F. M. Laigo. *Beekeeping in the Philippines.* Farm Bull. no. 27. Los Baños: Dept. Ag. Communications, 1968.
Sammataro, D. *Lesson Plans for Beekeeping.* Washington, D.C.: U.S. Peace Corps, 1980.
Saubolle, B. R., and R. Bachmann. *Beekeeping: An Introduction to Modern Beekeeping in Nepal.* Kathmandu: Sahayogi Press, 1979.
Smith, F. G. *Beekeeping in the Tropics.* London: Longmans, 1960.
Spence, J. D., D. D. Schlacbach, and A. J. Cummings. *La Apicultura Guía Práctica.* Guatemala: Impreso en Ediciones Fenacoac (3a Avenida 17-40, Zona 1, Nebaj), 1980.
Taylor, F. *Beekeeping for the Beginner.* Bull. no. 199. Pretoria, South Africa: Dept. Ag. and Forestry, 1939.
Verma, L. R. *Honeybees in Mountain Agriculture.* Boulder, Colo.: Westview Press, 1992.
Weaver, N., and E. C. Weaver. Beekeeping with the stingless bee *Melipona beecheii,* by the Yucatcán Maya. *Bee World* 62(1):7–19 (1981).

Management of Bee Colonies

Arnott, J. H., and S. E. Bland. *Beekeeping in Saskatchewan.* Regina: Saskatchewan Dept. Ag., 1954.
Beekeeping. ARS, Insect Notes no. 1. Raleigh, N.C.: North Carolina State Univ., 1974.
Beekeeping for Beginners. Home and Garden bull. no. 158. Washington, D.C.: GPO, 1974.
Beekeeping in Alaska. Pub. no. 701. Juneau, Alaska: Univ. of Alaska, CES, n.d.
Bees . . . for Pleasure and Profit. Medina, Ohio: A. I. Root, n.d.
Boggs, N. *Beekeeping in Colorado.* Circ. no. 41. Fort Collins: Colorado Ag. College, 1923.
Bonney, R. E. *Hive Management.* Pownal, Vt.: Garden Way Publishing, 1990.
Breed, J. D. How honeybees recognize their nestmates. *Bee World* 66(3):113–118 (1985).
Brown, R. *Honey Bees: A Guide to Management.* North Pomfret, Vt.: Crowood Press, 1988.
Burgin, C. J. *Introduction to Beekeeping.* Extension Entomology, no. B-153. College Station: Texas A&M Univ., 1974.
Caron, D. M. *Ten Trips for Suburban Beekeepers.* Leaflet no. 75. College Park: Univ. of Maryland, CES, 1973.

Caron, D. M. *Beekeeping in Maryland.* Bull. no. 223. 1967. Reprint, College Park: Univ. of Maryland, CES, 1975.

Clark, W. W., Jr. *Pennsylvanian Beekeeping.* Circ. no. 544. University Park: Pennsylvania State Univ., Ext. Serv., 1971.

Crane, E., and P. Walker. *The Impact of Pest Management on Bees and Pollination.* London: Tropical Development and Research Institute, 1983.

Crane, E., P. Walker, and R. Day. *Directory of Important World Honey Sources.* London: IBRA, 1984.

Dyce, E. J., and R. A. Morse. *Wintering Honey Bees in New York State.* Bull. no. 1054. Ithaca, N.Y.: Cornell Univ., CES, 1960.

Eckert, J. A., and H. A. Bess. *Fundamentals of Beekeeping in Hawaii.* Bull, no. 55. Honolulu: Univ. of Hawaii, Ext. Serv., 1952.

Farrar, C. L. *Life of the Honey Bee.* Extension bull. no. A-2279. Madison: Univ. of Wisconsin, Ext. Serv., 1967.

Haydak, M. H. *Beekeeping in Minnesota.* Bull. no. 204. St. Paul: Univ. of Minnesota, Ext. Serv., 1968.

Jaycox, E. R. *Beekeeping in Illinois.* Circ. no. 1000. Champaign-Urbana: Univ. of Illinois, CES, 1969.

Jaycox, E. R. *Destroying Bees and Wasps.* Circ. no. 1011. Champaign-Urbana: Univ. of Illinois, CES, 1969.

Jaycox, E. R. *Beekeeping in the Midwest.* Champaign-Urbana: Univ. of Illinois, CES, 1976.

Johansson, C. *Beekeeping.* Bull. no. 79. Corvallis, Ore.: Oregon State Univ., 1974.

Johansson, T. S. K., and M. P. Johansson. Wintering. *Bee World* 50(3):89–100 (1969).

Johansson, T. S. K., and M. P. Johansson. *Some Important Operations in Bee Management.* London: IBRA, 1978.

Johansson, T. S. K., and M. P. Johansson. The honeybee colony in winter. *Bee World* 60:155–170 (1979).

Kissinger, W. A. *Beekeeping in Montana.* Manuscript. Helena, Mont.: Dept. Ag., n.d.

Little, L. H., and L. D. Wallace. *A Bee Book for Beginners.* Nashville, Tenn.: Dept. Ag., 1972.

Martin, E. C. *Basic Beekeeping.* Farm Science Series no. E625. 1971. Revised, East Lansing: Michigan State Univ., 1975.

McCutcheon, D. M. *Indoor wintering of hives.* IBRA reprint no. M113. *Bee World* 65:19–37 (1984).

Moeller, F. E. *Overwintering of Honey Bee Colonies.* Products Research Report no. 169. Madison, Wis.: USDA-ARS and Univ. of Wisconsin, 1978.

Morse, R. A. *The Honeybee and Its Relatives.* Ithaca, N.Y.: Cornell Univ., CES, 1966.

Morse, R. A. *Florida Beekeeping.* Vol. 2, Bull. no. 10. Gainesville, Fla.: State Plant Board, 1956.

Mussen, E. C., L. Foote, et al. *Beekeeping in California.* Pub. no. 21422. Oakland: Univ. of California, CES, 1987.

O'Dell, W. T. *Beekeeping in South Carolina.* Clemson: Univ. of South Carolina, Ext. Serv., n.d.

Owens, C. D. *The Thermology of Wintering Honey Bee Colonies.* Bull. no. 1429. Washington, D.C.: USDA-ARS, 1971.

Phillips, E. F. *The Preparation of Bees for Outdoor Wintering.* USDA farmers' bull. no. 1012. Washington, D.C.: GPO, 1921.

Phillips, E. F. *Beekeeping in the Buckwheat Region.* USDA farmers' bull. no. 1216. Washington, D.C.: GPO, 1922.

Phillips, E. F. *Beekeeping in the Tulip-Tree Region.* USDA farmers' bull. no. 1222. Washington, D.C.: GPO, 1922.

Pike, H. A. *Beekeeping in Rhode Island.* Providence, R.I.: Dept. Ag. and Conservation, 1947.

Scheibner, R. A. *Beginning Beekeeping for Kentuckians.* No. 361. Lexington: Univ. of Kentucky, n.d.

Scott, H. E., R. C. Hillmann, and H. F. Greene. *Honey Bees in North Carolina.* Pub. no. 512. Raleigh: Univ. of North Carolina, Ag. Ext. Serv., 1975.

Shade and Water for the Honey Bee Colony. Leaflet no. 530. Washington, D.C.: GPO, 1964.

Shaw, F. R. *Beekeeping.* Pub. no. 148. Amherst: Univ. of Massachusetts, CES, 1963.

Shaw, F. R. *Bee Management through the Year.* USDA, Dept. Ent. Insect Information. Amherst: Univ. of Massachusetts, 1955.

Smith, M. V. *Caring for Bees in Schools.* Pub. no. 169. Toronto, Ontario: Dept. Ag., 1971.

Standifer, L. N. *Beekeeping in the United States.* Ag. handbook no. 335. 1967. Revised, Washington, D.C.: USDA, 1980.

Stephen, W. A. *Ohio Bee Lines.* Bull. no. 450. Columbus: Ohio State Univ., CES, 1971.

Taylor, R. *The Best of "Bee Talk."* Medina, Ohio: A. I. Root, 1988.

Thurber, P. F. *Bee Chats, Tips, and Gadgets.* Hamilton, Ill.: Hamilton Press, 1986.

Townsend, G. F., and P. W. Burke. *Beekeeping in Ontario.* Pub. no. 490. Guelph: Ministry of Ag. and Food, n.d.

Walstrom, W. T., B. H. Kantack, and W. L. Berndt. *Beekeeping in South Dakota.* Bull. no. EC565. Brookings: South Dakota State Univ., CES, n.d.

Williams, H. *Beekeeping in Tennessee.* Pub. no. 697. Knoxville: Univ. of Tennessee, ARS, 1975.

Wilson, W. T., and J. W. Brewer. *Beekeeping in the Rocky Mountain Region.* Fort Collins, Colo.: CES, 1974.

Miscellaneous Hive Products (excluding honey and beeswax)

Andrews, S. W. *All about Mead.* West Yorkshire, United Kingdom: Northern Bee Books, 1982.

Beck, B. F. *Bee Venom Therapy.* New York: D. Appleton-Century, 1935.

Broadman, J. *Bee Venom: The Natural Curative for Arthritis and Rheumatism.* Putnam, 1962.

Ghisalberti, E. L. *Propolis.* Reprint no. M99. London: IBRA, 1979.

Hocking, B., and F. Matsumura. Bee brood and food. *Bee World* 41(5):113–120 (1960).

Johansson, T. S. Royal jelly. *Bee World* 36:3–13 (1955).

Malone, F. *Bees Don't Get Arthritis.* New York: Dutton, 1974.

Marcucci, M. C. Propolis: Chemistry, composition, biological properties, and therapeutic activity. *Apidologie* 26:83–99 (1995).

Morse, R. A. *Making Mead.* Ithaca, N.Y.: Wicwas Press, 1980.

Morse, R. A. *General Information on Making Honey Wine at Home.* Ithaca, N.Y.: Cornell Univ., Dept. Ent., 1966.

Mraz, C. *Health and the Honeybee.* Burlington, Vt.: Queen City Publications, 1995.

Murat, F. *Propolis: The Eternal Natural Healer.* Miami, Fla.: 1982.

Parkhill, J. *Wonderful World of Pollen.* Berryville, Ark.: County Bazaar Publishing, 1982.

Rose, A. *Bee in Balance.* Bethesda, Md.: Starpoint Enterprize, 1994.

Simics, M. A. *Review of Bee Venom Collecting and More.* Calgary, Alberta: Apitronic Services, 1994.

Stein, I. *Royal Jelly.* Northamptonshire, United Kingdom: Thorsons Pub., 1989.

Taylor, R. L., and B. J. Carater. *Entertaining with Insects.* Yorba Linda, Calif.: Salutek Publishing, 1992.

Yoirish, N. *Curative Properties of Honey and Bee Venom.* San Francisco: New Glide Publishing, 1959.

Mites

Acarapis woodi in the United States. *ABJ* 124:805–806 (1984).

Apimondia: Program and summaries of reports, pp. 88–109. Thirty-fourth International Apicultural Congress, Lausanne, Switzerland, 1995.

Bienefeld, K., J. Radtke, and F. Zautke. Influence of thermoregulation within honeybee colonies on the reproduction success of *Varroa jacobsoni* Oud. *Apidologie* 26:329–330 (1995).

Calderone, N. W., and M. Spivak. Plant extracts for control of the parasite mite *Varroa jacobsoni* (Acari: Varroidae) in colonies of the western honey bee (Hymenoptera: Apidae). *J. Econ. Ent.* 88(5):1211–1215 (1995).

Cavalloro, R., ed. Varroa jacobsoni Oud. *Affecting Honey Bees.* Rotterdam, Netherlands: A. A. Balkema (POB 1675, 3000 BR), 1983.

Cox, R. L., J. O. Moffett, et al. Effects of late spring and summer menthol treatment on colony strength, honey production, and tracheal mite infestation levels. *ABJ* 129(8):547–549 (1989).

Danka, R. G., J. D. Villa, et al. Field test of resistance to *Acarapis woodi* (Acari: Tarsonemidae) and of colony production by four stocks of honey bees (Hymenoptera: Apidae). *J. Econ. Ent.* 88:584–591 (1995).

De Jong, D., R. A. Morse, and G. C. Eickwort. Mite pests of honey bees. *Ann. Rev. Ent.,* 27:229–252 (1982).

Delaplane, K. S. Controlling tracheal mites (Acari: Tarsonemidae) in colonies of honey bees (Hymenoptera: Apidae) with vegetable oil and menthol. *J. Econ. Ent.* 85(5):2118–2124 (1992).

Delfinado-Baker, M. *Acarapis woodi* in the United States. *ABJ* 124(11):805–806 (1984).

Dietz, A., and H. R. Hermann. *Biology, Detection, and Control of Varroa jacobsoni: A parasitic mite on honey bees.* Athens: University of Georgia, Dept. Ent., 1988.

Donzé, G., and P. M. Guerin. Behavioral attributes and parental care of Varroa mites parasitizing honeybee brood. *Behavioral Ecology and Sociobiology* 34:305–319 (1994).

Donzé, G., et al. Effect of mating frequency and brood cell infestation rates on the reproductive success of the honeybee parasite *Varroa jacobsoni. Ecological Entomology* 21:17–26 (1996).

Eckert, J. E. Acarapis mites of the honey bee (*Apis mellifera* Linnaeus). *J. Insect Pathology* 3:409–425 (1961).

Eickwort, G. C. Evolution and life-history patterns of mites associated with bees. *In* M. A. Houck, ed., *Mites: Ecological and Evolutionary Analyses of Life-History Patterns,* pp. 218–251. Ithaca, N.Y.: Chapman and Hall, 1993.

Eischen, F. A., W. T. Wilson, et al. Cultural practices that reduce populations of *Acarapis woodi* (Rennie). *ABJ* 128(3):209–211 (1988).

Engels, W. Varroa control by hyperthermia. *In* A. Matheson, ed., *New Perspectives on Varroa,* pp. 115–119. Cardiff, United Kingdom: IBRA, 1995.

Gamber, W. R. Fluvalinate scare should serve as a warning. *ABJ* 130:629 (1990).

Gerson, U., R. Mozes-Koch, and E. Cohen. Enzyme levels used to monitor pesticide resistance in *Varroa jacobsoni. JAR* 30:17–20 (1991).

Giordani, G. Laboratory research work on *Acarapis woodi* Rennie, the causative agent of acarine disease of *Apis mellifera* L. Note no. 5. *JAR* 6:147–157 (1967).

Harnaj, V., ed. *Varroasis: A Honey Bee Disease.* Bucharest, Romania: Apimondia Publishing House, 1977.

Hirst, S. On the mites *Acarapis woodi* (Rennie) associated with Isle of Wight bee disease. *Annals of the Magazine of Natural History* 7:509–519 (1921).

Hoopingarner, R. The time of fall treatment with Apistan® and winter survival of honey bee colonies. *ABJ* 135:535–536 (1995).

Hoppe, H., W. Ritter, and E. W. C. Stephen. The control of parasitic bee mites: *Varroa jacobsoni, Acarapis woodi,* and *Tropilaelaps clareae* with formic acid. *ABJ* 129(11):739–742 (1989).

Imdorf, A., S. Bogdanov, et al. Apilife var: A new varroacide with thymol as the main ingredient. *Bee World* 76(2):77–83 (1995).

Kjer, D. M., D. W. Ragsdale, and B. Furgala. A retrospective and prospective overview of the honey bee tracheal mite, *Acarapis woodi* R. *ABJ* 129: part 1, 25–28; part 2, 112–115 (1989).

Komeili, A. B., and J. T. Ambrose. Electron microscope studies of the tracheae and flight muscles of noninfested, *Acarapis woodi* infested, and crawling honey bees (*Apis mellifera*). *ABJ* 131(4):253–257 (1991).

Kraus, B., and R. E. Page Jr. Effect of vegetable oil on *Varroa jacobsoni* and honey bee colonies. *BeeScience* 3:157–161 (1995).

Lodesani, M., M. Colombo, and M. Spreafico. Ineffectiveness of Apistan® treatment against the mite *Varroa jacobsoni* Oud. in several districts of Lombardy (Italy). *Apidologie* 26:67–72 (1995).

Mattheson, A., ed. *New Perspectives on Varroa.* Cardiff, United Kingdom: IBRA, 1994.

Möbus, B., and L. Connor. *The Varroa Handbook.* Mytholmroyd, Hebden Bridge, United Kingdom: Northern Bee Books, 1988.

Moffett, J. O., R. L. Cox, et al. Menthol reduces winter populations of tracheal mites, Acarapis disease. *Bee World* 11:49–50 (1989).

Nazzi, F., and N. Milani. A technique for reproduction of *Varroa jacobsoni* Oud. under laboratory conditions. *Apidologie* 25:579–584 (1994).

Pettis, J. S., and W. T. Wilson. Life history of the honey bee tracheal mite (Acari: Tarsonemidae). *Annals of the Entomology Society of America* 89:368–374 (1995).

Phelan, L. P., A. W. Smith, and G. R. Needham. Mediation of host selection by cuticular hydrocarbons in the honey bee tracheal mite *Acarapis woodi* (Rennie). *J. Chemical Ecology* 17(2):463–473 (1991).

Sammataro, D. Tracking tracheal mites. *Gleanings Bee Culture* 118:206–208 (1990).

Sammataro, D., S. Cobey, et al. Controlling tracheal mites (Acari: Tarsonemidae) in honey bees (Hymenoptera: Apidae) with vegetable oil. *J. Econ. Ent.* 87:910–916 (1994).

Sammataro, D., and G. R. Needham. Host-seeking behaviour of tracheal mites (Acari: Tarsonemidae) on honey bees (Hymenoptera: Apidae). *Exp. Appl. Acarol.* 20:121–136 (1996).

Smith, A. W., and G. R. Needham. A new technique for the rapid removal of tracheal mites from honey bees for biological studies and diagnosis. *In* G. R. Needham et al., eds., *Africanized Honey Bees and Bee Mites,* pp. 530–534. Chichester, United Kingdom: Ellis Horwood, 1988.

Smith, A. W., R. E. Page Jr., and G. R. Needham. Vegetable oil disrupts the dispersal of tracheal mites,

Acarapis woodi (Rennie), to young host bees. *ABJ* 131(1):44–46 (1991).
Sugden, E. A., K. R. Williams, and D. Sammataro. IXth International Congress of Acarology: A honey bee mite round table. *Bee Culture* 123(2):80–81 (1995).
Wilson, W. T., J. R. Elliott, and J. J. Lackett. Antibiotic treatments last longer. *ABJ* 110:348,351 (1970).

Non-Apis Bee Pollinators

Batra, S. W. T. Solitary bees. *Scientific American* 250:120–127 (1984).
Batra, S. W. T. Japanese hornfaced bees: Gentle and efficient new pollinators. *Pomona* 22:3–5 (1989).
Bohart, G. E. Management of wild bees for the pollination of crops. *Ann. Rev. Ent.* 17:287–312 (1972).
Bohart, G. E., and F. E. Todd. Pollination of seed crops by insects. *USDA Yearbook of Agriculture* 240–246 (1961).
Buchmann, S. L. The ecology of oil flowers and their bees. *Annual Review of Ecology and Systematics* 18:343–369 (1987).
Buchmann, S. L., and G. P. Nabhan. *Forgotten Pollinators.* Washington, D.C.: Island Press, 1996.
Free, J. B., and C. G. Butler. *Bumblebees.* New York: Macmillan, 1959.
Frohlich, D. R. On the nesting biology of *Osmia (Chenosmia) bruneri* (Hymenoptera: Megachilidae). *J. Kans. Ent. Soc.* 56:123–130 (1983).
Griffin, B. L. *The Orchard Mason Bee.* Bellingham, Wash.: Knox Cellars Publishers, 1993.
Hurd, P. D. Jr., E. G. Linsley, and A. E. Michelbacher. Ecology of the squash and gourd bee, *Peponapis pruinosa,* on cultivated cucurbits in California (Hymenoptera: Apoidae). *Smithsonian Contributions to Zoology* (1964).
Kevan, P., ed. Alternative Pollinators for Ontario Crops. *Proceeding of the Entomological Society* Ontario 118 (1987).
Løken, A. Flower-visiting insects and their importance as pollinators. *Bee World* 62:130–140 (1981).
Michelbacher, A. E., P. D. Hurd Jr., and E. G. Linsley. Experimental introduction of squash bee (*Peponapis*) to improve yields of squashes, gourds, and pumpkins. *Bee World* 52:157–166 (1971).
O'Dell, C. R. Improve profits with adequate pollination. *American Fruit Grower* 115:16–17 (1995).
Orchard Bees. Auburn, IN 46706.
O'Toole, C., and A. Raw. *Bees of the World.* London: Blandford Books, 1991.
Parker, F. D., and D. R. Frohlich. Hybrid sunflower pollination by a manageable composite specialist: The sunflower leafcutter bee (Hymenoptera: Megachilidae). *Environmental Entomology* 12:576–581 (1983).
Plowright, R. C., and T. M. Laverty. The ecology and sociobiology of bumble bees. *Ann. Rev. Ent.* 29:175–199 (1984).
Southwick, E. E. Estimating the economic value of honey bees (Hymenoptera: Apidae) as agricultural pollinators in the United States. *J. Econ. Ent.* 85:621–633 (1992).
Stubbs, C. S., F. A. Drummond, and E. A. Osgood. *Osmia ribifloris biedermannii* and *Megachile rotundata* (Hymenoptera: Megachilidae) introduced into the lowbush Blueberry agroecosystem in Maine. *J. Kans. Ent. Soc.* 67:173–185 (1994).
Tepedino, V. J. The pollination efficiency of the Squash bee (*Peponapis pruinosa*) and the honey bee (*Apis mellifera*) on summer squash (*Cucurbita pepo*). *J. Kans. Ent. Soc.* 54:359–377 (1981).
Torchio, P. F. *Osmia ribifloris,* a native bee species developed as a commercially managed pollinator of Highbush Blueberry (Hymenoptera: Megachilidae). *J. Kans. Ent. Soc.* 63:427–436 (1990).
Willis, D. S. Foraging dynamics of *Peponapis pruinosa* (Hymenoptera: Anthophoridae) on pumpkin (*Cucurbita pepo*) in southern Ontario. *Canadian Entomology* 127:167–175 (1995).

Pesticides

Erickson, E. H. Jr., B. J. Erickson, and P. K. Flottum. Honey bees and pesticides. Four-part series. *ABJ* 123(10):724, 726–730; 797–800, 802–805, 814; 860–867 (1983); 124(1):42–45, 50 (1984).
Flottum, P. K., E. H. Erickson Jr., and B. J. Hanny. The honey bee–sweet corn relationship. *ABJ* 123:293–299 (1983).
Johansen, C., and D. R. Mayer. *Pollinator Protection: A Bee and Pesticide Handbook.* Cheshire, Conn.: Wicwas Press, 1990.
Protecting Honey Bees from Pesticides. Pub. no. 544. Washington, D.C.: USDA-ARS, 1972.
Wedberg, J. F., and E. H. Erickson. *Protecting Bees from Pesticides in Wisconsin.* Pub. no. A-3086. Madison: Univ. of Wisconsin, Ext. Serv., 1980.

Pollen and Pollination

Barth, F. G. *Insects and Flowers.* Princeton, N.J.: Princeton Univ. Press, 1991.
Beekeeping by Orchardists in Central Washington. Pub. no. EM-2607. Pullman: Washington State Univ., CES, 1975.
Chagnon, M. Complementary aspects of strawberry pollination by honey and indigenous bees (Hymenoptera). *J. Econ. Ent.* 86:416–420 (1993).
Collison, C. H., and E. C. Martin. Behaviour of honeybees foraging on male and female flower of *Cucumis sativus. JAR* 18:184–190 (1979).
Connor, L. J. *Bee Pollination of Crops in Ohio.* Bull. no. 559. Columbus: Ohio State Univ., CES.
Currie, R. W. Effect of synthetic queen mandibular pheromone sprays on honey bee (Hymenoptera: Apidae) pollination of berry crops. *J. Econ. Ent.* 85: 1300–1306 (1992).
DeGrandi-Hoffman, G. Influence of honey bee (Hymenoptera: Apidae) in-hive pollen transfer on cross-pollination and fruit set in apple. *Environmental Entomology* 15:723–725 (1986).
Delaplane, K. S., P. A. Thomas, and W. J. McLaurin. *Bee Pollination of Georgia Crop Plants.* CES Bull. no. 1106. Athens: Univ. of Georgia, 1994.
Dyce, E. J. *Honeybees and the Pollination Problem.* Ithaca, N.Y.: Cornell Univ., CES, 1960.
Evaluating Honey Bee Colonies for Pollination. A Pacific NW Ext. Pub. PNW 245, 1984.
Free, J. B. *Insect Pollination of Crops.* 2d ed. New York: Academic Press, 1993.
A Guide to Managing Bees for Crop Pollination. Canadian Assn. Professional Apiculturists, 1995.
Higo, H. A. Mechanisms by which honey bee (Hymenoptera: Apidae) queen pheromone sprays enhance pollination. *Annals of the Entomology Society of America* 88:366–373, 1995.
Hodges, D. *The Pollen Loads of the Honeybee.* 1974. Reprint, London: IBRA, 1994.

Jay, S. C. Spatial management of honey bees on crops. *Ann. Rev. Ent.* 31:49–65 (1986).
Kearns, C. A., and D. W. Inouye. *Techniques for Pollination Biologists.* Niwot: University Press of Colorado, 1993.
Lieux, M. H. Minor honeybee plants of Louisiana indicated by pollen analysis. *Economic Botany* 32:418–432 (1978).
Mayer, D. F. Honey bee foraging behavior on ornamental crabapple pollenizers and commercial apple cultivars. *HortScience* 24:510–512 (1989).
Mayer, D. F., C. A. Johansen and D. M. Burgett. *Bee Pollination of Tree Fruits.* Pacific NW Cooperative Extension, 1986.
McGregor, S. E. *Insect Pollination of Crops.* Washington, D.C.:USDA-ARS, 1976.
Meeuse, B. J. D. *The Story of Pollination.* New York: Ronald Press, 1961.
Myers, S. C. Manage pollination for early bloom apples. *American Fruit Grower* 113:26–27 (1993).
Nepi, M. Pollination, pollen viability, and pistil receptivity in *Cucurbita pepo. Annals of Botony* 72:527–536 (1993).
NeSmith, D. S. Variation in the onset of flowering of summer squash as a function of days and heat units. *J. Am. Soc. Hort. Sci.* 119:249–252 (1994).
O'Dell, C.R. Improve profits with adequate pollination. *American Fruit Grower* 115:16–17 (1995).
Proctor, M., P. Yeo, and A. Lack. *The Natural History of Pollination.* Portland, Oreg.: Timber Press, 1996.
Richards, A. J., ed. *The Pollination of Flowers by Insects.* New York: Academic Press, 1978.
Robinson, W. S., R. Nowogrodzik, and R. A. Morse. The value of honey bees as pollinators of U.S. crops. *ABJ* 129(6/7):411–423, 477–487 (1989).
Skinner, J. A. *Making a Pollination Contract.* Pub. no. 1516. Knoxville: Univ. of Tennessee, Ag. Ext. Serv., n.d.
Whynott, D. *Following the Bloom: Across American with the Migratory Beekeepers.* Boston: Beacon Press, 1991.

Queens

Doolittle, G. M. *Scientific Queen Rearing.* Hamilton, Ill.: *ABJ*, 1888.
Harp, E. P. *A Method of Holding Large Numbers of Honey Bee Queens in Laying Condition.* Madison: Univ. of Wisconsin and USDA-ARS, 1969.
Instrumental Insemination of Queen Bees. USDA handbook no. 390. Washington, D.C.: USDA, 1970.
Johansson, T. S. K., and M. D. Johansson. *Queen Introduction.* Flushing: City Univ. of New York, 1971.
Laidlaw, H. H. Jr. Organization and operation of a bee breeding program. *Proceedings of the 10th International Congress of Entomology* 4:1067–1078 (1958).
Laidlaw, H. H. *Instrumental Insemination of Honey Bee Queens.* Hamilton, Ill.: Dadant and Sons, 1977.
Laidlaw, H. H. Jr. *Contemporary Queen Rearing.* Hamilton, Ill.: Dadant and Sons, 1979.
Laidlaw, H. H. Jr., and J. E. Eckert. *Queen Rearing.* Berkeley: Univ. of California Press, 1962.
Mackensen, O., and K. W. Tucker. *Instrumental Insemination of Queen Bees.* Ag. handbook no. 390. Washington, D.C.: USDA, 1970.
Morse, R. A. *Rearing Queen Honey Bees.* 2d ed. Ithaca, N.Y.: Wicwas Press, 1994.
Page, R. E., and H. H. Laidlaw. Closed population honeybee breeding program. *Bee World* 66(2):63–72 (1985).
Queen Management. Medina, Ohio: A. I. Root, 1995.
Rinderer, T. E., ed. *Bee Genetics and Breeding.* New York: Academic Press, 1986.
Rothenbuhler, W. C. Behaviour genetics of nest cleaning in honey bees. I. Responses of four inbred lines to disease-killed brood. *Animal Behavior* 12:578–583 (1964a).
Rothenbuhler, W. C. Behaviour genetics of nest cleaning in honey bees. IV. Responses of F_1 and backcross generations to disease-killed brood. *American Zoologist* 4:111–128 (1964b).
Smailes, R. *Raise Your Own Queens.* Leaflet no. 10. 2d ed. Derby, United Kingdom: British Isles Bee Breeders Assn., 1977.
Spivak, M., and G. S. Reuter. *Successful Queen Rearing.* St. Paul: Univ. of Minnesota, Minnesota Ext. Serv., 1994.
Taber, S. *Breeding Super Bees.* Medina, Ohio: A. I. Root, 1987.

Social Insects

Free, J. B. *The Social Organization of Honeybees.* 1st ed. London: E. Arnold, 1977.
Free, J. B. *Pheromones of social bees.* London: Chapman and Hall, 1987.
Krebs, J. R., and N. B. Davis, eds. *Behavioural Ecology: An Evolutionary Approach.* 3d ed. Oxford, United Kingdom: Blackwell Scientific, 1993.
Michener, C. D. *The Social Behavior of the Bees.* Cambridge: Harvard Univ. Press, Belknap Press, 1974.
Moritz, R. F. A., and E. E. Southwick. *Bees as Superorganisms: An Evolutionary Reality.* New York: Springer-Verlag, 1992.
Papaj, D. R., and A. C. Lewis, eds. *Insect Learning.* New York: Chapman and Hall, 1993.
Ribbands, C. R. *Behavior and Social Life of Honeybees.* New York: Dover, 1964.
Seeley, T. D. *Honeybee Ecology: A Study of Adaptation in Social Life.* Princeton, N.J.: Princeton Univ. Press, 1985.
Seeley, T. D. *The Wisdom of the Hive.* Cambridge: Harvard Univ. Press, 1995.
Von Frisch, K. *Dance Language and Orientation of Bees.* Cambridge: Harvard Univ. Press, Belknap Press, 1967.
Von Frisch, K. *Bees: Their Vision, Chemical Senses, and Language.* Ithaca, N.Y.: Cornell Univ. Press, 1971.
Von Frisch, K. *Animal Architecture.* New York: Harcourt, Brace, Jovanovich, 1973.
Wenner, A. M. *The Bee Language Controversy: An Experience in Science.* Boulder, Colo.: Educational Programs Improvement, 1971.
Wenner, A. M., and P. H. Wells. *Anatomy of a Controversy: The Question of a "Language" among Bees.* New York: Columbia Univ. Press, 1990.
Wilson, E. O. *The Insect Societies.* Cambridge: Harvard Univ. Press, Belknap Press, 1971.
Wilson, E. O. *Sociobiology.* Cambridge: Harvard Univ. Press, Belknap Press, 1975.

Suppliers, Foreign

Chr. Graze K. G., 7057 Endersbach bei, Stuttgart, Germany.
Exeter Bee Supplies, Merrivale Rd., Exeter Rd. Ind. Estate, Okehampton, Devon EX20 1UD, United Kingdom.
Steele and Brodie, Beehive Works, Kilmany Td. Wormit, Newport-on-Tay, Fife DD6 8PG, United Kingdom.

E. H. Thorne, Beehive Works, Wragby, Lincolnshire, LN3 5LA, United Kingdom.

Suppliers, Major U.S.

A. I. Root Co., Box 706, Medina, OH 44258.
Better Bee, Rt. 4, Box 4070, Greenwich, NY 12834-9998.
Brushy Mountain Bee Farm, Rt. 1, Box 135, Moraview Falls, NC 28654.
Dadant and Sons, Inc., Hamilton, IL 62341.
Maxant Industries, Inc., POB 454, Ayer, MA 01432.
Mid-Continent Agrimarketing, Inc., Overland Park, KS 66204.
Walter T. Kelley Co., Clarkson, KY 42726.
Note: Check bee magazines for other suppliers.

Videos

James E. Tew, Extension Honey Bee Lab, OARDC/Dept. Entomology, 1680 Madison Ave., Wooster, OH 44691-4096.
University of Georgia Center for Continuing Education. <http://www.gactr.uga.edu/VideoCatalog/bees.html>.
University of Guelph, Independent Study/OAC Access. Main page: <http://www.uoguelph.ca/istudy/index.htm>. Bee-related information: <http://www.uoguelph.ca/istudy/foodpro.htm>.
Note: See bee journals for advertisments from various companies on bee videos.

Wintering

Barker, R. G. Indoor wintering of honeybee colonies in Manitoba. *Canadian Beekeeper* 5:36–37, 43–44 (1975).
Fingler, B., and D. Small. Indoor wintering in Manitoba. *Manitoba Beekeeper* (Fall):7–19 (1982).
Konrad, J. *Inside Wintering of Bees.* Centennial Report. Winnipeg, Manitoba: Red River Apiarists Assn., 1970.
McCutcheon, D. M. Indoor wintering of hives. *Bee World* 65(1):19–37 (1984).
Nelson, D. L. *Indoor Wintering: Outline of Basic Requirements.* NRG Pub. no. 82-1. Beaverlodge, Alberta: Research Station, Canada Agriculture, 1982.
Nelson, D. L., and G. D. Henn. Indoor wintering research highlights: Research Station, Beaverlodge, Alberta. *Canadian Beekeeper* 7:7–12 (1977).
O'Regan, F. J. *An Economic Analysis of Controlled Environment Storage for Honey Bees.* Working Paper. Kentville, Nova Scotia: AgCanada, n.d.
Pirker, H. J. Here's how Pirker produces our first package bees. *Alberta Bee Culture* 1(2):1, 18–19 (1976).
Pirker, H. J. Package bee production in Northern Canada. *Canadian Beekeeper* 7:17, 20–21 (1978).
Pirker, H. J. Steering factor humidity. *Canadian Beekeeper* 7:102–106 (1978).
Pirker, H. J. Brood rearing in the winter: Factors and methods. *Canadian Beekeeper* 8:69–71 (1980).
Specht, H. B. *Controlled Environment Storage for Honey Bees.* AFDA Project Report no. 06. AgCanada, 1987.
Wrubleski, E. M., and S. E. Bland. Honeybee Overwintering: Management and facilities. Saskatchewan, Canada: Dept. Ag., 1976.

Internet Resources

The following is a list of addresses for resources on the Internet. Because such addresses can change very quickly, these will serve just to get you started. A keyword search (on such words as "bee" or "honey") in any of the large search engines should give you a number of links to explore. Netscape maintains a list of search engines at <http://home.netscape.com/home/internet-search.html>.

American Apitherapy Society. *BeeOnline*—The Online Publication of the American Apitherapy Society. <http://www.beesting.com/#AAS>.
American Association of Professional Apiculturists (AAPA). University of Nebraska Extension—Beekeeping/Apiculture. http://ianrwww.unl.edu/ianr/entomol/beekpg/beekpg.htm> (all the AAPA links are at the bottom of the page). To get information on the AAPA's database or to order a copy, go to <http://gnv.ifas.ufl.edu/~ent1/software/det_bees.htm>.
Candles You Can Burn, Ltd. Candles and Candlemaking. <http://homepage.interaccess.com/~bmolo/>.
Cook, Forrest. The Mead Maker's Page. <http://www.atd.ucar.edu/rdp/gfc/mead/mead.html>.
Dick, Allen. Bees Page. <http://www.internode.net/Honeybee/beespage.htm>. This page includes a number of bee discussion links, including one to subscribe to Mr. Dick's listserve mailing list, Bee-L (address for subscriptions: <http://www.cuug.ab.ca:8001/~dicka/bee-1/html>).
National Honey Board Database. In text form: <http://sunsite.unc.edu/pub/academic/agriculture/entomology/beekeeping/general/National_Honey_Board/honref.txt>. Or zipped: http://sunsite.unc.edu/pub/academic/agriculture/entomology/beekeeping/general/National_Honey_Board/honref.zip>.
Ramel, G. J. L. The Insects Home Page. <http://www.ex.ac.uk/~gjlramel/six01.html>. One of the other pages at this site <http://www.ex.ac.uk/~gjlramel/entolink.html> is a list of other entomological Web sites.
sci.agriculture.beekeeping. Usenet news group on beekeeping.
Texas A&M Agricultural Communications—News, Public Affairs. Various articles on Africanized honey bees. <http://ag.arizona.edu/AES/mac/ahb/ahbhome.html>. Use "Africanized" as the keyword for the search.
U.S. Code, Title 7, Chapter <http://www.law.cornell.edu/uscode/7/ch77.html> and 77 <http://www.law.cornell.edu/uscode/7/ch77.html>. U.S. law concerning honey bees.
University of Michigan Museum of Zoology, Insect Division. Ento[mological] Suppliers. <http://insects.ummz.lsa.umich.edu/entostuff.html>.

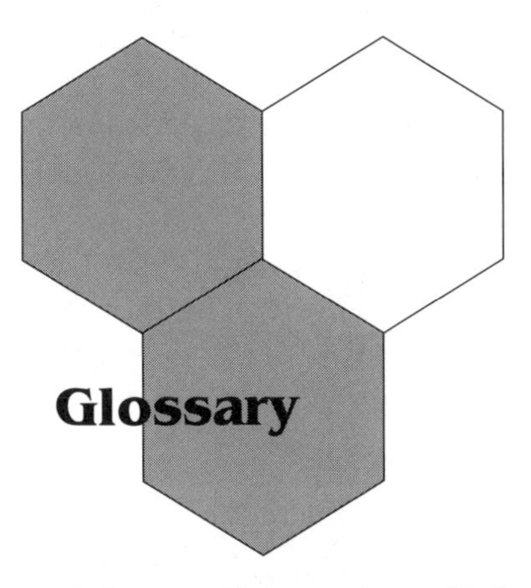

Glossary

abdomen The posterior or third region of the body of the bee that encloses the honey stomach, stomach proper, intestines, sting, and reproductive organs.

abscond To abandon a hive because of wax moth, excessive heat or water, mites, lack of food, or other causes.

acarine disease The name of the disease later known to be caused by the mite *Acarapis woodi* (Rennie) infesting thoracic tracheae of adult bees; *see* **tracheal mite.**

acarology The study of mites.

Africanized honey bee A race of honey bee *Apis mellifera scutellata* that was introduced into South America in the early 1950s to improve the genetics of the European honey bees imported earlier. It was hoped to create a hybrid able to perform better in tropical conditions.

afterswarms Swarms which may leave the hive after the first, or prime, swarm has departed and which are usually led by a virgin queen. These swarms are also known as secondary or tertiary swarms.

alarm odor or alarm pheromone A chemical substance released from the vicinity of a worker bee's sting that alerts other bees to danger. Isopentyl acetate is the principal alarm odor; 2-heptanone, found in worker bee mandibular glands, is a secondary alarm odor.

allergic reaction A systemic or general reaction to bee venom.

American foulbrood A brood disease of bees that affects the late larval and prepupal stages and is caused by the bacterium *Bacillus larvae* (*larvae*).

amino acid An organic compound made up of an amine group and a carboxyl group (generic formula is $H_2N-CHR-COOH$, where R = 20 residual or side groups). Proteins are made up of about 20 different amino acids.

amoeba A unicellular animal in the phylum Sarcodina. The amoeba that affects honey bees is *Malpighamoeba mellificae.*

antenna (pl. antennae) One of two long, segmented, sensory filaments extending from the head of insects; includes taste and odor receptors.

antibody A protein molecule, called immunoglobulin, produced by lymphocytes that reacts with foreign proteins (antigens) in the blood of animals. Antibodies bind to the antigen to mark them for destruction by other elements of the immune system.

antigen A foreign protein entering a body and stimulating the body's immune defense system (lymphocyte cells) to secrete antibodies that will bind to a foreign material, labeling it so the antibodies will recognize it.

apiary The location and sum total of honey bee colonies in a single location; a beeyard.

apiculture The science and art of raising honey bees for economic benefit.

Apis cerana Indian or Asian honey bees. This species is found throughout Asia, and in certain countries it is the honey bee of commerce. It is the natural host of the varroa mite. Closely related species, recently found, are *A. koschevnikovi* and *A. nigrocincta.*

Apis dorsata The second largest of the species of honey bees, it is often referred to as the giant bee; this species is found in Southeast Asia only. *A. laboriosa* is now the largest bee known.

Apis florea The smallest of the species of honey bees, it is often called the dwarf bee. This species is found in Asia, though it is more Western in distribution than the other Asian species. A newly discovered related species is *A. andreniformis.*

Apis mellifera The European honey bee found throughout the Western world though originally Near Eastern in origin. Has been carried from Europe to all continents of the world except the Arctic and Antarctica.

Apistan strips Plastic strips impregnated with the pesticide fluvalinate, used to control varroa mites.

apitherapy Refers to the application of products from the bee hive, including honey, propolis, wax, pollen, and bee venom, for therapeutic purposes.

Bacillus larvae (larvae) The organism (bacterium) that causes American foulbrood disease.

balling a queen An attack on a queen by a number of worker bees intent on killing her by pulling at her legs and wings, suffocating, and overheating her. In this process the bees form a small cluster or ball around the queen.

banking a queen To place a caged queen or queens into a colony made queenless in order to hold them until introduced into another queenless colony.

bee bread Pollen gathered by bees, mixed with various liquids, and sealed in the comb for later use as a protein source for larvae and young bees.

bee cellar A building made especially for overwintering bee colonies in climates where winters are severe.

bee escape A device that permits bees to pass only one way, preventing their return. It is used to remove bees from honey supers.

Bee Go A chemical (usually benzaldehyde) repellent used with a fume board to remove bees from honey supers.

bee louse A flightless fly, *Braula coeca* (Diptera: Braulidae) found only on honey bees. They steal food from a bee's mouth but are not considered a serious problem. They can be controlled with Apistan strips.

bee paralysis Describes the condition in which adult bees are unable to fly or work normally. May be caused by virus induced by parasitic bee mites, such as varroa.

bee space A space which permits free passage for a bee but which is too small to encourage comb building and too large to induce propolization by bees. It measures from ¼ to ⅜ inch (6.4–9.5 mm).

bee suit Coveralls, usually white, with attached veils, that fit over regular clothing and are used when working bees.

bee veil A cloth or wire netting worn with a hat or helmet for protecting the head and neck from stinging insects.

beehive A box or receptacle for housing a colony of bees. Modern beehives are made of wood or plastic and adhere to the bee space dimensions.

beeswax A complex mixture of organic compounds secreted by four pairs of glands on the ventral or underside of the worker bee's abdomen and used by bees for building comb. Its melting point is 143.6–147.2°F (62–64°C). Its flash point is 490–525°F (254–274°C).

bottom board The floor of a beehive.

brace comb or burr comb Small pieces of comb made as connecting links between two frames or between a comb and the hive itself. Burr comb often does not connect to any other part but is an extension of the comb, frequently built during a honeyflow.

brood Immature stages of the bees not yet emerged from their cells (eggs, larvae, and pupae).

brood chamber That part of the hive interior in which brood is reared; the brood chamber includes one or more hive bodies and the combs within.

brood foundation Heavier wax foundation sheets, usually wired with vertical wire and used in broodnest frames.

brood rearing The raising of young bees from eggs to adults.

broodnest The part of the hive interior in which brood is reared; this is the warmest part of a colony, because the eggs and larvae must be incubated at around 95°F (35°C).

Buckfast Hybrid A strain of bees developed by the late Brother Adam, in England; bred for resistance to tracheal mites, disinclination to swarm, winter hardiness, honey production, and good temperament.

bulk comb Refers to frames of freshly capped honey that is sold in the frame. Today, bulk comb is mostly produced for display at state or county fairs.

candied honey Crystallized, granulated, or solidified honey; *see* **Dyce process.**

capped brood Larval cells that have been capped over with a brown wax cover; this allows the larvae to spin their cocoons and turn into pupae.

cappings The thin, light-colored wax that covers cells full of honey. Once cut from honey frames, these wax covers are referred to as cappings and make the finest grade of beeswax.

cappings scratcher A forklike device used to scratch honey cappings before extraction so as to pull off the wax covers; also used to pull up capped brood to test for the presence of varroa mites.

Carniolan bee A race of honey bee, *A. m. carnica,* named for Carniola, Austria, but originating in the Balkan region. It is gray black, very gentle, conserves honey well, and is the source of the ARS Y-C-1 bees.

caste The two types of female bees: workers and queens. Drones are male bees and are therefore not a caste.

Caucasian bee A race of honey bee, *A. m. caucasica,* originating in the Caucasus Mountains; grayish in color, they tend to propolize more than other races.

cell The hexagonal unit compartment of a comb.

cell bar A wooden strip on which queen cups are placed for rearing queens.

cellar wintering To winter bees by placing them in an unheated cellar or special building during the winter months.

chalkbrood A fungus disease caused by *Ascosphera apis,* which turns larvae into white and then gray or black mummies.

chilled brood Immature bees that have died from being too cold, frequently because the field force of a colony has been killed off by pesticides or by a late-season cold snap.

chorion Membrane or "shell" covering a bee egg.

chunk comb honey Comb honey cut from the comb and placed with liquid honey in glass containers.

clarifying The removal of any foreign material or wax from extracted honey usually by allowing the honey to settle, either at room or higher temperatures. Wax and most other material will float to the top and can be skimmed off.

cleansing flight Bees flying out of the hive in early spring or after a long confinement to defecate.

cluster The form or arrangement in which bees cling together after swarming or at cold temperatures.

colony An aggregate of worker bees, drones, and a queen living together in a hive or any other dwelling as one social unit.

comb A back-to-back arrangement of two series of hexagonal cells made of beeswax to hold eggs, brood, pollen, and honey. There are approximately five worker cells to the linear inch and about four drone cells to the inch.

comb honey Honey in the comb usually produced in small wooden sections of 4¼ inch × 4¼ inch, in plastic rings (Ross Rounds), or plastic boxes (Half-Comb Cassettes).

compound eyes The two large lateral eyes of the adult honey bee composed of many lens elements called ommatidia.

conical escape An escape board made with cones set in a frame to permit a one-way exit for bees; used to clear honey supers of bees.

cremed honey *See* **Dyce process.**

crystallization Or granulation of honey; when honey as a supersaturated solution has candied, or become solid, instead of liquid. Creamed honey is a soft form of granulated honey and is labeled as "cremed" honey.

cut-comb honey Honeycomb cut to fit small plastic boxes or wrapped individually for retail sale.

cuticle Waxy, outermost layer of an insect; the exoskeleton.

cyst Resistant, thick-walled cell formed by some organisms, such as amoebae.

dance Or bee dance; a movement made by a forager bee that conveys information to other hive mates. Distance, direction, and the kind of food to look for are indicated. Other components of the dance include scent and vibration.

dearth A period of time when no forage is available for bees to collect, because of weather conditions (drought or flood, unusually cold or hot temperatures) or time of year.

deep super Hive furniture that holds standard, full-depth frames; the usual depth is 9½ or 9⅝ inches.

Demaree The surname of the beekeeper who invented a once popular method of swarm control; also used as a verb ("to Demaree") in describing this method. It consists of separating the queen from most of the brood.

dequeen To remove a queen from a colony.

dextrose *See* **glucose.**

diastase An enzyme in honey.

disease-resistant bees Strain of bees selected from stock showing high survival despite the presence of diseases or mites.

dividing Separating the bees and furniture of a colony in such a way as to form two or more colonies.

division board Also called a *dummy board* or *follower* *board.* A thin vertical board of the same dimensions as a frame. It is used to reduce the size of the brood chamber (to a few frames) or to fill up the gap in a hive body using only nine frames. It is also used to divide the hive permanently into two or more parts.

division board feeder A plastic container the shape of a frame, hung in a hive and filled with syrup to feed bees.

division screen A screened board the size of an inner cover that is used to separate two parts of a colony, for rearing queens above an established colony. Also called a Snelgrove board.

drawn combs Honeycomb having the cell walls built up by honey bees from a sheet of foundation base.

drifting The movement of bees that have lost their location and enter other hives. Young bees drift more than older bees, and drones drift more than workers. Common where hives are placed in long rows: bees drift to ends of rows.

drone The male honey bee, which develops from unfertilized eggs (haploid, or having half the number of chromosomes).

drone brood Brood that is reared in larger cells and produces drone bees. Drone pupae have dome-shaped tops.

drone comb Comb having cells measuring about four cells per linear inch. Drone comb has about 18½ cells per square inch.

drone congregating areas (DCA) A specific area to which drones fly year after year, waiting for virgin queens to pass. DCAs are located in areas with easily discernible landmarks, horizons, or other geographic features.

drone layer A queen that lays only unfertilized eggs, resulting in only drone offspring, because she is old, is low on sperm, was improperly mated, was not mated at all, or is diseased or injured.

drumming Pounding on the sides of a hive (or log) to drive the bees upward. The method is used in transferring bees from bee trees into a beehive.

dwindling The rapid dying off of old bees in the spring. Sometimes called *spring dwindling,* it can be caused by nosema and viral diseases or mite infestation or simply by the lack of new bees being produced to replace bees dying of old age.

Dyce process A patented process involving pasteurization and controlled granulation to produce a finely granulated honey product that spreads easily at room temperature. It produces cremed or granulated honey.

dysentery A condition of adult bees similar to diarrhea in humans. It usually occurs during winter, caused by unfavorable wintering conditions and low-quality food, and can be confused with nosema disease, which is caused by a protozoan. It is recognized by an accumulation of feces on the inner or outer surface of hive furniture. Dysentery is first detected by small spots of feces around the entrance and within the hive and during extreme conditions will darken the snow around affected hives.

egg The first stage in the bee's life cycle, usually laid by the queen. It is cylindrical, 1⁄16 inch (1.6 mm) long, and enclosed with a flexible shell, or chorion.

embed To force wire into wax foundation by heat, pressure, or both in order to strengthen the finished comb.

embedder A device, either mechanical or electrical, that forces wire into wax foundation by heat, pressure, or both.

emergency queen cells Queen cells found on the surface of the comb, which are all of the same age and built when the queen is suddenly lost.

emerging brood Young or teneral bees chewing their way out of the capped brood cells.

entrance reducer A strip of wood notched with different-sized holes to regulate the size of the hive entrance and hence the flow of bees into or out of a hive.

enzyme A protein that has particular chemical groups on its surface, turning it into a catalyst for quickening chemical reactions.

escape board A board having one or more one-way bee escapes in it; used to remove bees from honey supers.

European foulbrood A bacterial brood disease of bees caused by *Melissococcus (= Streptococcus) pluton.*

eusocial Characterized by overlapping generations of members of the same nest that cooperate in rearing young and in which there is a division of labor—that is, a worker or sterile caste.

extender patties Vegetable shortening and sugar patties made with the antibiotic Terramycin added. This patty is used to suppress American and European foulbrood diseases.

extracted honey Honey removed from combs by means of a centrifugal extractor.

Extraction The process of uncapping, spinning, and recovering the liquid honey.

extractor A machine used for removing honey from combs by spinning the frames and throwing the honey against the extractor walls; the combs remain intact. Extracting is the act of spinning honey in an extractor.

eyelets Metal pieces that fit in the holes of frame end bars; used to keep reinforcing wires from cutting into the wood and thus slackening the taut wires.

feces Excreta of bees.

feeders Various types of appliances and containers for feeding bees sugar syrup.

fermentation A chemical breakdown of high-moisture honey by yeasts. In honey, fermentation is caused only by the genus *Zygosaccharomyces,* a yeast able to tolerate the high sugar content of honey.

fertile queen A queen inseminated instrumentally or naturally with drone spermatozoa, which is stored in her spermatheca. She is capable of laying fertilized eggs.

fertilization Usually refers to eggs laid by queen bees that become fertilized when sperm stored in a spermatheca is mixed with the egg while it is being laid. Can also refer to the process in which a pollen grain grows down a flower's female organ, or stigma, to fertilize eggs in the ovary, producing seed and fruit.

festooning The activity of young bees, engorged with honey and hanging on to each other, secreting beeswax scales.

field bees, or foragers Worker bees, usually 16 or more days old, that work in the field to collect nectar (or rob honey from other hives), pollen, water, and propolis.

flight path Refers to the direction bees fly leaving their hive. If obstructed by a beekeeper standing in front of the hive entrance, bees will pile up behind the obstruction and may become aggressive.

food chamber A hive body filled with honey; used as extra food or for winter stores.

foulbrood Malignant, contagious bacterial diseases

of honey bee brood. *See* **American foulbrood** and **European foulbrood.**

foundation A thin sheet of beeswax embossed or stamped with the base of a normal worker cell on which the bees will construct complete or drawn comb.

frame Four pieces of wood that form a rectangle, which is designed to hold honeycomb. A frame consists of one top bar, a bottom bar (one or two pieces), and two end bars.

fructose A simple sugar (or monosaccharide), also called levulose (fruit sugar); it is usually the predominant carbohydrate in honey.

fumagillin An antibiotic, known by several trade names, to control nosema disease.

fume board A cloth-coated wooden frame with a metal cover, sprinkled with a bee repellent such as Bee Go, used to remove bees from honey supers.

Fumidil-B A trade name for fumagillin.

***Galleria mellonella* L.** The scientific name for the greater wax moth. The larvae of this moth are used for fish bait and for scientific research.

glucose Or dextrose (grape sugar), a simple sugar. It is one of the two main sugars found in honey and forms most of the solid phase in granulated honey.

grafting A process of removing newly hatched worker larvae from their cells and placing them in artificial queen cups to rear them into queens.

granulation A term applied to crystallized, candied, creamed, or solidified honey.

grease patty A patty made with vegetable shortening and sugar and used to control tracheal mites. If mixed with Terramycin, it is called an *extender patty* and is used to control foulbrood diseases.

head The front (anterior) part of an insect containing the eyes, antennae, and mouthparts.

2-heptanone A chemical substance which is produced in the mandibular glands of worker honey bees and which elicits an alarm reaction.

hive (n) A home for bees provided by humans—that is, a wooden hive box. (v) To place a swarm into a hive box.

hive body A wooden box that encloses the frames.

hive scale A device used to weigh beehives to determine the amount of weight gained or lost over a certain time.

hive stand A structure that serves as a base support for the hive body to keep it off the ground.

hive tool A metal device with a curled (or L-shaped) scraping surface at one end and a flat blade at the other end. Used to separate hive furniture when inspecting bees, to scrape frames, and to remove frames from the hive.

homeostasis Maintaining a steady state, such as a constant nest temperature, by means of feedback responses. For example, if the outside temperature drops, bees respond to the change by clustering and shivering to bring the nest temperature back to where it belongs.

honey A sweet, viscous liquid composed of sugars. Made from flower nectar gathered by the bees, it is ripened or evaporated into honey and stored in the combs. Well-cured honey contains about 17 percent water, the rest being solids.

honey bee The common name for *Apis mellifera* (honey bearer). The word is correctly written as two words, although some bee journals and texts cling to the use of "honey bee" as one word. An arthropod in the class Insecta, order Hymenoptera, and superfamily Apoidea: a social, honey-collecting insect living in perennial colonies.

honey house A building used for extracting and packaging honey, storing supers, and so forth.

honey stomach or sac An enlargement of the back or posterior end of the bee's esophagus that lies in the front part of the abdomen. Can hold a large amount of liquid as a result of its invaginated walls.

honeydew Liquid excreted by insects that suck plant sap, such as aphids, which contains sugars; sometimes collected by bees and stored as honeydew honey.

honeyflow Loosely, a time when there is a plentiful supply of nectar that bees can collect. Its signs include fresh, white wax and combs filled with liquid. It is a time when bees produce and store surplus honey.

hybrid queen The result of crossing different bee races or lines to produce a queen of superior qualities.

hygroscopic Refers to the ability of honey to absorb water from the air.

hypopharyngeal glands A pair of organs that produce brood food and royal jelly; they are located in the head of a worker bee.

increase To add to the number of existing colonies in an apiary by dividing those already on hand, by hiving swarms, or with packages.

infertile Not able to reproduce.

inner cover A lightweight cover with an oblong hole in its center; used under a standard telescoping outer cover on a beehive.

insecticides Chemicals that kill insects.

instrumental insemination The introduction of drone spermatozoa into the genital organs of a virgin queen by means of special instruments; sometimes called II or artificial insemination (AI).

introduction, or queen, cage A small box made of wire screen and wood or of plastic, used in shipping queens and introducing a new queen to a colony.

invertase An enzyme that speeds the transformation of sucrose (a complex sugar) into the monosaccharides (or simple sugars) fructose and glucose.

Isle of Wight disease An early and little-used name for acarine disease, the infestation of bees by tracheal mites.

isopentyl acetate The alarm odor in honey bees that is produced in the sting chamber. Also called isoamyl acetate.

Italian bee (*A. m. ligustica*) The most common race of European bees used commercially. Introduced from Italy in the 1860s, workers have brown and yellow bands on their abdomen; queens have brown or orange abdomens with few or no stripes.

larva (pl. larvae) The second stage in the development of an insect, such as the honey bee, that undergoes complete metamorphosis. Comparable to a caterpillar (or eating) stage of a moth or butterfly. Capped larvae are brood that have spun their cocoons before turning into pupae.

laying worker A worker bee that lays eggs, which will develop into drones because they are unfertilized. Laying workers usually develop in colonies that have been queenless for a long time.

levulose *See* **fructose**.

mandibles The jaws of an insect. In the honey bee and most insects, the mandibles move in a horizontal rather than in a vertical plane. They are used by bees to form honeycomb, to scrape pollen, or to pick up hive debris.

mating flight The flight taken by a virgin queen, during which she mates in the air with one or more drones. Typically, queens mate with up to 20 drones during several mating flights.

mead A wine made with honey as the sweetener.

menthol crystals Crystalline form of the essential oil of the mint plant *Mentha arvensis*; used to control tracheal mites.

metamorphosis The four-stage development process of most insects: egg, larva, pupa, and adult; also called *complete metamorphosis.*

migratory beekeeping Moving colonies from one locality to another during a single season, to pollinate different crops or to take advantage of more than one honeyflow.

mite Eight-legged creatures or acarines (like ticks and spiders). At least two species are parasitic on bees in the United States—tracheal and varroa mites.

movable frame A wooden frame containing honeycomb that is constructed on the principle of the bee space. When placed in a hive, it remains unattached (by pieces of brace comb or heavy deposits of propolis) to its surroundings and is therefore easily removed for inspection.

Nasanoff gland (also spelled Nassonoff [original spelling], Nasanov, or Nasonov) The name given to the gland associated with the seventh abdominal tergite of the worker honey bee. This gland is commonly called the scent gland, because its odor attracts bees to gather in a cluster.

nectar A sweet plant exudation containing sugars that is secreted by special nectary glands. It is found chiefly in the flowers or reproductive organ of plants. Extrafloral nectaries are found outside flowers (on leaves or stems).

nectaries Specialized tissues which are contained in organs or glands of plants and which secrete nectar.

Nosema apis The scientific name of a protozoan, a microsporidian parasite of honey bees, causing nosema disease.

nosema disease An abnormal condition of adult bees resulting from the presence of *Nosema apis* in their intestines. The abdomen of the bee swells, and the midgut is distended. Should be treated with the antibiotic fumagillin.

nucleus hive A small colony of bees often used in queen rearing and called a *nuc.* Comes in sizes of from three to five frames.

nurse bees Young worker bees with fully functional food glands whose duty is to feed larvae and the queen and to perform particular hive duties. Generally, nurse bees are 3–10 days old.

observation hive A hive made largely of glass or Plexiglas to permit observation of the bees at work from inside a building.

ocellus (pl. ocelli) A simple eye with a single lens. The honey bee has three ocelli on the top of its head, with which it distinguishes light from dark.

ommatidium (pl. ommatidia) One of the visual units or lenses that makes up the compound eye.

osmophilic yeasts Yeasts which occur naturally in honey and which are responsible for fermentation in honey that has more than 18 percent water. These yeasts belong to the genus *Zygosaccharomyces.*

out apiary An apiary kept at a distance from the home of the beekeeper; also called an *outyard.*

outer cover The top cover that fits on a hive to protect it from the weather; the two most common covers are migratory (flush with the hive edge) and telescoping (telescopes over the edge).

outyard Also called an out apiary; an apiary that is kept some distance from the home apiary.

ovary The egg-producing organ of a plant or animal.

oxytetracycline The only registered antibiotic (sold as Terramycin) to control American and European foulbrood.

package A quantity of bees from 2 to 5 pounds, shipped in a special wire-screened box with or without a queen.

paradichlorobenzene (PDB) A crystalline chemical also called *moth crystals,* used to control wax moths.

parthenogenesis The development of young from unfertilized eggs. In honey bees, the unfertilized eggs produce only drone bees, which are laid by virgin queens, laying workers, and old (or young) mated queens.

pesticide Any chemical applied to a crop to protect it from weeds, pests, or diseases.

pheromone A chemical substance released externally by one insect (or animal) that stimulates a response in other insects (or animals) of the same species.

piping A series of sounds made by queens: a loud, shrill tone followed by shorter ones. Usually made by a newly emerged virgin queen to elicit "quacking" from queens still in their cell, which enables her to locate and destroy them.

play flights Short flights taken in front of the hive and in its vicinity to acquaint young bees with their immediate surroundings and hive location. These are sometimes mistaken for robbing or preparation for swarming, but play flights are common in the late afternoon. They are also called *orientation flights.*

pollen The dustlike male reproductive cell bodies of flowers formed on the anthers and collected by bees. It provides the protein part of the honey bee's diet.

pollen basket (corbicula) A flattened depression surrounded by curved spines located on the outside of the tibiae of the bee's hind legs. It is used to carry pollen gathered from flowers back to the hive; the pollen is packed as "pellets."

pollen insert A device inserted into the entrance of a colony into which hand-collected pollen is placed. As the bees leave the hive and pass through the trap, some of the pollen adheres to their bodies. This allows the pollen to be carried to the target blossoms, resulting in cross-pollination.

pollen patty A cake or patty made of pollen pellets and sugar syrup. Patties are fed to stimulate brood rearing.

pollen substitute A food material used to substitute wholly for pollen in the bee's diet; commonly made from soy flour or other products or a special mixture available from bee supply companies. *See also* **wheast**.

pollen supplement A bee food that is mixed with pollen to augment the bees' diet. It can contain brewer's yeast (distiller's soluble from a brewer), soybean flour, natural pollen, and other ingredients formulated in different ways to be digestible to bees.

pollen trap A device for removing pollen from the pollen baskets of bees as they return to their hives.

pollination The transfer of pollen from the anthers (male part) to the stigma (female part) of flowers.

pollinator The agent that transmits the pollen.

pollinizer The plant that furnishes pollen. Crabapples are often pollinizers in apple orchards.

prime swarm The first swarm to issue from the parent colony; usually contains the old queen.

proboscis The tongue of a bee, which combines the maxillae and labium mouthparts.

propolis, or bee glue A gluey or resinous material collected from trees and plants by the bees and used to strengthen the comb or to seal cracks; it has antimicrobial activity.

protein Naturally occurring, complex organic macromolecule that contains amino acids. Pollen contains protein, which is an important nutrient to developing bees.

pupa (pl. pupae) The third stage in the development of an insect in which the insect is encapsulated in a cocoon. In this stage, the organs of the larva are replaced by those that will be used as an adult.

queen A fully developed, mated female bee. The queen is larger and longer than a worker bee and is recognized by workers because of special pheromones.

queen cage A small cage in which a queen and five or six worker bees may be confined. It is used for shipping queens and typically contains a candy plug.

queen cell A special, elongated cell resembling a peanut shell, in which the queen is reared. It is usually an inch or more in length and hangs down from the comb in a vertical position.

queen cup A cup-shaped cell hanging vertically on the comb. They are commercially made of beeswax or plastic. It becomes a queen cell when an egg is placed in it; the bees feed and add wax to it as the larva grows.

queen excluder A device made of wire, wood and wire, plastic, or punched metal, having openings from about 0.163 to 0.167 inch (0.414 to 0.424 cm). This device permits workers to pass through but excludes queens and drones. It is used to confine the queen to a specific part of the hive—usually the brood chamber.

queen substance A glandular secretion from the mandibular glands of a queen bee ((E)-9-oxo-2-decenoic acid, or 9-ODA), which the attendant worker bees collect and pass around the entire colony. If the supply of this substance is inadequate or missing, the colony will initiate queen cell construction. It is also the sex attractant that entices drones to virgin queens.

queenright A bee colony with a laying queen.

QueenTab A small Apistan strip formulated for a queen cage.

rabbet A recessed ridge on the upper inside ends of a beehive on which frames hang. Some rabbets are cut so that resting frames will be at the right bee space to the top of the box; others are lower, requiring a metal strip to correct the bee space. Also refers to a piece of metal that fits over this ridge as a frame rest.

rendering wax The process of melting combs and cappings to separate wax from its impurities and thus to refine the beeswax; the first step is usually done by means of a hot water tank or solar wax melter. To refine wax further, hot water, or steam, and pressure are needed.

requeen To place a new queen into a hive intentionally made queenless.

reversing To exchange places of different hive bodies of the same colony, usually to expand the nest. Done in the spring, the upper hive box full of bees and brood is reversed with the lower, emptier box on the bottom board.

robbing Refers to the behavior of bees that steal honey and nectar from other colonies.

round sections Section comb honey made in plastic rings instead of square wooden or plastic boxes. Also called Ross Rounds.

royal jelly A highly nutritious glandular secretion of young bees. It is used to feed the queen and the young queen larvae.

sacbrood A brood disease of bees caused by a filterable virus.

scent gland *See* **Nasanoff gland.**

section comb Comb honey produced in special equipment: wooden section boxes, round plastic sections, or plastic cassettes.

sex attractant *See* **queen substance.**

shallow super Any one of several super sizes less than the depth of a deep super. Commonly, shallow supers vary from 4¼ inches to 6¼ inches in depth.

side bars Also called end bars, the wooden sides of a frame.

skep A dome-shaped beehive without movable frames, usually made of twisted straw. The use of skeps for keeping bees is illegal in most states in the United States.

slumgum The refuse from melted comb and cappings after the wax has been rendered or removed.

smoker A metal container with attached bellows that burns organic fuels to generate smoke. Used to control defensive behavior of bees during routine colony inspections.

solar wax melter A glass-covered, insulated box in which one melts wax from combs and cappings by using the heat of the sun.

spermatheca A small sac connected with the oviduct of the queen, in which is stored the spermatozoa received during mating with drones.

spermatozoa Male reproduction cells (gametes) that fertilize eggs.

spiracles Openings in the body wall connected to the tracheal tubes through which the bee breathes.

split (v) To divide components of a hive and its population of bees to form a new colony. (n) A colony divided, thus increasing the number of hives.

spring dwindling A term that refers to the weakened condition of a colony in the spring. Can be the result of poor overwintering ability (genetics), inadequate or poor food stores, diseases, or mite predation.

Starline hybrid An Italian strain, crossbred for vigor, honey production, and prolific populations of bees.

sting An organ of defense of workers and queen bees. It is an egg-laying device (or ovipositor) modified to form a piercing shaft through which painful organic venom is delivered.
sucrose Also known as table, cane, or beet sugar, sucrose is a white granular substance fed to bees. It is a complex sugar or disaccharide, made by the combination of one molecule of glucose and one of fructose.
super A piece of hive furniture in which bees store surplus honey, so called because it is placed over or above the brood chamber.
supering The act of placing supers on a colony in anticipation of a honey crop.
supersede To rear a young queen that will replace the mother queen in the same hive. Shortly after the young queen starts to lay eggs, the old queen usually disappears.
supersedure cells Queen cells, few in number and of different ages, located in the middle of the broodnest.
surplus honey The honey removed from a hive, in excess of what bees need for their own use, such as winter food stores.
swarm The aggregate of worker bees, drones, and a queen or queens that leave the original colony to establish a new one. Swarming is the natural propagation method to form a new bee colony.
swarming season The time of year, usually late spring, when swarms normally occur.

teneral bee A recently molted, pale, soft-bodied individual (e.g., a young bee about to emerge from the cell). Callow bees are newly emerged, pale (i.e., not yet tanned) young bees that have hardened cuticles.
Terramycin The trade name of the antibiotic (oxytetracycline) used to combat European foulbrood and American foulbroud disease. It is the only antibiotic registered in the United States for suppressing these diseases.
thin super foundation A thin, wireless, wax foundation that is used for section comb or chunk honey production.
thixotropic Refers to the gelatinous state of some honey as a result of particular nectar proteins. The most commonly known honey that shows this property is heather honey; it is pressed out of the comb instead of extracted.
thorax The central or second region, between the head and the abdomen, of the bee's body that supports the wings and legs.
top bar The top part of a frame.
tracheae The breathing tubes of an insect opening into the spiracles.
tracheal mite A mite, *Acarapis woodi* (Rennie) (Acari: Tarsonemidae), that inhabits the tracheae of adult honey bees; it can be controlled with grease patties and menthol crystals.
travel stain The darkened or stained surface of comb honey caused by bees walking on its surface over a long time; usually from propolis and pollen.
trophallaxis Food exchange between nestmates. Older bees not only transfer food to younger bees but also pheromones from the queen. This exchange also conveys other information, such as colony or queen condition, and information on the external environment, via incoming food.
Tropilaelaps clareae One of two common mites associated with Asian honey bees; the other is *Euvarroa* spp. These mites are not currently present in the New World.

uncapped brood Brood from one to five days old and not covered by the brood capping.
uncapping knife A knife with a sharp blade used to shave or remove the cappings from combs of sealed honey before extracting. The knives can be heated by steam, hot water, or electricity.
unfertilized egg An ovum (egg) that has not been united with the sperm. Insect eggs not fertilized usually become males.
uniting Combining two or more weak colonies to form a large one. To prevent fighting between colonies, a sheet of newspaper is placed between the colonies until they become familiar with each other's scent.
unripe honey Honey that is more like nectar, containing over 18 percent water.
unsealed brood or open brood Brood or larval bees not yet capped over with wax seals; the term can include eggs.

***Varroa jacobsoni* Oudemans** Varroa mite; a destructive, parasitic mite (eight-legged), related to ticks, that feeds on (drone) brood but is carried by adult bees. Introduced into the United States in 1986. It is controlled with Apistan strips.
ventriculus The stomach of the bee, located in the abdomen between the honey stomach and the hindgut; also called the midgut.
virgin queen An unmated queen. If she begins to lay without mating, she can lay only unfertilized eggs, which will become drones.

wasp A close relative of the honey bee, usually in the genus *Vespula*. They are carnivorous, and some species prey on bees.
wax bloom A powdery coating that forms on the surface of beeswax. It is made up of volatile components of beeswax that over time migrate to and accumulate on the surface of the wax.
wax glands The eight glands that secrete beeswax. They are located in pairs on the last four visible ventral abdominal segments (sternites) of young workers.
wax moth A moth (*Galleria mellonella* L.) whose larvae eat comb, pollen, and pupae.
wax scale A drop of liquid beeswax, produced by and expelled from the wax gland, that hardens into a scale on contact with air; from this form it is shaped into honeycomb.
wheast A dried yeast grown in cottage cheese whey. It is 57 percent protein by weight and is fed to bees as a pollen substitute. *See* **pollen substitute**.
windbreaks Specially constructed fences or natural barriers used to reduce the force of the wind in an apiary during cold weather.
winter cluster The arrangement or organization of adult bees within the hive during the winter period.
winter hardiness The ability of some strains of bees to survive long winters through frugal use of honey stores and low bee populations.
wired foundation Wax foundation containing embedded vertical wires to prevent the finished drawn comb from sagging.
worker bee A female bee whose organs of reproduction are only partially developed. Workers are re-

sponsible for carrying on all the routine tasks of a bee colony.

worker jelly A diet fed to worker larvae by nurse bees, composed of 20–40 percent white component (from the mandibular glands, containing lipids) and 60–80 percent of the clear one for the first two days of larvae life. Gradually, the mandibular gland liquid is withdrawn and the clear component from the hypopharyngeal glands (containing proteins) is fed; this is called *modified worker jelly*. A mixture of pollen and honey is fed for the last two days.

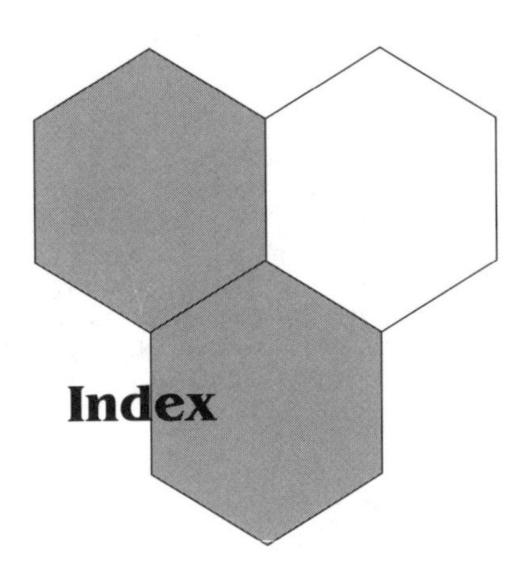

Index